Information Security and Cryptography
Texts and Monographs

T0207254

Hans Delfs
Helmut Knebl

Introduction
to Cryptography

Principles and Applications

Second Edition

Springer

Authors

Prof. Dr. Hans Delfs
Georg-Simon-Ohm University
of Applied Sciences Nürnberg
Department of Computer Science
Keßlerplatz 12
90489 Nürnberg
Germany

Hans.Delfs@fh-nuernberg.de

Prof. Dr. Helmut Knebl
Georg-Simon-Ohm University
of Applied Sciences Nürnberg
Department of Computer Science
Keßlerplatz 12
90489 Nürnberg
Germany

Helmut.Knebl@fh-nuernberg.de

Series Editor

Prof. Dr. Ueli Maurer
Inst. für Theoretische Informatik
ETH Zürich, 8092 Zürich
Switzerland

ACM Computing Classification: E.3

ISSN 1619-7100
ISBN 978-3-642-08040-1   e-ISBN 978-3-540-49244-3

Springer is a part of Springer Science+Business Media
springer.com

© Springer-Verlag Berlin Heidelberg 2007
Softcover reprint of the hardcover 2nd edition 2007

Cover design: KünkelLopka, Heidelberg

Preface to the Second, Extended Edition

New topics have been included in the second edition. They reflect recent progress in the field of cryptography and supplement the material covered in the first edition. Major extensions and enhancements are the following.

- A complete description of the Advanced Encryption Standard AES is given in Chapter 2 on symmetric encryption.
- In Appendix A, there is a new section on polynomials and finite fields. There we offer a basic explanation of finite fields, which is necessary to understand the AES.
- The description of cryptographic hash functions in Chapter 3 has been extended. It now also includes, for example, the HMAC construction of message authentication codes.
- Bleichenbacher's 1-Million-Chosen-Ciphertext Attack against schemes that implement the RSA encryption standard PKCS#1 is discussed in detail in Chapter 3. This attack proves that adaptively-chosen-ciphertext attacks can be a real danger in practice.
- In Chapter 9 on provably secure encryption we have added typical security proofs for public-key encryption schemes that resist adaptively-chosen-ciphertext attacks. Two prominent examples are studied – Boneh's simple-OAEP, or SAEP for short, and Cramer-Shoup's public key encryption.
- Security proofs in the random oracle model are now included. Full-domain-hash RSA signatures and SAEP serve as examples.

Furthermore, the text has been updated and clarified at various points. Errors and inaccuracies have been corrected.

We thank our readers and our students for their comments and hints, and we are indebted to our colleague Patricia Shiroma-Brockmann and Ronan Nugent at Springer for proof-reading the English copy of the new and revised chapters.

Nürnberg, December 2006   Hans Delfs, Helmut Knebl

Preface

The rapid growth of electronic communication means that issues in information security are of increasing practical importance. Messages exchanged over worldwide publicly accessible computer networks must be kept confidential and protected against manipulation. Electronic business requires digital signatures that are valid in law, and secure payment protocols. Modern cryptography provides solutions to all these problems.

This book originates from courses given for students in computer science at the Georg-Simon-Ohm University of Applied Sciences, Nürnberg. It is intended as a course on cryptography for advanced undergraduate and graduate students in computer science, mathematics and electrical engineering.

In its first part (Chapters 1–4), it covers – at an undergraduate level – the key concepts from symmetric and asymmetric encryption, digital signatures and cryptographic protocols, including, for example, identification schemes, electronic elections and digital cash. The focus is on asymmetric cryptography and the underlying modular algebra. Since we avoid probability theory in the first part, we necessarily have to work with informal definitions of, for example, one-way functions and collision-resistant hash functions.

It is the goal of the second part (Chapters 5–10) to show, using probability theory, how basic notions like the security of cryptographic schemes and the one-way property of functions can be made precise, and which assumptions guarantee the security of public-key cryptographic schemes such as RSA. More advanced topics, like the bit security of one-way functions, computationally perfect pseudorandom generators and the close relation between the randomness and security of cryptographic schemes, are addressed. Typical examples of provably secure encryption and signature schemes and their security proofs are given.

Though particular attention is given to the mathematical foundations and, in the second part, precise definitions, no special background in mathematics is presumed. An introductory course typically taught for beginning students in mathematics and computer science is sufficient. The reader should be familiar with the elementary notions of algebra, such as groups, rings and fields, and, in the second part, with the basics of probability theory. Appendix A contains an exposition of the results from algebra and number theory necessary for an understanding of the cryptographic methods. It includes proofs and covers, for example, basics like Euclid's algorithm and the Chinese Remainder Theorem, but also more advanced material like Legendre and Jacobi symbols and probabilistic prime number tests. The concepts and results from probability and information theory that are applied in the second part of the book are given in full in Appendix B. To keep the mathematics easy, we do not address elliptic curve cryptography. We illustrate the key concepts of public-key cryptography by the classical examples like RSA in the quotient rings $\mathbb{Z}_n$ of the integers $\mathbb{Z}$.

The book starts with an introduction into classical symmetric encryption in Chapter 2. The principles of public-key cryptography and their use for encryption and digital signatures are discussed in detail in Chapter 3. The famous and widely used RSA, ElGamal's methods and the digital signature standard, Rabin's encryption and signature schemes serve as the outstanding examples. The underlying one-way functions – modular exponentiation, modular powers and modular squaring – are used throughout the book, also in the second part.

Chapter 4 presents typical cryptographic protocols, including key exchange, identification and commitment schemes, electronic cash and electronic elections.

The following chapters focus on a precise definition of the key concepts and the security of public-key cryptography. Attacks are modeled by probabilistic polynomial algorithms (Chapter 5). One-way functions as the basic building blocks and the security assumptions underlying modern public-key cryptography are studied in Chapter 6. In particular, the bit security of the RSA function, the discrete logarithm and the Rabin function is analyzed in detail (Chapter 7). The close relation between one-way functions and computationally perfect pseudorandom generators meeting the needs of cryptography is explained in Chapter 8. Provable security properties of encryption schemes are the central topic of Chapter 9. It is clarified that randomness is the key to security. We start with the classical notions of provable security originating from Shannon's work on information theory. Typical examples of more recent results on the security of public-key encryption schemes are given, taking into account the computational complexity of attacking algorithms. A short introduction to cryptosystems, whose security can be proven by information-theoretic methods without any assumptions on the hardness of computational problems ("unconditional security approach"), supplements the section. Finally, we discuss in Chapter 10 the levels of security of digital signatures and give examples of signature schemes, whose security can be proven solely under standard assumptions like the factoring assumption, including a typical security proof.

Each chapter (except Chapter 1) closes with a collection of exercises. Answers to the exercises are provided on the Web page for this book: www.informatik.fh-nuernberg.de/DelfsKnebl/Cryptography.

We thank our colleagues and students for pointing out errors and suggesting improvements. In particular, we express our thanks to Jörg Schwenk, Harald Stieber and Rainer Weber. We are grateful to Jimmy Upton for his comments and suggestions, and we are very much indebted to Patricia Shiroma-Brockmann for proof-reading the English copy. Finally, we would like to thank Alfred Hofmann at Springer-Verlag for his support during the writing and publication of this book.

Nürnberg, December 2001   Hans Delfs, Helmut Knebl

Contents

Notation

Page

1. Introduction

Cryptography is the science of keeping secrets secret. Assume a sender referred to here and in what follows as *Alice* (as is commonly used) wants to send a message m to a receiver referred to as *Bob*. She uses an insecure communication channel. For example, the channel could be a computer network or a telephone line. There is a problem if the message contains confidential information. The message could be intercepted and read by an eavesdropper. Or, even worse, the adversary, as usual referred to here as *Eve*, might be able to modify the message during transmission in such a way that the legitimate recipient Bob does not detect the manipulation.

One objective of cryptography is to provide methods for preventing such attacks. Other objectives are discussed in Section 1.2.

1.1 Encryption and Secrecy

The fundamental and classical task of cryptography is to provide *confidentiality* by *encryption methods*. The message to be transmitted – it can be some text, numerical data, an executable program or any other kind of information – is called the *plaintext*. Alice *encrypts* the plaintext m and obtains the *ciphertext* c. The ciphertext c is transmitted to Bob. Bob turns the ciphertext back into the plaintext by *decryption*. To *decrypt*, Bob needs some secret information, a secret *decryption key*.[1] Adversary Eve still may intercept the ciphertext. However, the encryption should guarantee secrecy and prevent her from deriving any information about the plaintext from the observed ciphertext.

Encryption is very old. For example, *Caesar's shift cipher*[2] was introduced more than 2000 years ago. Every encryption method provides an encryption algorithm E and a decryption algorithm D. In classical encryption schemes, both algorithms depend on the same secret key k. This key k is used for both encryption and decryption. These encryption methods are therefore called

---

[1] Sometimes the terms *encipher* and *decipher* are used instead of encrypt and decrypt.

[2] Each plaintext character is replaced by the character 3 to the right modulo 26, i.e., a is replaced by d, b by e,..., x by a, y by b and z by c.

symmetric. For example, in Caesar's cipher the secret key is the offset 3 of the shift. We have

$$D(k, E(k, m)) = m \text{ for each plaintext } m.$$

Symmetric encryption and the important examples DES (data encryption standard) and AES (advanced encryption standard) are discussed in Chapter 2.

In 1976, W. Diffie and M.E. Hellman published their famous paper, *New Directions in Cryptography* ([DifHel76]). There they introduced the revolutionary concept of *public-key cryptography*. They provided a solution to the long standing problem of key exchange and pointed the way to digital signatures. The *public-key encryption* methods (comprehensively studied in Chapter 3) are *asymmetric*. Each recipient of messages has his personal key $k = (pk, sk)$, consisting of two parts: pk is the encryption key and is made public, sk is the decryption key and is kept secret. If Alice wants to send a message m to Bob, she encrypts m by use of Bob's publicly known encryption key pk. Bob decrypts the ciphertext by use of his decryption key sk, which is known only to him. We have

$$D(sk, E(pk, m)) = m.$$

Mathematically speaking, public-key encryption is a so-called *one-way function* with a *trapdoor*. Everyone can easily encrypt a plaintext using the public key pk, but the other direction is difficult. It is practically impossible to deduce the plaintext from the ciphertext, without knowing the secret key sk (which is called the trapdoor information).

Public-key encryption methods require more complex computations and are less efficient than classical symmetric methods. Thus symmetric methods are used for the encryption of large amounts of data. Before applying symmetric encryption, Alice and Bob have to agree on a key. To keep this key secret, they need a secure communication channel. It is common practice to use public-key encryption for this purpose.

1.2 The Objectives of Cryptography

Providing confidentiality is not the only objective of cryptography. Cryptography is also used to provide solutions for other problems:

1. *Data integrity.* The receiver of a message should be able to check whether the message was modified during transmission, either accidentally or deliberately. No one should be able to substitute a false message for the original message, or for parts of it.
2. *Authentication.* The receiver of a message should be able to verify its origin. No one should be able to send a message to Bob and pretend to

be Alice (*data origin authentication*). When initiating a communication, Alice and Bob should be able to identify each other (*entity authentication*).

3. *Non-repudiation.* The sender should not be able to later deny that she sent a message.

If messages are written on paper, the medium – paper – provides a certain security against manipulation. Handwritten personal signatures are intended to guarantee authentication and non-repudiation. If electronic media are used, the medium itself provides no security at all, since it is easy to replace some bytes in a message during its transmission over a computer network, and it is particularly easy if the network is publicly accessible, like the Internet.

So, while encryption has a long history,[3] the need for techniques providing data integrity and authentication resulted from the rapidly increasing significance of electronic communication.

There are symmetric as well as public-key methods to ensure the integrity of messages. Classical symmetric methods require a secret key k that is shared by sender and receiver. The message m is augmented by a *message authentication code* (MAC). The code is generated by an algorithm and depends on the secret key. The augmented message $(m, MAC(k, m))$ is protected against modifications. The receiver may test the integrity of an incoming message $(m, \overline{m})$ by checking whether

$$MAC(k, m) = \overline{m}.$$

Message authentication codes may be implemented by keyed hash functions (see Chapter 3).

Digital signatures require public-key methods (see Chapter 3 for examples and details). As with classical handwritten signatures, they are intended to provide authentication and non-repudiation. Note that non-repudiation is an indispensable feature if digital signatures are used to sign contracts. Digital signatures depend on the secret key of the signer – they can be generated only by him. On the other hand, anyone can check whether a signature is valid, by applying a publicly known verification algorithm *Verify*, which depends on the public key of the signer. If Alice wants to sign the message m, she applies the algorithm *Sign* with her secret key sk and gets the signature $Sign(sk, m)$. Bob receives a signature s for message m, and may then check the signature by testing whether

$$Verify(pk, s, m) = ok,$$

with Alice's public key pk.

It is common not to sign the message itself, but to apply a *cryptographic hash function* (see Section 3.4) first and then sign the hash value. In schemes

---

[3] For the long history of cryptography, see [Kahn67].

like the famous RSA (named after its inventors: Rivest, Shamir and Adleman), the decryption algorithm is used to generate signatures and the encryption algorithm is used to verify them. This approach to digital signatures is therefore often referred to as the "hash-then-decrypt" paradigm (see Section 3.4.5 for details). More sophisticated signature schemes, like the probabilistic signature scheme (PSS), require more steps. Modifying the hash value by pseudorandom sequences turns signing into a probabilistic procedure (see Section 3.4.5).

Digital signatures depend on the message. Distinct messages yield different signatures. Thus, like classical message authentication codes, digital signatures can also be used to guarantee the integrity of messages.

1.3 Attacks

The primary goal of cryptography is to keep the plaintext secret from eavesdroppers trying to get some information about the plaintext. As discussed before, adversaries may also be active and try to modify the message. Then, cryptography is expected to guarantee the integrity of the messages. Adversaries are assumed to have complete access to the communication channel.

Cryptanalysis is the science of studying attacks against cryptographic schemes. Successful attacks may, for example, recover the plaintext (or parts of the plaintext) from the ciphertext, substitute parts of the original message, or forge digital signatures. Cryptography and cryptanalysis are often subsumed by the more general term *cryptology*.

A fundamental assumption in cryptanalysis was first stated by A. Kerkhoff in the nineteenth century. It is usually referred to as *Kerkhoff's Principle*. It states that the adversary knows all the details of the cryptosystem, including algorithms and their implementations. According to this principle, the security of a cryptosystem must be entirely based on the secret keys.

Attacks on the secrecy of an encryption scheme try to recover plaintexts from ciphertexts, or even more drastically, to recover the secret key. The following survey is restricted to passive attacks. The adversary, as usual we call her Eve, does not try to modify the messages. She monitors the communication channel and the end points of the channel. So she may not only intercept the ciphertext, but (at least from time to time) she may be able to observe the encryption and decryption of messages. She has no information about the key. For example, Eve might be the operator of a bank computer. She sees incoming ciphertexts and sometimes also the corresponding plaintexts. Or she observes the outgoing plaintexts and the generated ciphertexts. Perhaps she manages to let encrypt plaintexts or decrypt ciphertexts of her own choice.

The possible attacks depend on the actual resources of the adversary Eve. They are usually classified as follows:

1. *Ciphertext-only attack.* Eve has the ability to obtain ciphertexts. This is likely to be the case in any encryption situation. Even if Eve cannot perform the more sophisticated attacks described below, one must assume that she can get access to encrypted messages. An encryption method that cannot resist a ciphertext-only attack is completely insecure.

2. *Known-plaintext attack.* Eve has the ability to obtain plaintext-ciphertext pairs. Using the information from these pairs, she attempts to decrypt a ciphertext for which she does not have the plaintext. At first glance, it might appear that such information would not ordinarily be available to an attacker. However, it very often is available. Messages may be sent in standard formats which Eve knows.

3. *Chosen-plaintext attack.* Eve has the ability to obtain ciphertexts for plaintexts of her choosing. Then she attempts to decrypt a ciphertext for which she does not have the plaintext. While again this may seem unlikely, there are many cases in which Eve can do just this. For example, she sends some interesting information to her intended victim which she is confident he will encrypt and send out. This type of attack assumes that Eve must first obtain whatever plaintext-ciphertext pairs she wants and then do her analysis, without any further interaction. This means that she only needs access to the encrypting device once.

4. *Adaptively-chosen-plaintext attack.* This is the same as the previous attack, except now Eve may do some analysis on the plaintext-ciphertext pairs, and subsequently get more pairs. She may switch between gathering pairs and performing the analysis as often as she likes. This means that she has either lengthy access to the encrypting device or can somehow make repeated use of it.

5. *Chosen- and adaptively-chosen-ciphertext attack.* These two attacks are similar to the above plaintext attacks. Eve can choose ciphertexts and gets the corresponding plaintexts. She has access to the decryption device.

1.4 Cryptographic Protocols

Encryption and decryption algorithms, cryptographic hash functions or *pseudorandom generators* (see Section 2.1, Chapter 8) are the basic building blocks (also called cryptographic primitives) for solving problems involving secrecy, authentication or data integrity.

In many cases a single building block is not sufficient to solve the given problem: different primitives must be combined. A series of steps must be executed to accomplish a given task. Such a well-defined series of steps is called a *cryptographic protocol.* As is also common, we add another condition: we require that two or more parties are involved. We only use the term protocol if at least two people are required to complete the task.

As a counter example, take a look at digital signature schemes. A typical scheme for generating a digital signature first applies a cryptographic hash function h to the message m and then, in a second step, computes the signature by applying a public-key decryption algorithm to the hash value $h(m)$. Both steps are done by one person. Thus, we do not call it a protocol.

Typical examples of protocols are protocols for user identification. There are many situations where the identity of a user Alice has to be verified. Alice wants to log in to a remote computer, for example, or to get access to an account for electronic banking. Passwords or PIN numbers are used for this purpose. This method is not always secure. For example, anyone who observes Alice's password or PIN when transmitted might be able to impersonate her. We sketch a simple *challenge-and-response* protocol which prevents this attack (however, it is not perfect; see Section 4.2.1).

The protocol is based on a public-key signature scheme, and we assume that Alice has a key $k = (pk, sk)$ for this scheme. Now, Alice can prove her identity to Bob in the following way.

1. Bob randomly chooses a "challenge" c and sends it to Alice.
2. Alice signs c with her secret key, $s := Sign(sk, c)$, and sends the "response" s to Bob.
3. Bob accepts Alice's proof of identity, if $Verify(pk, s, c) = ok$.

Only Alice can return a valid signature of the challenge c, because only she knows the secret key sk. Thus, Alice proves her identity, without showing her secret. No one can observe Alice's secret key, not even the verifier Bob.

Suppose that an eavesdropper Eve observed the exchanged messages. Later, she wants to impersonate Alice. Since Bob selects his challenge c at random (from a huge set), the probability that he uses the same challenge twice is very small. Therefore, Eve cannot gain any advantage by her observations.

The parties in a protocol can be friends or adversaries. Protocols can be attacked. The attacks may be directed against the underlying cryptographic algorithms or against the implementation of the algorithms and protocols. There may also be attacks against a protocol itself. There may be passive attacks performed by an eavesdropper, where the only purpose is to obtain information. An adversary may also try to gain an advantage by actively manipulating the protocol. She might pretend to be someone else, substitute messages or replay old messages.

Important protocols for key exchange, electronic elections, digital cash and interactive proofs of identity are discussed in Chapter 4.

1.5 Provable Security

It is desirable to design cryptosystems that are *provably secure*. Provably secure means that mathematical proofs show that the cryptosystem resists certain types of attacks. Pioneering work in this field was done by C.E. Shannon. In his information theory, he developed measures for the amount of information associated with a message and the notion of perfect secrecy. A *perfectly secret* cipher perfectly resists all ciphertext-only attacks. An adversary gets no information at all about the plaintext, even if his resources in computing power and time are unlimited. *Vernam's one-time pad* (see Section 2.1), which encrypts a message m by XORing it bitwise with a truly random bit string, is the most famous perfectly secret cipher. It even resists all the passive attacks mentioned. This can be mathematically proven by Shannon's theory. Classical information-theoretic security is discussed in Section 9.1; an introduction to Shannon's information theory may be found in Appendix B. Unfortunately, Vernam's one-time pad and all perfectly secret ciphers are usually impractical. It is not practical in most situations to generate and handle truly random bit sequences of sufficient length as required for perfect secrecy.

More recent approaches to provable security therefore abandon the ideal of perfect secrecy and the (unrealistic) assumption of unbounded computing power. The computational complexity of algorithms is taken into account. Only attacks that might be *feasible* in practice are considered. Feasible means that the attack can be performed by an *efficient algorithm*. Of course, here the question about the right notion of efficiency arises. Certainly, algorithms with non-polynomial running time are inefficient. Vice versa algorithms with polynomial running time are often considered as the efficient ones. In this book, we also adopt this notion of efficiency.

The way a cryptographic scheme is attacked might be influenced by random events. Adversary Eve might toss a coin to decide which case she tries next. Therefore, *probabilistic algorithms* are used to model attackers. Breaking an encryption system, for example by a ciphertext-only attack, means that a probabilistic algorithm with polynomial running time manages to derive information about the plaintext from the ciphertext, with some non-negligible probability. Probabilistic algorithms can toss coins, and their control flow may be at least partially directed by these random events. By using random sources, they can be implemented in practice. They must not be confused with non-deterministic algorithms. The notion of probabilistic (polynomial) algorithms and the underlying probabilistic model are discussed in Chapter 5.

The security of a public-key cryptosystem is based on the hardness of some computational problem (there is no efficient algorithm for solving the problem). For example, the secret keys of an RSA scheme could be easily figured out if computing the prime factors of a large integer were possible.[4]

---

[4] What "large" means depends on the available computing power. Today, a 1024-bit integer is considered as large.

However, it is believed that factoring large integers is infeasible.[5] There are no mathematical proofs for the hardness of the computational problems used in public-key systems. Therefore, security proofs for public-key methods are always conditional: they depend on the validity of the underlying assumption.

The assumption usually states that a certain function f is one way; i.e., f can be computed efficiently, but it is infeasible to compute x from $f(x)$. The assumptions, as well as the notion of a one-way function, can be made very precise by the use of probabilistic polynomial algorithms. The probability of successfully inverting the function by a probabilistic polynomial algorithm is negligibly small, and negligibly small means that it is asymptotically less than any given polynomial bound (see Chapter 6, Definition 6.12). Important examples, like the factoring, discrete logarithm and quadratic residuosity assumptions, are included in this book (see Chapter 6).

There are analogies to the classical notions of security. Shannon's perfect secrecy has a computational analogy: *ciphertext indistinguishability* (or *semantic security*). An encryption is perfectly secret if and only if an adversary cannot distinguish between two plaintexts, even if her computing resources are unlimited: if adversary Eve knows that a ciphertext c is the encryption of either m or m', she has no better chance than $1/2$ of choosing the right one. Ciphertext indistinguishability – also called *polynomial-time indistinguishability* – means that Eve's chance of successfully applying a probabilistic polynomial algorithm is at most negligibly greater than $1/2$ (Chapter 9, Definition 9.14).

As a typical result, it is proven in Section 9.4 that *public-key one-time pads* are ciphertext-indistinguishable. This means, for example, that the RSA public-key one-time pad is ciphertext-indistinguishable under the sole assumption that the RSA function is one way. A public-key one-time pad is similar to Vernam's one-time pad. The difference is that the message m is XORed with a pseudorandom bit sequence which is generated from a short truly random seed, by means of a one-way function.

Thus, one-way functions are not only the essential ingredients of public-key encryption and digital signatures. They also yield computationally perfect pseudorandom bit generators (Chapter 8). If f is a one-way function, it is not only impossible to compute x from $f(x)$, but certain bits (called hard-core bits) of x are equally difficult to deduce. This feature is called the bit security of a one-way function. For example, the least-significant bit is a hard-core bit for the RSA function $x \mapsto x^e \bmod n$. Starting with a truly random seed, repeatedly applying f and taking the hard-core bit in each step, you get a pseudorandom bit sequence. These bit sequences cannot be distinguished from truly random bit sequences by an efficient algorithm, or, equivalently (Yao's Theorem, Section 8.2), it is practically impossible to predict the next bit from the previous ones. So they are really computationally perfect.

---

[5] It is not known whether breaking RSA is easier than factoring the modulus. See Chapters 3 and 6 for a detailed discussion.

The bit security of important one-way functions is studied in detail in Chapter 7 including an in-depth analysis of the probabilities involved.

Randomness and the security of cryptographic schemes are closely related. There is no security without randomness. An encryption method provides secrecy only if the ciphertexts appear random to the adversary Eve. Vernam's one-time pad is perfectly secret, because, due to the truly random key string k, the encrypted message $m \oplus k$ [6] is a truly random bit sequence for Eve. The public-key one-time pad is ciphertext-indistinguishable, because if Eve applies an efficient probabilistic algorithm, she cannot distinguish the pseudorandom key string and, as a consequence, the ciphertext from a truly random sequence.

Public-key one-time pads are secure against passive eavesdroppers, who perform a ciphertext-only attack (see Section 1.3 above for a classification of attacks). However, active adversaries, who perform adaptively-chosen-ciphertext attacks, can be a real danger in practice – as demonstrated by Bleichenbacher's 1-Million-Chosen-Ciphertext Attack (Section 3.3.3). Therefore, security against such attacks is also desirable. In Section 9.5, we study two examples of public-key encryption schemes which are secure against adaptively-chosen-ciphertext attacks, and their security proofs. One of the examples, Cramer-Shoup's public key encryption scheme, was the first practical scheme whose security proof is based solely on a standard number-theoretic assumption and a standard assumption of hash functions (collision-resistance).

The ideal cryptographic hash function is a *random function*. It yields hash values which cannot be distinguished from randomly selected and uniformly distributed values. Such a random function is also called a *random oracle*. Sometimes, the security of a cryptographic scheme can be proven in the *random oracle model*. In addition to the assumed hardness of a computational problem, such a proof relies on the assumption that the hash functions used in the scheme are truly random functions. Examples of such schemes include the public-key encryption schemes OAEP (Section 3.3.4) and SAEP (Section 9.5.1), the above mentioned signature scheme PSS and full-domain-hash RSA signatures (Section 3.4.5). We give the random-oracle proofs for SAEP and full-domain-hash signatures.

Truly random functions can not be implemented, nor even perfectly approximated in practice. Therefore, a proof in the random oracle model can never be a complete security proof. The hash functions used in practice are constructed to be good approximations to the ideal of random functions. However, there were surprising errors in the past (see Section 3.4).

We distinguished different types of attacks on an encryption scheme. In a similar way, the attacks on signature schemes can be classified and different levels of security can be defined. We introduce this classification in Chapter 10 and give examples of signature schemes whose security can be proven solely under standard assumptions (like the factoring or the strong RSA as-

---

[6] $\oplus$ denotes the bitwise XOR operator, see page 13.

sumption). No assumptions on the randomness of a hash function have to be made, in contrast, for example, to schemes like PSS. A typical security proof for the highest level of security is included. For the given signature scheme, we show that not a single signature can be forged, even if the attacker Eve is able to obtain valid signatures from the legitimate signer, for messages she has chosen adaptively.

The security proofs for public-key systems are always conditional and depend on (widely believed, but unproven) assumptions. On the other hand, Shannon's notion of perfect secrecy and, in particular, the perfect secrecy of Vernam's one-time pad are unconditional. Although perfect unconditional security is not reachable in most practical situations, there are promising attempts to design practical cryptosystems which provably come close to perfect information-theoretic security. The proofs are based on classical information-theoretic methods and do not depend on unproven assumptions. The security relies on the fact that communication channels are noisy or on the limited storage capacity of an adversary. Some results in this approach are reviewed in the chapter on provably secure encryption (Section 9.6).

2. Symmetric-Key Encryption

In this chapter, we give an introduction to symmetric-key encryption. We explain the notions of stream and block ciphers. The operation modes of block ciphers are studied and, as prominent examples for block ciphers, DES and AES are described.

Symmetric-key encryption provides secrecy when two parties, say Alice and Bob, communicate. An adversary who intercepts a message should not get any significant information about its content.

To set up a secure communication channel, Alice and Bob first agree on a key k. They keep their shared key k secret. Before sending a message m to Bob, Alice encrypts m by using the encryption algorithm E and the key k. She obtains the ciphertext $c = E(k, m)$ and sends c to Bob. By using the decryption algorithm D and the same key k, Bob decrypts c to recover the plaintext $m = D(k, c)$.

We speak of symmetric encryption, because both communication partners use the same key k for encryption and decryption. The encryption and decryption algorithms E and D are publicly known. Anyone can decrypt a ciphertext, if he or she knows the key. Thus, the key k has to be kept secret.

A basic problem in a symmetric scheme is how Alice and Bob can agree on a shared secret key k in a secure and efficient way. For this key exchange, the methods of public-key cryptography are needed, which we discuss in the subsequent chapters. There were no solutions to the key exchange problem, until the revolutionary concept of public-key cryptography was discovered 30 years ago.

We require that the encrypted plaintext m can be uniquely recovered from the ciphertext c. This means that for a fixed key k, the encryption map must be bijective. Mathematically, symmetric encryption may be considered as follows.

Definition 2.1. A *symmetric-key encryption scheme* consists of a map

$$E : K \times M \longrightarrow C,$$

such that for each $k \in K$, the map

$$E_k : M \longrightarrow C, \ m \longmapsto E(k, m)$$

is invertible. The elements $m \in M$ are the *plaintexts* (also called *messages*). C is the set of *ciphertexts* or *cryptograms*, the elements $k \in K$ are the *keys*. E_k is called the *encryption function* with respect to the key k. The inverse function $D_k := E_k^{-1}$ is called the *decryption function*. It is assumed that efficient algorithms to compute E_k and D_k exist.

The key k is shared between the communication partners and kept secret. A basic security requirement for the encryption map E is that, without knowing the key k, it should be impossible to successfully execute the decryption function D_k. Important examples of symmetric-key encryption schemes – Vernam's one-time pad, DES and AES – are given below.

Among all encryption algorithms, symmetric-key encryption algorithms have the fastest implementations in hardware and software. Therefore, they are very well-suited to the encryption of large amounts of data. If Alice and Bob want to use a symmetric-key encryption scheme, they first have to exchange a secret key. For this, they have to use a secure communication channel. Public-key encryption methods, which we study in Chapter 3, are often used for this purpose. Public-key encryption schemes are less efficient and hence not suitable for large amounts of data. Thus, symmetric-key encryption and public-key encryption complement each other to provide practical cryptosystems.

We distinguish between *block ciphers* and *stream ciphers*. The encryption function of a block cipher processes plaintexts of fixed length. A stream cipher operates on streams of plaintext. Processing character by character, it encrypts plaintext strings of arbitrary length. If the plaintext length exceeds the block length of a block cipher, various *modes of operation* are used. Some of them yield stream ciphers. Thus, block ciphers may also be regarded as building blocks for stream ciphers.

2.1 Stream Ciphers

Definition 2.2. Let K be a set of keys and M be a set of plaintexts. In this context, the elements of M are called *characters*.
A stream cipher

$$E^* : K^* \times M^* \longrightarrow C^*, E^*(k, m) := c := c_1 c_2 c_3 \ldots$$

encrypts a stream $m := m_1 m_2 m_3 \ldots \in M^*$ of plaintext characters $m_i \in M$ as a stream $c := c_1 c_2 c_3 \ldots \in C^*$ of ciphertext characters $c_i \in C$ by using a *key stream* $k := k_1 k_2 k_3 \ldots \in K^*, k_i \in K$.
The plaintext stream $m = m_1 m_2 m_3 \ldots$ is encrypted character by character. For this purpose, there is an encryption map

$$E : K \times M \longrightarrow C,$$

which encrypts the single plaintext characters m_i with the corresponding key character k_i:

$$c_i = E_{k_i}(m_i) = E(k_i, m_i), i = 1, 2, \ldots.$$

Typically, the characters in M and C and the key elements in K are binary digits or bytes.

Of course, encrypting plaintext characters with E_{k_i} must be a bijective map, for every key character $k_i \in K$. Decrypting a ciphertext stream $c :=$ $c_1 c_2 c_3 \ldots$ is done character by character by applying the decryption map D with the same key stream $k = k_1 k_2 k_3 \ldots$ that was used for encryption:

$$c = c_1 c_2 c_3 \ldots \mapsto D(k, c) := D_{k_1}(c_1) D_{k_2}(c_2) D_{k_3}(c_3) \ldots.$$

A necessity for stream ciphers comes, for example, from operating systems, where input and output is done with so-called streams.

Of course, the key stream in a stream cipher has to be kept secret. It is not necessarily the secret key which is shared between the communication partners; the key stream might be generated from the shared secret key by a pseudorandom generator (see below).

Notation. In most stream ciphers, the binary exclusive-or operator XOR for bits $a, b \in \{0, 1\}$ – considered as truth values – is applied. We have a XOR $b = 1$, if $a = 0$ and $b = 1$ or $a = 1$ and $b = 0$, and a XOR $b = 0$, if $a = b = 0$ or $a = b = 1$. XORing two bits a and b means to add them modulo 2, i.e., we have a XOR $b = a + b \bmod 2$. As is common practice, we denote the XOR-operator by $\oplus$, $a \oplus b := a$ XOR b, and we use $\oplus$ also for the binary operator that bitwise XORs two bit strings. If $a = a_1 a_2 \ldots a_n$ and $b = b_1 b_2 \ldots b_n$ are bit strings, then

$$a \oplus b := (a_1 \text{ XOR } b_1)(a_2 \text{ XOR } b_2) \ldots (a_n \text{ XOR } b_n).$$

Vernam's One-Time Pad. The most famous example of a stream cipher is *Vernam's one-time pad* (see [Vernam19] and [Vernam26]). It is easy to describe. Plaintexts, keys and ciphertexts are bit strings. To encrypt a message $m := m_1 m_2 m_3 \ldots$, where $m_i \in \{0, 1\}$, a key stream $k := k_1 k_2 k_3 \ldots$, with $k_i \in \{0, 1\}$, is needed. Encryption and decryption are given by bitwise XORing with the key stream:

$$E^*(k, m) := k \oplus m \text{ and } D^*(k, c) := k \oplus c.$$

Obviously, encryption and decryption are inverses of each other. Each bit in the key stream is chosen at random and independently, and the key stream is used only for the encryption of one message m. This fact explains the name "one-time pad". If a key stream k were used twice to encrypt m and $\overline{m}$, we could derive $m \oplus \overline{m}$ from the cryptograms c and $\overline{c}$ and thus obtain information about the plaintexts, by computing $c \oplus \overline{c} = m \oplus k \oplus \overline{m} \oplus k = m \oplus \overline{m} \oplus k \oplus k = m \oplus \overline{m}$.

There are obvious disadvantages to Vernam's one-time pad. Truly random keys of the same length as the message have to be generated and securely transmitted to the recipient. There are very few situations where this is practical. Reportedly, the hotline between Washington and Moscow was encrypted with a one-time pad; the keys were transported by a trusted courier.

Nevertheless, most practical stream ciphers work as Vernam's one-time pad. The difference is that a pseudorandom key stream is taken instead of the truly random key stream. A pseudorandom key stream looks like a random key stream, but actually the bits are generated from a short (truly) random seed by a deterministic algorithm. In practice, such *pseudorandom generators* can be based on specific operation modes of block ciphers or on feedback shift registers. We study the operation modes of block ciphers in Section 2.2.3 (e.g. see the cipher and output feedback modes). Feedback shift registers can be implemented to run very fast on relatively simple hardware. This fact makes them especially attractive. More about these generators and stream ciphers can be found, for example, in [MenOorVan96]. There are also public-key stream ciphers, in which the pseudorandom key stream is generated by using public-key techniques. We discuss these pseudorandom bit generators and the resulting stream ciphers in detail in Chapters 8 and 9.

Back to Vernam's one-time pad. Its advantage is that one can prove that it is secure – an adversary observing a cryptogram does not have the slightest idea what the plaintext is. We discuss this point in the simplest case, where the message m consists of a single bit. Alice and Bob want to exchange one of the messages yes $= 1$ or no $= 0$. Previously, they exchanged the key bit k, which was the outcome of an unbiased coin toss.

First, we assume that each of the two messages yes and no is equally likely. The adversary, we call her Eve, intercepts the cryptogram c. Since the key bit is truly random, Eve can only derive that c encrypts yes or no with probability $1/2$. Thus, she has not the slightest idea which of the two is encrypted. Her only chance of making a decision is to toss a coin. She can do this, however, without seeing the cryptogram c.

If one of the two messages has a greater probability, Eve also cannot gain any advantage by intercepting the cryptogram c. Assume, for example, that the probability of a 1 is $3/4$ and the probability of a 0 is $1/4$. Then the cryptogram c encrypts 0 with probability $1/4$ and 1 with probability $3/4$, irrespective of whether $c = 0$ or $c = 1$. Thus, Eve cannot learn more from the cryptogram than she has learned a priori about the distribution of the plaintexts.

Our discussion for one-bit messages may be transferred to the general case of n-bit messages. The amount of information an attacker may obtain is made precise by information theory. The level of security we achieve with the one-time pad is called perfect secrecy (see Chapter 9 for details). Note that we have to assume that all messages have the same length n (if necessary, they are padded out). Otherwise, some information – the length of the message – would leak to the attacker.

The Vernam one-time pad not only resists a ciphertext-only attack as proven formally in Chapter 9, but it resists all the attacks defined in Chapter 1. Each cryptogram has the same probability. Eve does not learn anything, not even about the probabilities of the plaintexts, if she does not know them a priori. For each message, the key is chosen at random and independently from the previous ones. Thus, Eve cannot gain any advantage by observing plaintext-ciphertext pairs, not even if she has chosen the plaintexts adaptively.

The Vernam one-time pad ensures the confidentiality of messages, but it does not protect messages against modifications. If someone changes bits in the cryptogram and the decrypted cryptogram makes sense, the receiver will not notice it.

2.2 Block Ciphers

Definition 2.3. A *block cipher* is a symmetric-key encryption scheme with $M = C = \{0,1\}^n$ and key space $K = \{0,1\}^r$:

$$E : \{0,1\}^r \times \{0,1\}^n \longrightarrow \{0,1\}^n, \ (k,m) \longmapsto E(k,m).$$

Using a secret key k of binary length r, the encryption algorithm E encrypts plaintext blocks m of a fixed binary length n and the resulting ciphertext blocks $c = E(k,m)$ also have length n. n is called the *block length* of the cipher.

Typical block lengths are 64 (as in DES) or 128 (as in AES), typical key lengths are 56 (as in DES) or 128, 192 and 256 (as in AES).

Let us consider a block cipher E with block length n and key length r. There are 2^n plaintext blocks and 2^n ciphertext blocks of length n. For a fixed key k, the encryption function $E_k : m \mapsto E(k,m)$ maps $\{0,1\}^n$ bijectively to $\{0,1\}^n$ – it is a permutation[1] of $\{0,1\}^n$. Thus, to choose a key k, means to select a permutation E_k of $\{0,1\}^n$, and this permutation is then used to encrypt the plaintext blocks. The 2^r permutations E_k, with k running through the set $\{0,1\}^r$ of keys, form an almost negligibly small subset in the tremendously large set of all permutations of $\{0,1\}^n$, which consists of $2^n!$ elements. So, when we randomly choose an r-bit key k for E, then we restrict our selection of the encryption permutation to an extremely small subset.

From these considerations, we conclude that we cannot have the ideal block cipher with perfect secrecy in practice. Namely, in the preceding Section 2.1, we discussed a stream cipher with perfect secrecy, the Vernam one-time pad. Perfect secrecy results from a maximum amount of randomness: for each

---

[1] A map $f : D \longrightarrow D$ is called a *permutation* of D, if f is bijective.

message bit, a random key bit is chosen (we will prove in Chapter 9 that less randomness in key generation destroys perfect secrecy, see Theorem 9.6). We conclude that the maximal level of security in a block cipher also requires a maximum of randomness, and this in turn means that – when choosing a key – we would have to select a random element from the set of all permutations of $\{0,1\}^n$. Unfortunately, this turns out to be completely impractical. We could try to enumerate all permutations π of $\{0,1\}^n$, $\pi_1, \pi_2, \pi_3, \ldots$, and then randomly select one by randomly selecting an index (this index would be the key). Since there are $2^n!$ permutations, we need $\log_2(2^n!)$-bit numbers to encode the indexes. By Stirling's approximation formula $k! \approx \sqrt{2\pi k}\left(k/e\right)^k$, we derive that $\log_2(2^n!) \approx (n - 1.44)2^n$. This is a huge number. For a block length n of 64 bits, we would need approximately 2^{67} bytes to store a single key. There is no storage medium with such capacity.

Thus, in a real block cipher, we have to restrict ourselves to much smaller keys and choose the encryption permutation E_k for a key k from a much smaller set of 2^r permutations, with r typically in the range of 56 to 256. Nevertheless, the designers of a block cipher try to approximate the ideal. The idea is to get an encryption function which behaves like a randomly chosen function from the very huge set of all permutations.

2.2.1 DES

The *data encryption standard* (DES), originally specified in [FIPS46 1977], was previously the most widely used symmetric-key encryption algorithm. Governments, banks and applications in commerce took the DES as the basis for secure and authentic communication.

We give a high-level description of the DES encryption and decryption functions. The DES algorithm takes 56-bit keys and 64-bit plaintext messages as inputs and outputs a 64-bit cryptogram:[2]

$$\text{DES} : \{0,1\}^{56} \times \{0,1\}^{64} \longrightarrow \{0,1\}^{64}$$

If the key k is chosen, we get

$$\text{DES}_k : \{0,1\}^{64} \longrightarrow \{0,1\}^{64}, \ x \longmapsto \text{DES}(k, x).$$

An encryption with DES_k consists of 16 major steps, called rounds. In each of the 16 rounds, a 48-bit round key k_i is used. The 16 round keys $k_1, \ldots, k_{16}$ are computed from the 56-bit key k by using an algorithm which is studied in Exercise 1 at the end of this chapter.

In the definition of DES, one of the basic building blocks is a map

$$f : \{0,1\}^{48} \times \{0,1\}^{32} \longrightarrow \{0,1\}^{32},$$

---

[2] Actually, the 56 bits of the key are packed with 8 bits of parity.

which transforms a 32-bit message block x with a 48-bit round key $\tilde{k}$. f is composed of a substitution S and a permutation P:

$$f(\tilde{k}, x) = P(S(E(x) \oplus \tilde{k})).$$

The 32 message bits are extended to 48 bits, $x \mapsto E(x)$ (some of the 32 bits are used twice), and XORed with the 48-bit round key $\tilde{k}$. The resulting 48 bits are divided into eight groups of 6 bits, and each group is substituted by 4 bits. Thus, we get 32 bits which are then permuted by P. The cryptographic strength of the DES function depends on the design of f, especially on the design of the eight famous *S-boxes* which handle the eight substitutions (for details, see [FIPS46 1977]).

We define for $i = 1, \ldots, 16$

$$\phi_i : \{0,1\}^{32} \times \{0,1\}^{32} \longrightarrow \{0,1\}^{32} \times \{0,1\}^{32}, \ (x, y) \longmapsto (x \oplus f(k_i, y), y).$$

ϕ_i transforms 64-bit blocks and for this transformation, a 64-bit block is split into two 32-bit halves x and y. We have

$$\phi_i \circ \phi_i(x, y) = \phi_i(x \oplus f(k_i, y), y) = (x \oplus f(k_i, y) \oplus f(k_i, y), y) = (x, y).^3$$

Hence, ϕ_i is bijective and $\phi_i^{-1} = \phi_i.^4$ The fact that ϕ_i is bijective does not depend on any properties of f.

The DES$_k$ function is obtained by composing $\phi_1, \ldots, \phi_{16}$ and the map

$$\mu : \{0,1\}^{32} \times \{0,1\}^{32} \longrightarrow \{0,1\}^{32} \times \{0,1\}^{32}, \ (x, y) \longmapsto (y, x),$$

which interchanges the left and the right half of a 64-bit block (x, y).

Namely,

$$\text{DES}_k : \{0,1\}^{64} \longrightarrow \{0,1\}^{64},$$

$$\text{DES}_k(x) := IP^{-1} \circ \phi_{16} \circ \mu \circ \phi_{15} \circ \ldots \mu \circ \phi_2 \circ \mu \circ \phi_1 \circ IP(x).$$

Here, IP is a publicly known permutation without cryptographic significance.

We see that a DES cryptogram is obtained by 16 encryptions of the same type using 16 different round keys that are derived from the original 56-bit key. ϕ_i is called the encryption of round i. After each round, except the last one, the left and the right half of the argument are interchanged. A block cipher that is computed by iteratively applying a round function to the plaintext is called an *iterated cipher*. If the round function has the form of the DES round function ϕ_i, the cipher is called a *Feistel cipher*. H. Feistel developed the *Lucifer* algorithm, which was a predecessor of the DES algorithm. The idea of using an alternating sequence of permutations and substitutions to get an iterated cipher can be attributed to C.E. Shannon (see [Shannon49]).

---

[3] $g \circ h$ denotes the composition of maps: $g \circ h(x) := g(h(x))$.
[4] As usual, if $f : D \longrightarrow R$ is a bijective map, we denote the inverse map by f^{-1}.

Notation. We also write $\text{DES}_{k_1 \ldots k_{16}}$ for DES_k to indicate that the round keys, derived from k, are used in this order for encryption.

The following Proposition 2.4 means that the DES encryption function may also be used for decryption. For decryption, the round keys $k_1 \ldots k_{16}$ are supplied in reverse order.

Proposition 2.4. *For all messages* $x \in \{0,1\}^{64}$

$$\text{DES}_{k_{16} \ldots k_1}(\text{DES}_{k_1 \ldots k_{16}}(x)) = x.$$

In other words,

$$\text{DES}_{k_{16} \ldots k_1} \circ \text{DES}_{k_1 \ldots k_{16}} = \text{id}.$$

Proof. Since $\phi_i = \phi_i^{-1}$ (see above) and, obviously, $\mu = \mu^{-1}$, we get

$$\text{DES}_{k_{16} \ldots k_1} \circ \text{DES}_{k_1 \ldots k_{16}}$$
$$= \text{IP}^{-1} \circ \phi_1 \circ \mu \circ \phi_2 \circ \ldots \mu \circ \phi_{16} \circ \text{IP} \circ \text{IP}^{-1} \circ \phi_{16} \circ \mu \circ \phi_{15} \circ \ldots \mu \circ \phi_1 \circ \text{IP}$$
$$= \text{id}.$$

This proves the proposition. $\qquad\qquad\qquad\qquad\qquad\qquad\qquad\qquad\qquad\qquad\qquad\Box$

Shortly after DES was published, W. Diffie and M.E. Hellman criticized the short key size of 56 bits in [DifHel77]. They suggested using DES in multiple encryption mode. In triple encryption mode with three independent 56-bit keys k_1, k_2 and k_3, the cryptogram c is computed by $\text{DES}_{k_3}(\text{DES}_{k_2}(\text{DES}_{k_1}(m)))$. This can strengthen the DES because the set of DES_k functions is not a group (i.e., $\text{DES}_{k_2} \circ \text{DES}_{k_1}$ is not a DES_k function), a fact which was shown in [CamWie92]. Moreover, it is shown there that 10^{2499} is a lower bound for the size of the subgroup generated by the DES_k functions in the symmetric group. A small order of this subgroup would imply a less secure multiple encryption mode.

The DES algorithm is well-studied and a lot of cryptanalysis has been performed. Special methods like linear and differential cryptanalysis have been developed and applied to attempt to break DES. However, the best practical attack known is an exhaustive key search. Assume some plaintext-ciphertext pairs (m_i, c_i), $i = 1, \ldots, n$, are given. An exhaustive key search tries to find the key by testing $\text{DES}(k, m_i) = c_i$, $i = 1, \ldots, n$, for all possible $k \in \{0,1\}^{56}$. If such a k is found, the probability that k is really the key is very high. Special computers were proposed to perform an exhaustive key search (see [DifHel77]). Recently a specially designed supercomputer and a worldwide network of nearly 100 000 PCs on the Internet were able to find out the key after 22 hours and 15 minutes (see [RSALabs]). This effort recovered one key. This work would need to be repeated for each additional key to be recovered.

The key size and the block size of DES have become too small to resist the progress in computer technology. The U.S. National Institute of Standards and Technology (NIST) had standardized DES in the 1970s. After more than 20 "DES years" the search for a successor, the AES, was started.

2.2.2 AES

In January 1997, the National Institute of Standards and Technology started an open selection process for a new encryption standard – *the advanced encryption standard*, or *AES* for short. NIST encouraged parties worldwide to submit proposals for the new standard. The proposals were required to support a block size of at least 128 bits, and three key sizes of 128, 192 and 256 bits.

The selection process was divided into two rounds. In the first round, 15 of the submitted 21 proposals were accepted as AES candidates. The candidates were evaluated by a public discussion. The international cryptographic community was asked for comments on the proposed block ciphers. Five candidates were chosen for the second round: MARS (IBM), RC6 (RSA), Rijndael (Daemen and Rijmen), Serpent (Anderson, Biham and Knudsen) and Twofish (Counterpane). Three international "AES Candidate Conferences" were held, and in October 2000 NIST selected the *Rijndael cipher* to be the AES (see [NIST2000]).

Rijndael (see [DaeRij02]) was developed by J. Daemen and V. Rijmen. It is an iterated block cipher and supports different block and key sizes. Block and key sizes of 128, 160, 192, 224 and 256 bits can be combined independently.

The only difference between Rijndael and AES is that AES supports only a subset of Rijndael's block and key sizes. The AES fixes the block length to 128 bits, and uses the three key lengths 128, 192 and 256 bits.

Besides encryption, Rijndael (like many block ciphers) is suited for other cryptographic tasks, for example, the construction of cryptographic hash functions (see Section 3.4.2) or pseudorandom bit generators (see Section 2.2.3). Rijndael can be implemented efficiently on a wide range of processors and on dedicated hardware.

Structure of Rijndael. Rijndael is an iterated block cipher. The iterations are called rounds. The number of rounds, which we denote by N_r, depends on the block length and the key length. In each round except the final round, the same round function is applied, each time with a different round key. The round function of the final round differs slightly. The round keys $key_1, \ldots, key_{N_r}$ are derived from the secret key k by using the key schedule algorithm, which we describe below.

We use the terminology of [DaeRij02] in our description of Rijndael. A byte, as usual, consists of 8 bits, and by a *word* we mean a sequence of 32 bits or, equivalently, 4 bytes.

Rijndael is byte-oriented. Input and output (plaintext block, key, ciphertext block) are considered as one-dimensional arrays of 8-bit-bytes. Both block length and key length are multiples of 32 bits. We denote by N_b the block length in bits divided by 32 and by N_k the key length in bits divided by 32. Thus, a Rijndael block consists of N_b words (or $4 \cdot N_b$ bytes), and a Rijndael key consists of N_k words (or $4 \cdot N_k$ bytes).

The following table shows the number of rounds N_r as a function of N_k and N_b:

	N_b				
N_k	4	5	6	7	8
4	10	11	12	13	14
5	11	11	12	13	14
6	12	12	12	13	14
7	13	13	13	13	14
8	14	14	14	14	14

In particular, AES with key length 128 bits (and the fixed AES block length of 128 bits) consists of 10 rounds.

The round function of Rijndael, and its steps, operate on an intermediate result, called the *state*. The *state* is a block of N_b words (or $4 \cdot N_b$ bytes). At the beginning of an encryption, the variable *state* is initialized with the plaintext block, and at the end, *state* contains the ciphertext block.

The intermediate result *state* is considered as a 4-row matrix of bytes with N_b columns. Each column contains one of the N_b words of *state*.

The following table shows the state matrix in the case of block length 192 bits. We have 6 state words. Each column of the matrix represents a state word consisting of 4 bytes.

$a_{0,0}$	$a_{0,1}$	$a_{0,2}$	$a_{0,3}$	$a_{0,4}$	$a_{0,5}$
$a_{1,0}$	$a_{1,1}$	$a_{1,2}$	$a_{1,3}$	$a_{1,4}$	$a_{1,5}$
$a_{2,0}$	$a_{2,1}$	$a_{2,2}$	$a_{2,3}$	$a_{2,4}$	$a_{2,5}$
$a_{3,0}$	$a_{3,1}$	$a_{3,2}$	$a_{3,3}$	$a_{3,4}$	$a_{3,5}$

The Rijndael Algorithm. An encryption with Rijndael consists of an initial round key addition, followed by applying the round function $(N_r - 1)$-times, and a final round with a slightly modified round function. The round function is composed of the SubBytes, ShiftRows and MixColumns steps and an addition of the round key (see next section). In the final round, the Mix-Columns step is omitted. A high level description of the Rijndael algorithm follows:

Algorithm 2.5.
   byteString *Rijndael*(byteString *plaintextBlock, key*)

    1   *InitState*(*plaintextBlock, state*)
    2   *AddKey*(*state, key₀*)
    3   for $i \leftarrow 1$ to $N_r - 1$ do
    4   *SubBytes*(*state*)
    5   *ShiftRows*(*state*)
    6   *MixColumns*(*state*)
    7   *AddKey*(*state, key$_i$*)
    8   *SubBytes*(*state*)
    9   *ShiftRows*(*state*)
   10 *AddKey*(*state, key$_{N_r}$*)
   11 return *state*;

    The input and output blocks of the Rijndael algorithm are byte strings of $4 \cdot N_b$ bytes. In the beginning, the state matrix is initialized with the plaintext block. The matrix is filled column by column. The ciphertext is taken from the state matrix after the last round. Here, the matrix is read column by column.

    All steps of the round function – SubBytes, ShiftRows, MixColumns, AddKey – are invertible. Therefore, decrypting with Rijndael means to apply the inverse functions of SubBytes, ShiftRows, MixColumns and AddKey, in the reverse order.

The Round function. We describe now the steps – SubBytes, ShiftRows, MixColumns and AddKey – of the round function. The Rijndael algorithm and its steps are byte-oriented. They operate on the bytes of the state matrix. In Rijndael, bytes are usually considered as elements of the finite field $\mathbb{F}_{2^8}$ with 2^8 elements, and $\mathbb{F}_{2^8}$ is constructed as an extension of the field $\mathbb{F}_2$ with 2 elements by using the irreducible polynomial $X^8 + X^4 + X^3 + X + 1$ (see Appendix A.5.3). Then adding (which is the same as bitwise XORing) and multiplying bytes means to add and multiply them as elements of the field $\mathbb{F}_{2^8}$.

The SubBytes Step. SubBytes is the only non-linear transformation of Rijndael. It substitutes the bytes of the state matrix byte by byte, by applying the function S_{RD}[5] to each element of the matrix *state*. The function S_{RD} is also called the S-box; it does not depend on the key. The same S-box is used for all byte positions. The S-box S_{RD} is composed of two maps, f and g. First f and then g is applied:

$$S_{RD}(x) = g \circ f(x) = g(f(x)) \quad (x \text{ a byte}).$$

    Both maps, f and g, have a simple algebraic description.

    To understand f, we consider a byte x as an element of the finite field $\mathbb{F}_{2^8}$. Then f simply maps x to its multiplicative inverse x^{-1}:

---

[5] Rijmen and Daemen's S-box.

$$f : \mathbb{F}_{2^8} \longrightarrow \mathbb{F}_{2^8}, \; x \longmapsto \begin{cases} x^{-1} & \text{if } x \neq 0, \\ 0 & \text{if } x = 0. \end{cases}$$

To understand g, we consider a byte x as a vector of 8 bits or, more precisely, as a vector of length 8 over the field $\mathbb{F}_2$ with 2 elements[6]. Then g is the $\mathbb{F}_2$-affine map

$$g : \mathbb{F}_2^8 \longrightarrow \mathbb{F}_2^8, \; x \longmapsto Ax + b,$$

composed of a linear map $x \mapsto Ax$ and a translation with vector b. The matrix A of the linear map and b are given by

$$A := \begin{pmatrix} 1&0&0&0&1&1&1&1 \\ 1&1&0&0&0&1&1&1 \\ 1&1&1&0&0&0&1&1 \\ 1&1&1&1&0&0&0&1 \\ 1&1&1&1&1&0&0&0 \\ 0&1&1&1&1&1&0&0 \\ 0&0&1&1&1&1&1&0 \\ 0&0&0&1&1&1&1&1 \end{pmatrix} \quad \text{and} \quad b := \begin{pmatrix} 1 \\ 1 \\ 0 \\ 0 \\ 0 \\ 1 \\ 1 \\ 0 \end{pmatrix}.$$

The S-box $\mathrm{S_{RD}}$ operates on each of the state bytes of the state matrix independently. For a block length of 128 bits, we have:

$a_{0,0}$	$a_{0,1}$	$a_{0,2}$	$a_{0,3}$
$a_{1,0}$	$a_{1,1}$	$a_{1,2}$	$a_{1,3}$
$a_{2,0}$	$a_{2,1}$	$a_{2,2}$	$a_{2,3}$
$a_{3,0}$	$a_{3,1}$	$a_{3,2}$	$a_{3,3}$

$\longmapsto$

$\mathrm{S_{RD}}(a_{0,0})$	$\mathrm{S_{RD}}(a_{0,1})$	$\mathrm{S_{RD}}(a_{0,2})$	$\mathrm{S_{RD}}(a_{0,3})$
$\mathrm{S_{RD}}(a_{1,0})$	$\mathrm{S_{RD}}(a_{1,1})$	$\mathrm{S_{RD}}(a_{1,2})$	$\mathrm{S_{RD}}(a_{1,3})$
$\mathrm{S_{RD}}(a_{2,0})$	$\mathrm{S_{RD}}(a_{2,1})$	$\mathrm{S_{RD}}(a_{2,2})$	$\mathrm{S_{RD}}(a_{2,3})$
$\mathrm{S_{RD}}(a_{3,0})$	$\mathrm{S_{RD}}(a_{3,1})$	$\mathrm{S_{RD}}(a_{3,2})$	$\mathrm{S_{RD}}(a_{3,3})$

Both maps f and g are invertible. We even have $f = f^{-1}$. Thus the S-box $\mathrm{S_{RD}}$ is invertible and $\mathrm{S_{RD}}^{-1} = f^{-1} \circ g^{-1} = f \circ g^{-1}$.

The ShiftRows Step. The ShiftRows transformation performs a cyclic left shift of the rows of the state matrix. The offsets are different for each row and depend on the block length N_b.

N_b	1. row	2. row	3. row	4. row
4	0	1	2	3
5	0	1	2	3
6	0	1	2	3
7	0	1	2	4
8	0	1	3	4

For a block length of 128 bits ($N_b = 4$), as in AES, ShiftRows is the map

---

[6] Recall that the field $\mathbb{F}_2$ with 2 elements consists of the residues modulo 2, i.e., $\mathbb{F}_2 = \mathbb{Z}_2 = \{0, 1\}$.

a	b	c	d
e	f	g	h
i	j	k	l
m	n	o	p

$\longmapsto$

a	b	c	d
f	g	h	e
k	l	i	j
p	m	n	o

Obviously, ShiftRows is invertible. The inverse operation is obtained by cyclic right shifts with the same offsets.

The MixColumns Step. The MixColumns transformation operates on each column of the state matrix independently. We consider a column $a = (a_0, a_1, a_2, a_3)$ as a polynomial $a(X) = a_3 X^3 + a_2 X^2 + a_1 X + a_0$ of degree ≤ 3, with coefficients in $\mathbb{F}_{2^8}$.

Then MixColumns transforms a column a by multiplying it with the fixed polynomial

$$c(X) := 03\,X^3 + 01\,X^2 + 01\,X + 02$$

and taking the residue of the product modulo $X^4 + 1$:

$$a(X) \mapsto a(X) \cdot c(X) \bmod (X^4 + 1).$$

The coefficients of c are elements of $\mathbb{F}_{2^8}$. Hence, they are represented as bytes, and a byte is given by two hexadecimal digits, as usual.

The transformations of MixColumns, multiplying by $c(X)$ and taking the residue modulo $X^4 + 1$, are $\mathbb{F}_{2^8}$-linear maps. Hence MixColumns is a linear map of vectors of length 4 over $\mathbb{F}_{2^8}$. It is given by the following 4×4-matrix over $\mathbb{F}_{2^8}$:

$$\begin{pmatrix} 02\ 03\ 01\ 01 \\ 01\ 02\ 03\ 01 \\ 01\ 01\ 02\ 03 \\ 03\ 01\ 01\ 02 \end{pmatrix}.$$

Again, bytes are represented by two hexadecimal digits.

MixColumns transforms each column of the state matrix independently. For a block length of 128 bits, as in AES, we get

$a_{0,0}$	$a_{0,1}$	$a_{0,2}$	$a_{0,3}$
$a_{1,0}$	$a_{1,1}$	$a_{1,2}$	$a_{1,3}$
$a_{2,0}$	$a_{2,1}$	$a_{2,2}$	$a_{2,3}$
$a_{3,0}$	$a_{3,1}$	$a_{3,2}$	$a_{3,3}$

$\longmapsto$

$b_{0,0}$	$b_{0,1}$	$b_{0,2}$	$b_{0,3}$
$b_{1,0}$	$b_{1,1}$	$b_{1,2}$	$b_{1,3}$
$b_{2,0}$	$b_{2,1}$	$b_{2,2}$	$b_{2,3}$
$b_{3,0}$	$b_{3,1}$	$b_{3,2}$	$b_{3,3}$

where

$$\begin{pmatrix} b_{0,j} \\ b_{1,j} \\ b_{2,j} \\ b_{3,j} \end{pmatrix} = \begin{pmatrix} 02\ 03\ 01\ 01 \\ 01\ 02\ 03\ 01 \\ 01\ 01\ 02\ 03 \\ 03\ 01\ 01\ 02 \end{pmatrix} \cdot \begin{pmatrix} a_{0,j} \\ a_{1,j} \\ a_{2,j} \\ a_{3,j} \end{pmatrix}, \quad j = 0, 1, 2, 3.$$

The polynomial $c(X)$ is relatively prime to $X^4 + 1$. Therefore $c(X)$ is a unit modulo $X^4 + 1$. Its inverse is

$$d(X) = 0B\,X^3 + 0D\,X^2 + 09\,X + 0E,$$

i.e., $c(X){\cdot}d(X) \bmod (X^4{+}1) = 1$. This implies that MixColumns is invertible. The inverse operation is to multiply each column of the state matrix by $d(X)$ modulo $X^4 + 1$.

AddKey. The operation AddKey is the only operation in Rijndael that depends on the secret key k, which is shared by the communication partners. It adds a round key to the intermediate result *state*. The round keys are derived from the secret key k by applying the key schedule algorithm, which is described in the next section. Round keys are bit strings and, as the intermediate results *state*, they have block length, i.e., each round key is a sequence of N_b words. AddKey simply bitwise XORs the *state* with the *roundkey* to get the new value of *state*:

$$(state, roundkey) \mapsto state \oplus roundkey.$$

Since we arrange *state* as a matrix, a round key is also represented as a round key matrix of bytes with 4 rows and N_b columns. Each of the N_b words of the round key yields a column. Then the corresponding entries of the state matrix and the round key matrix are bitwise XORed by AddKey to get the new state matrix. Note that bitwise XORing two bytes means to add two elements of the field $\mathbb{F}_{2^8}$.

Obviously, AddKey is invertible. It is inverse to itself. To invert it, you simply apply AddKey a second time with the same round key.

The Key Schedule. The secret key k consists of N_k 4-byte-words. The Rijndael algorithm needs a round key for each round and one round key for the initial key addition. Thus we have to generate $N_r + 1$ round keys (as before, N_r is the number of rounds). A round key consists of N_b words. If we concatenate all the round keys, we get a string of $N_b(N_r + 1)$ words. We call this string the expanded key.

The expanded key is derived from the secret key k by the key expansion procedure, which we describe below. The round keys

$$key_0, key_1, key_2, \ldots, key_{N_r}$$

are then selected from the expanded key $ExpKey$: key_0 consists of the first N_b words of $ExpKey$, key_1 consists of the next N_b words of $ExpKey$, and so on.

To explain the key expansion procedure, we use functions f_j, defined for multiples j of N_k, and a function g. All these functions map words (x_0, x_1, x_2, x_3), which each consist of 4 bytes, to words.

g simply applies the S-box S_{RD} (see SubBytes above) to each byte:

$$(x_0, x_1, x_2, x_3) \mapsto (S_{RD}(x_0), S_{RD}(x_1), S_{RD}(x_2), S_{RD}(x_3)).$$

If j is a multiple of N_k, i.e., $j \equiv 0 \bmod N_k$, we define f_j by

$$(x_0, x_1, x_2, x_3) \mapsto (\text{S}_{\text{RD}}(x_1) \oplus RC\left[j/N_k\right], \text{S}_{\text{RD}}(x_2), \text{S}_{\text{RD}}(x_3), \text{S}_{\text{RD}}(x_0)).$$

Here, so-called round constants $RC[i]$ are used. They are defined as follows. First, recall that in our representation, the elements of $\mathbb{F}_{2^8}$ are the residues of polynomials with coefficients in $\mathbb{F}_2$ modulo $P(X) = X^8 + X^4 + X^3 + X + 1$. Now, the round constant $RC[i] \in \mathbb{F}_{2^8}$ is defined by $RC[i] := X^{i-1} \bmod P(X)$.

Relying on the non-linear S-box S_{RD}, the functions f_j and g are also non-linear.

We are ready to describe the key expansion. We denote by

$$ExpKey[j], \ 0 \leq j < N_b(N_r + 1),$$

the words of the expanded key. The first N_k words are initialized with the secret key k. The following words are computed recursively. $ExpKey[j]$ depends on $ExpKey[j - N_k]$ and on $ExpKey[j - 1]$.

Depending on the key length N_k, there are two versions of the key expansion procedure, one for $N_k \leq 6$, the other for $N_k > 6$. We have for $N_k \leq 6$:

$$ExpKey[j] := \begin{cases} ExpKey[j - N_k] \oplus f_j(ExpKey[j-1]) & \text{if } j \equiv 0 \bmod N_k, \\ ExpKey[j - N_k] \oplus ExpKey[j-1] & \text{if } j \not\equiv 0 \bmod N_k. \end{cases}$$

If $N_k > 6$, we have:

$$ExpKey[j] := \begin{cases} ExpKey[j - N_k] \oplus f_j(ExpKey[j-1]) & \text{if } j \equiv 0 \bmod N_k, \\ ExpKey[j - N_k] \oplus g(ExpKey[j-1]) & \text{if } j \not\equiv 0 \bmod N_k \\ & \text{and } j \equiv 4 \bmod N_k, \\ ExpKey[j - N_k] \oplus ExpKey[j-1] & \text{else.} \end{cases}$$

2.2.3 Modes of Operation

Block ciphers need some extension, because in practice most of the messages have a size that is distinct from the block length. Often the message length exceeds the block length. Modes of operation handle this problem. They were first specified in conjunction with DES, but they can be applied to any block cipher.

We consider a block cipher E with block length n. We fix a key k and, as usual, we denote the encryption function with this key k by

$$E_k : \{0,1\}^n \longrightarrow \{0,1\}^n,$$

for example, $E_k = \text{DES}_k$. To encrypt a message m that is longer than n bits we apply a mode of operation: The message m is decomposed into blocks of some fixed bit length r, $m = m_1 m_2 \ldots m_l$, and then these blocks are encrypted iteratively. The length r of the blocks m_i is not in all modes of operation equal to the block length n of the cipher. There are modes of operation, where r can be smaller than n, for example, the cipher feedback and the output feedback modes below. In electronic code book mode and cipher-block chaining mode, which we discuss first, the block length r is equal to the block length n of the block cipher.

If the block length r does not divide the length of our message, we have to complete the last block. The last block is padded out with some bits. After applying the decryption function, the receiver must remove the padding. Therefore, he must know how many bits were added. This can be achieved, for example, by storing the number of padded bits in the last byte of the last block.

Electronic Codebook Mode. The electronic code book mode is the straightforward mode. The encryption is deterministic – identical plaintext blocks result in identical ciphertext blocks. The encryption works like a codebook. Each block of m is encrypted independently of the other blocks. Transmission bit errors in a single ciphertext block affect the decryption only of that block.

In this mode, we have $r = n$. The electronic codebook mode is implemented by the following algorithm:

Algorithm 2.6.
> bitString $ecbEncrypt$(bitString m)
> 1   divide m into $m_1 \dots m_l$
> 2   for $i \leftarrow 1$ to l do
> 3 $c_i \leftarrow E_k(m_i)$.
> 4   return $c_1 \dots c_l$

For decryption, the same algorithm can be used with the decryption function E_k^{-1} in place of E_k.

If we encrypt many blocks, partial information about the plaintext is revealed. For example, an eavesdropper Eve detects whether a certain block repeatedly occurs in the sequence of plaintext blocks, or, more generally, she can figure out how often a certain plaintext block occurs. Therefore, other modes of operation are preferable.

Cipher-Block Chaining Mode. In this mode, we have $r = n$. Encryption in the cipher-block chaining mode is implemented by the following algorithm:

Algorithm 2.7.
> bitString $cbcEncrypt$(bitString m)
> 1   select $c_0 \in \{0,1\}^n$ at random
> 2   divide m into $m_1 \dots m_l$
> 3   for $i \leftarrow 1$ to l do
> 4 $c_i \leftarrow E_k(m_i \oplus c_{i-1})$
> 5   return $c_0 c_1 \dots c_l$

Choosing the initial value c_0 at random prevents almost with certainty that the same initial value c_0 is used for more than one encryption. This is important for security. Suppose for a moment that the same c_0 is used for two messages m and m'. Then, an eavesdropper Eve can immediately detect whether the first l blocks of m and m' coincide, because in this case the first l ciphertext blocks are the same.

If a message is encrypted twice, then, with a very high probability, the initial values are different, and hence the resulting ciphertexts are distinct. The ciphertext depends on the plaintext, the key and a randomly chosen initial value. We obtain a *randomized encryption algorithm*.

Decryption in cipher-block chaining mode is implemented by the following algorithm:

Algorithm 2.8.
    bitString *cbcDecrypt*(bitString c)
    1   divide c into $c_0 c_1 \ldots c_l$
    2   for $i \leftarrow 1$ to l do
    3        $m_i \leftarrow E_k^{-1}(c_i) \oplus c_{i-1}$
    4   return $m_1 \ldots m_l$

The cryptogram $c = c_0 c_1 \ldots c_l$ has one block more than the plaintext. The initial value c_0 needs not be secret, but its integrity must be guaranteed in order to decrypt c_1 correctly.

A transmission bit error in block c_i affects the decryption of the blocks c_i and c_{i+1}. The block recovered from c_i will appear random (here we assume that even a small change in the input of a block cipher will produce a random-looking output), while the plaintext recovered from c_{i+1} has bit errors precisely where c_i did. The block c_{i+2} is decrypted correctly. The cipher-block chaining mode is self-synchronizing, even if one or more entire blocks are lost. A lost ciphertext block results in the loss of the corresponding plaintext block and errors in the next plaintext block.

In both the electronic codebook mode and cipher-block chaining mode, E_k^{-1} is applied for decryption. Hence, both modes are also applicable with public-key encryption methods, where the computation of E_k^{-1} requires the recipient's secret, while E_k can be easily computed by everyone.

Cipher Feedback Mode. Let lsb_l denote the l least significant (rightmost) bits of a bit string, msb_l the l most significant (leftmost) bits of a bit string, and let $\parallel$ denote the concatenation of bit strings.

In the cipher feedback mode, we have $1 \leq r \leq n$ (recall that the plaintext m is divided into blocks of length r). Let $x_1 \in \{0,1\}^n$ be a randomly chosen initial value. The cipher feedback mode is implemented by the following algorithm:

Algorithm 2.9.

bitString $cfbEnCrypt$(bitString m, x_1)
1   divide m into $m_1 \ldots m_l$
2   for $i \leftarrow 1$ to l do
3       $c_i \leftarrow m_i \oplus \mathrm{msb}_r(E_k(x_i))$
4       $x_{i+1} \leftarrow \mathrm{lsb}_{n-r}(x_i)\|c_i$
5   return $c_1 \ldots c_l$

We get a stream cipher in this way. The key stream is computed by using E_k, and depends on the key underlying E_k, on an initial value x_1 and on the ciphertext blocks already computed. Actually, x_{i+1} depends on the first $\lceil n/r \rceil$ members[7] of the sequence $c_i, c_{i-1}, \ldots, c_1, x_1$. The key stream is obtained in blocks of length r. The message can be processed bit by bit and messages of arbitrary length can be encrypted without padding. If one block of the key stream is consumed, the next block is computed. The initial value x_1 is transmitted to the recipient. It does not need to be secret if E_k is the encryption function of a symmetric cryptosystem (an attacker does not know the key underlying E_k). The recipient can compute $E_k(x_1)$ – hence m_1 and x_2 – from x_1 and the cryptogram c_1, then $E_k(x_2), m_2$ and x_3, and so on.

For each encryption, a new initial value x_1 is chosen at random. This prevents almost with certainty that the same initial value x_1 is used for more than one encryption. As in every stream cipher, this is important for security. If the same initial value x_1 is used for two messages m and m', then an eavesdropper Eve immediately finds out whether the first l blocks of m and m' coincide. In this case, the first l blocks of the generated key stream, and hence the first l ciphertext blocks are the same for m and m'.

A transmission bit error in block c_i affects the decryption of that block and the next $\lceil n/r \rceil$ ciphertext blocks. The block recovered from c_i has bit errors precisely where c_i did. The next $\lceil n/r \rceil$ ciphertext blocks will be decrypted into random-looking blocks (again we assume that even a small change in the input of a block cipher will produce a random-looking output). The cipher feedback mode is self-synchronizing after $\lceil n/r \rceil$ steps, even if one or more entire blocks are lost.

Output Feedback Mode. As in the cipher feedback mode, we have $1 \leq r \leq n$. Let $x_1 \in \{0,1\}^n$ be a randomly chosen initial value. The output feedback mode is implemented by the following algorithm:

Algorithm 2.10.

bitString $ofbEnCrypt$(bitString m, x_1)
1   divide m into $m_1 \ldots m_l$
2   for $i \leftarrow 1$ to l do
3       $c_i \leftarrow m_i \oplus \mathrm{msb}_r(E_k(x_i))$
4       $x_{i+1} \leftarrow E_k(x_i)$
5   return $c_1 \ldots c_l$

---

[7] $\lceil x \rceil$ denotes the smallest integer $\geq x$.

There are two different output feedback modes discussed in the literature. The one we introduced is considered to have better security properties and was specified in [ISO/IEC 10116]. In the output feedback mode, plaintexts of arbitrary length can be encrypted without padding. As in the cipher feedback mode, the plaintext is considered as a bit stream and each bit is XORed with a bit of a key stream. The key stream depends only on an initial value x_1 and is iteratively computed by $x_{i+1} = E_k(x_i)$. The initial value x_1 is transmitted to the recipient. It does not need to be secret if E_k is the encryption function of a symmetric cryptosystem (an attacker does not know the key underlying E_k). For decryption, the same algorithm can be used.

It is essential for security that the initial value is chosen randomly and independently from the previous ones. This prevents almost with certainty that the same initial value x_1 is used for more than one encryption. If the same initial value x_1 is used for two messages m and m', then identical key streams are generated for m and m', and an eavesdropper Eve immediately computes the difference between m and m' from the ciphertexts: $m \oplus m' = c \oplus c'$. Thus, it is strongly recommended to choose a new random initial value for each message.

A transmission bit error in block c_i only affects the decryption of that block. The block recovered from c_i has bit errors precisely where c_i did. However, the output feedback mode will not recover from a lost ciphertext block – all following ciphertext blocks will be decrypted incorrectly.

Security. We mentioned before that the electronic codebook mode has some shortcomings. The question arises as to what amount the mode of operation weakens the cryptographic strength of a block cipher. A systematic treatment of this question can be found in [BelDesJokRog97]. First, a model for the security of block ciphers is developed, the so-called *pseudorandom function model* or *pseudorandom permutation model*.

As discussed in Section 2.2, ideally we would like to choose the encryption function of a block cipher from the huge set of all permutations on $\{0, 1\}^n$ in a truly random way. This approach might be called the "truly random permutation model". In practice, we have to follow the "pseudorandom permutation model": the encryption function is chosen randomly, but from a much smaller family $(F_k)_{k \in K}$ of permutations on $\{0, 1\}^n$, like DES.

In [BelDesJokRog97], the security of the cipher-block chaining mode is reduced to the security of the pseudorandom family $(F_k)_{k \in K}$. Here, security of the family $(F_k)_{k \in K}$ means that no efficient algorithm is able to distinguish elements randomly chosen from $(F_k)_{k \in K}$ from elements randomly chosen from the set of all permutations. This notion of security for pseudorandom function families is analogously defined as the notion of computationally perfect pseudorandom bit generators, which will be studied in detail in Chapter 8. [BelDesJokRog97] also consider a mode of operation similar to the output feedback mode, called the XOR scheme, and its security is also reduced to the security of the underlying pseudorandom function family.

Exercises

1. The following algorithm computes the round keys k_i, $i = 1, \ldots, 16$, for DES from the 64-bit key k. Only 56 of the 64 bits are used and permutated. This is done by a map PC1. The result $PC1(k)$ is divided into two halves, C_0 and D_0, of 28 bits.

   Algorithm 2.11.
   bitString *DESKeyGenerator*(bitString k)
   1   $(C_0, D_0) \leftarrow PC1(k)$
   2   for $i \leftarrow 1$ to 16 do
   3       $(C_i, D_i) \leftarrow (LS_i(C_{i-1}), LS_i(D_{i-1}))$
   4       $k_i \leftarrow PC2(C_i, D_i)$
   5   return $k_1 \ldots k_{16}$

   Here LS_i is a cyclic left shift by one position if $i = 1, 2, 9$ or 16, and by two positions otherwise. The maps

   $$PC1 : \{0,1\}^{64} \longrightarrow \{0,1\}^{56}, \quad PC2 : \{0,1\}^{56} \longrightarrow \{0,1\}^{48}$$

   are defined by the tables

PC1								PC2				
57	49	41	33	25	17	9	14	17	11	24	1	5
1	58	50	42	34	26	18	3	28	15	6	21	10
10	2	59	51	43	35	27	23	19	12	4	26	8
19	11	3	60	52	44	36	16	7	27	20	13	2
63	55	47	39	31	23	15	41	52	31	37	47	55
7	62	54	46	38	30	22	30	40	51	45	33	48
14	6	61	53	45	37	29	44	49	39	56	34	53
21	13	5	28	20	12	4	46	42	50	36	29	32

   The tables are read line by line and describe how to get the images, i.e.,

   $$PC1(x_1, \ldots, x_{64}) = (x_{57}, x_{49}, x_{41}, \ldots, x_{12}, x_4),$$
   $$PC2(x_1, \ldots, x_{56}) = (x_{14}, x_{17}, x_{11}, \ldots, x_{29}, x_{32}).$$

   The bits 8, 16, 24, 32, 40, 48, 56 and 64 of k are not used. They are defined in such a way that odd parity holds for each byte of k. A key k is defined to be *weak* if $k_1 = k_2 = \ldots = k_{16}$.
   Show that exactly four weak keys exist, and determine these keys.

2. In this exercise, $\overline{x}$ denotes the bitwise complement of a bit string x. Let DES : $\{0,1\}^{64} \times \{0,1\}^{56} \longrightarrow \{0,1\}^{64}$ be the DES function.

   a. Show that $DES(\overline{k}, \overline{x}) = \overline{DES(k, x)}$, for $k \in \{0,1\}^{56}, x \in \{0,1\}^{64}$.
   b. Let $(m, DES_k(m))$ be a plaintext-ciphertext pair. We try to find the unknown key k by an exhaustive key search. Show that the number of encryptions we have to compute can be reduced from 2^{56} to 2^{55} if the pair $(\overline{m}, DES_k(\overline{m}))$ is also known.

3. The key stream in the output feedback mode is periodic, i.e., there exists an $i \in \mathbb{N}$ such that $x_i = x_1$. The lowest positive integer with this property is called the *period* of the key stream. Let f be randomly chosen from the set of all permutations on $\{0, 1\}^n$. Show that the average period of the key stream is $2^{n-1} + 1/2$ if the initial value $x_1 \in \{0, 1\}^n$ is chosen at random.

3. Public-Key Cryptography

The basic idea of public-key cryptography are public keys. Each person's key is separated into two parts: a public key for encryption available to everyone and a secret key for decryption which is kept secret by the owner. In this chapter we introduce the concept of public-key cryptography. Then we discuss some of the most important examples of public-key cryptosystems, such as the RSA, ElGamal and Rabin cryptosystems. These all provide encryption and digital signatures.

3.1 The Concept of Public-Key Cryptography

Classical symmetric cryptography provides a secure communication channel to each pair of users. In order to establish such a channel, the users must agree on a common secret key. After establishing a secure communication channel, the secrecy of a message can be guaranteed. Symmetric cryptography also includes methods to detect modifications of messages and methods to verify the origin of a message. Thus, confidentiality and integrity can be accomplished using secret key techniques.

However, public key techniques have to be used for a secure distribution of secret keys, and at least some important forms of authentication and non-repudiation also require public-key methods, such as digital signatures. A digital signature should be the digital counterpart of a handwritten signature. The signature must depend on the message to be signed and a secret known only to the signer. An unbiased third party should be able to verify the signature without access to the signer's secret.

In a public-key encryption scheme, the communication partners do not share a secret key. Each user has a pair of keys: a *secret key sk* known only to him and a *public key pk* known to everyone.

Suppose Bob has such a key pair (pk, sk) and Alice wants to encrypt a message m for Bob. Like everyone else, Alice knows Bob's public key pk. She computes the ciphertext $c = E(pk, m)$ by applying the encryption function E with Bob's public key pk. As before, we denote encrypting with a fixed key pk by E_{pk}, i.e., $E_{pk}(m) := E(pk, m)$. Obviously, the encryption scheme can only be secure if it is practically infeasible to compute m from $c = E_{pk}(m)$.

But how can Bob then recover the message m from the ciphertext c? This is where Bob's secret key is used. The encryption function E_{pk} must have the property that the pre-image m of the ciphertext $c = E_{pk}(m)$ is easy to compute using Bob's secret key sk. Since only Bob knows the secret key, he is the only one who can decrypt the message. Even Alice, who encrypted the message m, would not be able to get m from $E_{pk}(m)$ if she lost m. Of course, efficient algorithms must exist to perform encryption and decryption.

We summarize the requirements of public-key cryptography. We are looking for a family of functions $(E_{pk})_{pk \in PK}$ such that each function E_{pk} is computable by an efficient algorithm. It should be practically infeasible to compute pre-images of E_{pk}. Such families $(E_{pk})_{pk \in PK}$ are called *families of one-way functions* or *one-way functions* for short. Here, PK denotes the set of available public keys.[1] For each function E_{pk} in the family, there should be some information sk to be kept secret which enables an efficient computation of the inverse of E_{pk}. This secret information is called the *trapdoor information*. One-way functions with this property are called *trapdoor functions*.

In 1976, W. Diffie and M.E. Hellman published the idea of public-key cryptography in their famous paper "New Directions in Cryptography" ([DifHel76]). They introduced a public-key method for key agreement which is in use to this day. In addition, they described how digital signatures would work, and proposed, as an open question, the search for such a function. The first public-key cryptosystem that could function as both a key agreement mechanism and as a digital signature was the RSA cryptosystem published in 1978 ([RivShaAdl78]). RSA is named after the inventors: R. Rivest, A. Shamir and L. Adleman. The RSA cryptosystem provides encryption and digital signatures and is the most popular and widely used public-key cryptosystem today. We shall describe the RSA cryptosystem in Section 3.3. It is based on the difficulty of factoring large numbers, which enables the construction of one-way functions with a trapdoor. Another basis for one-way functions is the difficulty of extracting discrete logarithms. These two problems from number theory are the foundations of most public-key cryptosystems used today.

Each participant in a public-key cryptosystem needs his personal key $k = (pk, sk)$, consisting of a public and a secret (also called private) part. To guarantee the security of the cryptosystem, it must be infeasible to compute the secret key sk from the public key pk, and it must be possible to randomly choose the keys k from a huge parameter space. An efficient algorithm must be available to perform this random choice. If Bob wants to participate in the cryptosystem, he randomly selects his key $k = (pk, sk)$, keeps sk secret and publishes pk. Now everyone can use pk in order to encrypt messages for Bob.

To discuss the basic idea of digital signatures, we assume that we have a family $(E_{pk})_{pk \in PK}$ of trapdoor functions and that each function E_{pk} is

---

[1] A rigorous definition of one-way functions is given in Definition 6.12.

bijective. Such a family of trapdoor permutations can be used for digital signatures. Let pk be Alice's public key. To compute the inverse E_{pk}^{-1} of E_{pk}, the secret key sk of Alice is required. So Alice is the only one who is able to do this. If Alice wants to sign a message m, she computes $E_{pk}^{-1}(m)$ and takes this value as signature s of m. Everyone can verify Alice's signature s by using Alice's public key pk and computing $E_{pk}(s)$. If $E_{pk}(s) = m$, Bob is convinced that Alice really signed m because only Alice was able to compute $E_{pk}^{-1}(m)$.

An important straightforward application of public-key cryptosystems is the distribution of session keys. A session key is a secret key used in a classical symmetric encryption scheme to encrypt the messages of a single communication session. If Alice knows Bob's public key, then she may generate a session key, encrypt it with Bob's public key and send it to Bob. Digital signatures are used to guarantee the authenticity of public keys by certification authorities. The certification authority signs the public key of each user with her secret key. The signature can be verified with the public key of the certification authority. Cryptographic protocols for user authentication and advanced cryptographic protocols, like bit commitment schemes, oblivious transfer and zero-knowledge interactive proof systems, have been developed. Today they are fundamental to Internet communication and electronic commerce.

Public-key cryptography is also important for theoretical computer science: theories of security were developed and the impact on complexity theory should be mentioned.

3.2 Modular Arithmetic

In this section, we give a brief overview of the modular arithmetic necessary to understand the cryptosystems we discuss in this chapter. Details can be found in Appendix A.

3.2.1 The Integers

Let $\mathbb{Z}$ denote the ordered set $\{\ldots, -3, -2, -1, 0, 1, 2, 3, \ldots\}$. The elements of $\mathbb{Z}$ are called integers or numbers. The integers greater than 0 are called natural numbers and are denoted by $\mathbb{N}$. The sum $n+m$ and the product $n \cdot m$ of integers are defined. Addition and multiplication satisfy the axioms of a commutative ring with a unit element. We call $\mathbb{Z}$ the ring of integers.

Addition, Multiplication and Exponentiation. Efficient algorithms exist for the addition and multiplication of numbers.[2] An efficient algorithm is an algorithm whose running time is bounded by a polynomial in the size of its

---

[2] Simplifying slightly, we only consider non-negative integers, which is sufficient for all our purposes.

input. The size of a number is the length of its binary encoding, i.e., the size of $n \in \mathbb{N}$ is equal to $\lfloor \log_2(n) \rfloor + 1$.[3] It is denoted by $|n|$. Let $a, b \in \mathbb{N}$, $a, b \leq n$, and $k := \lfloor \log_2(n) \rfloor + 1$. The number of bit operations for the computation of $a + b$ is $O(k)$, whereas for the multiplication $a \cdot b$ it is $O(k^2)$. Multiplication can be improved to $O(k \log_2(k))$ if a fast multiplication algorithm is used.

Exponentiation is also an operation that occurs often. The repeated squaring method (Algorithm A.26) yields an efficient algorithm for the computation of a^n. It requires at most $2 \cdot |n|$ modular multiplications. We compute, for example, a^{16} by $(((a^2)^2)^2)^2$, which are four squarings, in contrast to the 15 multiplications that are necessary for the naive method. If the exponent is not a power of 2, the computation has to be modified a little. For example, $a^{14} = ((a^2 a)^2 a)^2$ is computable by three squarings and two multiplications, instead of by 13 multiplications.

Division with Remainder. If m and n are integers, $m \neq 0$, we can divide n by m with a remainder. We can write $n = q \cdot m + r$ in a unique way such that $0 \leq r < \operatorname{abs}(m)$.[4] The number q is called the *quotient* and r is called the *remainder* of the division. They are unique. Often we denote r by $a \bmod b$.

An integer m divides an integer n if n is a multiple of m, i.e., $n = mq$ for an integer q. We say, m is a *divisor* or *factor* of n. The *greatest common divisor* $\gcd(m, n)$ of numbers $m, n \neq 0$ is the largest positive integer dividing m and n. $\gcd(0, 0)$ is defined to be zero. If $\gcd(m, n) = 1$, then m is called *relatively prime to* n, or *prime to* n for short.

The *Euclidean algorithm* computes the greatest common divisor of two numbers and is one of the oldest algorithms in mathematics:

Algorithm 3.1.
```
int gcd(int a, b)
  1   while b ≠ 0 do
  2       r ← a mod b
  3       a ← b
  4       b ← r
  5   return abs(a)
```

The algorithm computes $\gcd(a, b)$ for $a \neq 0$ and $b \neq 0$. It terminates because the non-negative number r decreases in each step. Note that $\gcd(a, b)$ is invariant in the while loop, because $\gcd(a, b) = \gcd(b, a \bmod b)$. In the last step, the remainder r becomes 0 and we get $\gcd(a, b) = \gcd(a, 0) = \operatorname{abs}(a)$.

Primes and Factorization. A natural number $p \neq 1$ is a *prime number*, or simply a *prime*, if 1 and p are the only divisors of p. If a number $n \in \mathbb{N}$ is not prime, it is called *composite*. Primes are essential for setting up the public-key cryptosystems we describe in this chapter. Fortunately, there are very fast algorithms (so-called probabilistic primality tests) for finding – at

---

[3] $\lfloor x \rfloor$ denotes the greatest integer less than or equal to x.
[4] $\operatorname{abs}(m)$ denotes the absolute value of m.

least with a high probability (though not with mathematical certainty) – the correct answer to the question whether a given number is prime or not (see Appendix A.8). Primes are the basic building blocks for numbers. This statement is made precise by the *Fundamental Theorem of Arithmetic*.

Theorem 3.2 *(Fundamental Theorem of Arithmetic). Let $n \in \mathbb{N}, n \geq 2$. There exist pairwise distinct primes $p_1, \ldots, p_k$ and exponents $e_1, \ldots, e_k \in \mathbb{N}, e_i \geq 1, i = 1, \ldots, k$, such that*

$$n = \prod_{i=1}^{k} p_i^{e_i}.$$

The primes $p_1, \ldots, p_k$ and exponents $e_1, \ldots, e_k$ are unique.

It is easy to multiply two numbers, but the design of an efficient algorithm for calculating the prime factors of a number is an old mathematical problem. For example, this was already studied by the famous mathematician C.F. Gauß about 200 years ago (see, e.g., [Riesel94] for details on Gauß' factoring method). However, to this day, we do not have a practical algorithm for factoring extremely large numbers.

3.2.2 The Integers Modulo n

The Residue Class Ring Modulo n. Let n be a positive integer. Let a and b be integers. Then a is *congruent to b modulo n*, written $a \equiv b \bmod n$, if a and b leave the same remainder when divided by n or, equivalently, if n divides $a - b$. We obtain an equivalence relation. The equivalence class of a is the set of all numbers congruent to a. It is denoted by $[a]$ and called the *residue class* of a modulo n. The set of residue classes $\{[a] \mid a \in \mathbb{Z}\}$ is called the set of *integers modulo n* and is denoted by $\mathbb{Z}_n$.

Each number is congruent to a unique number r in the range $0 \leq r \leq n - 1$. Therefore the numbers $0, \ldots, n - 1$ form a set of representatives of the elements of $\mathbb{Z}_n$. We call them the *natural representatives*.

The equivalence relation is compatible with addition and multiplication in $\mathbb{Z}$, i.e., if $a \equiv a' \bmod n$ and $b \equiv b' \bmod n$, then $a + b \equiv (a' + b') \bmod n$ and $a \cdot b \equiv (a' \cdot b') \bmod n$. Consequently, addition and multiplication on $\mathbb{Z}$ induce an addition and multiplication on $\mathbb{Z}_n$:

$$[a] + [b] := [a + b],$$
$$[a] \cdot [b] := [a \cdot b].$$

Addition and multiplication satisfy the axioms of a commutative ring with a unit element. We call $\mathbb{Z}_n$ the *residue class ring modulo n*.

Although we can calculate in $\mathbb{Z}_n$ as in $\mathbb{Z}$, there are some important differences. First, we do not have an ordering of the elements of $\mathbb{Z}_n$ which is compatible with addition and multiplication. For example, if we assume that we have such an ordering in $\mathbb{Z}_5$ and that $[0] < [1]$, then $[0] < [1]+[1]+[1]+[1]+[1] = [0]$, which is a contradiction. A similar calculation shows that the assumption $[1] < [0]$ also leads to a contradiction.

Another fact is that $[a] \cdot [b]$ can be $[0]$ for $[a] \neq [0]$ and $[b] \neq [0]$. For example, $[2] \cdot [3] = [0]$ in $\mathbb{Z}_6$. Such elements – $[a]$ and $[b]$ – are called *zero divisors*.

The Prime Residue Class Group Modulo n. In $\mathbb{Z}$, elements a and b satisfy $a \cdot b = 1$ if and only if both a and b are equal to 1, or both are equal to -1. We say that 1 and -1 have multiplicative inverse elements. In $\mathbb{Z}_n$, this can happen more frequently. In $\mathbb{Z}_5$, for example, every class different from $[0]$ has a multiplicative inverse element. Elements in a ring which have multiplicative inverses are called *units* and form a group under multiplication.

An element $[a]$ in $\mathbb{Z}_n$ has the multiplicative inverse element $[b]$, if $ab \equiv 1 \bmod n$ or, equivalently, n divides $1 - ab$. This means we have an equation $nm + ab = 1$, with suitable m. The equation implies that $\gcd(a, n) = 1$. On the other hand, if numbers a, n with $\gcd(a, n) = 1$ are given, an equation $nm + ab = 1$, with suitable b and m, can be derived from a and n by the extended Euclidean algorithm (Algorithm A.5). Hence, $[a]$ is a unit in $\mathbb{Z}_n$ and the inverse element is $[b]$. Thus, the elements of the group of units of $\mathbb{Z}_n$ are represented by the numbers prime to n.

$$\mathbb{Z}_n^* := \{[a] \mid 1 \leq a \leq n - 1 \text{ and } \gcd(a, n) = 1\}$$

is called the *prime residue class group modulo n*. The number of elements in $\mathbb{Z}_n^*$ (also called the order of $\mathbb{Z}_n^*$) is the number of integers in the interval $[1, n-1]$ which are prime to n. This number is denoted by $\varphi(n)$. The function φ is called the *Euler phi function* or the *Euler totient function*.

For every element a in a finite group G, we have $a^{|G|} = e$, with e being the neutral element of G.[5] This is an elementary and easy to prove feature of finite groups. Thus, we have for a number a prime to n

$$a^{\varphi(n)} \equiv 1 \bmod n.$$

This is called *Euler's Theorem* or, if n is a prime, *Fermat's Theorem*.

If $\prod_{i=1}^{k} p_i^{e_i}$ is the prime factorization of n, then the Euler phi function can be computed by the formula (see Corollary A.30)

$$\varphi(n) = n \prod_{i=1}^{k} \left(1 - \frac{1}{p_i}\right).$$

If p is a prime, then every integer in $\{1, \ldots, p - 1\}$ is prime to p. Therefore, every element in $\mathbb{Z}_p \setminus \{0\}$ is invertible and $\mathbb{Z}_p$ is a field. The group of units $\mathbb{Z}_p^*$ is a cyclic group with $p - 1$ elements, i.e., $\mathbb{Z}_p^* = \{g, g^2, \ldots, g^{p-1} = [1]\}$

---

[5] $|G|$ denotes the number of elements of G (called the *cardinality* or *order* of G).

for some $g \in \mathbb{Z}_{p-1}$. Such a g is called a *generator* of $\mathbb{Z}_p^*$. Generators are also called *primitive elements modulo p* or *primitive roots modulo p* (see Definition A.37).

We can now introduce three functions which may be used as one-way functions and which are hence very important in cryptography.

Discrete Exponentiation. Let p denote a prime number and g be a primitive root in $\mathbb{Z}_p$.

$$\text{Exp} : \mathbb{Z}_{p-1} \longrightarrow \mathbb{Z}_p^*, \ x \longmapsto g^x$$

is called the discrete exponential function. Exp is a homomorphism from the additive group $\mathbb{Z}_{p-1}$ to the multiplicative group $\mathbb{Z}_p^*$, i.e., $\text{Exp}(x+y) = \text{Exp}(x) \cdot \text{Exp}(y)$, and Exp is bijective. In other words, Exp is an isomorphism of groups. This follows immediately from the definition of a primitive root. The inverse function

$$\text{Log} : \mathbb{Z}_p^* \longrightarrow \mathbb{Z}_{p-1}$$

is called the *discrete logarithm function*. We use the adjective "discrete" to distinguish Exp and Log for finite groups from the classical functions defined for the reals.

Exp is efficiently computable, for example by the repeated squaring method (see Section 3.2.1), whereas no efficient algorithm is known to exist for computing the inverse function Log for sufficiently large primes p. This statement is made precise by the discrete logarithm assumption (see Definition 6.1).

Modular Powers. Let n denote the product of two distinct primes p and q and let e be prime to $\varphi(n)$.

$$\text{RSA}_e : \mathbb{Z}_n \longrightarrow \mathbb{Z}_n, \ x \longmapsto x^e$$

is called the *RSA function*.

Proposition 3.3. *Using the same notation as above, let d be a multiplicative inverse element of e modulo $\varphi(n)$ (note that d is also prime to $\varphi(n)$ and RSA_d is defined). Then*

$$\text{RSA}_d \circ \text{RSA}_e = \text{RSA}_e \circ \text{RSA}_d = \text{id}_{\mathbb{Z}_n}.$$

Proof. We show $x^{ed} = x$, for $x \in \mathbb{Z}_n$. First let $x \in \mathbb{Z}_n^*$. The group $\mathbb{Z}_n^*$ has order $\varphi(n)$, hence $x^{\varphi(n)} = [1]$ and therefore $x^{ed} = x^{ed \bmod \varphi(n)} = x$. In the case $x \notin \mathbb{Z}_n^*$, p or q is a factor of x. If both divide x, we have $x = 0$ and $x^{ed} = 0$. Thus the equalities hold. Observe that $\varphi(n) = (p-1)(q-1)$ (see Corollary A.30). If p divides x and q does not divide x, then $(x^e)^d \bmod p = 0$, $x \bmod p = 0$ and $(x^e)^d \equiv x^{ed \bmod (q-1)} \equiv x \bmod q$, because $ed \equiv 1 \bmod (q-1)$. This shows that $(x^e)^d \equiv x \bmod n$. The case where p does not divide x and q divides x

follows analogously. Thereby $(x^e)^d = x^{ed} = x$ for all $x \in \mathbb{Z}_n$, and we have proven our assertion.   □

We see that RSA_e is an (easily computable) permutation of $\mathbb{Z}_n$. Knowing d, it is also easy to compute the inverse, which is simply RSA_d. However, if d is a secret, it is believed to be infeasible to invert RSA_e (provided that p and q are very large).

Modular Squares. Let p and q denote distinct prime numbers and $n = pq$.

$$\text{Square} : \mathbb{Z}_n \longrightarrow \mathbb{Z}_n, \ x \longmapsto x^2$$

is called the *Square function*. Each element $y \in \mathbb{Z}_n, y \neq 0$, either has 0, 2 or 4 pre-images. If both p and q are primes $\equiv 3 \bmod 4$, then -1 is not a square modulo p and modulo q, and it easily follows that Square becomes a bijective map by restricting the domain and range to the subset of squares in $\mathbb{Z}_n^*$ (for details see Appendix A.6). If the factors of n are known, pre-images of Square (called *square roots*) are efficiently computable (see Proposition A.62). Again, without knowing the factors p and q, computing square roots is practically impossible for p, q sufficiently large.

On the Difficulty of Extracting Roots and Discrete Logarithms. Let
$g, k \in \mathbb{N}, g, k \geq 2$, and let

$$F : \mathbb{Z} \longrightarrow \mathbb{Z}$$

denote one of the maps $x \longmapsto x^2, x \longmapsto x^k$ or $x \longmapsto g^x$. Usually these maps are considered as real functions. Efficient algorithms for the computation of values and pre-images are known. These algorithms rely heavily on the ordering of the reals and the fact that F is, at least piecewise, monotonic. The map F is efficiently computable by integer arithmetic with a fast exponentiation algorithm (see Section 3.2.1). If we use such an algorithm to compute F, we get the following algorithm for computing a pre-image x of $y = F(x)$.

For simplicity, we restrict the domain to the positive integers. Then all three functions are monotonic and hence injective. It is easy to find integers a and b with $a \leq x \leq b$. If $F(a) \neq y$ and $F(b) \neq y$, we call $FInvers(y, a, b)$.

Algorithm 3.4.
```
   int FInvers(int y, a, b)
   1  repeat
   2      c ← (a + b) div 2
   3      if F(c) < y
   4          then a ← c
   5          else  b ← c
   6  until F(c) = y
   7  return c
```

F is efficiently computable. The repeat-until loop terminates after $O(\log_2(|b-a|))$ steps. Hence, *FInvers* is also efficiently computable, and we see that F considered as a function from $\mathbb{Z}$ to $\mathbb{Z}$ can easily be inverted in an efficient way.

Now consider the same maps modulo n. The function F is still efficiently computable (see above or Algorithm A.26). The algorithm *FInvers*, however, does not work in modular arithmetic. The reason is that in modular arithmetic, it does not make sense to ask in line 3 whether $F(c) < y$. The ring $\mathbb{Z}_n$ has no order which is compatible with the arithmetic operations. The best we could do to adapt the algorithm above for modular arithmetic is to test all elements c in $\mathbb{Z}_n$ until we reach $F(c) = y$. However, this leads to an algorithm with exponential running time (n is exponential in the binary length $|n|$ of n).

If the factors of n are kept secret, no efficient algorithm is known today to invert RSA and Square. The same holds for Exp. It is widely believed that no efficient algorithms exist to compute the pre-images. No one could prove, however, this statement in the past. These assumptions are defined in detail in Chapter 6. They are the basis for security proofs in public-key cryptography (Chapters 9 and 10).

3.3 RSA

The RSA cryptosystem is based on facts from elementary number theory which have been known for 250 years. To set up an RSA cryptosystem, we have to multiply two very large primes and make their product n public. n is part of the public key, whereas the factors of n are kept secret and are used as the secret key. The basic idea is that the factors of n cannot be recovered from n. In fact, the security of the RSA encryption function depends on the tremendous difficulty of factoring, but the equivalence is not proven.

We now describe in detail how RSA works. We discuss key generation, encryption and decryption as well as digital signatures.

3.3.1 Key Generation and Encryption

Key Generation. Each user Alice of the RSA cryptosystem has her own public and secret keys. The key generation algorithm proceeds in three steps (see also Section 6.4):

1. Choose large distinct primes p and q, and compute $n = p \cdot q$.
2. Choose e that is prime to $\varphi(n)$. The pair (n, e) is published as the public key.
3. Compute d with $ed \equiv 1 \bmod \varphi(n)$. (n, d) is used as the secret key.

Recall that $\varphi(n) = (p-1)(q-1)$ (Corollary A.30). The numbers n, e and d are referred to as the *modulus*, and *encryption* and *decryption exponents*, respectively. To decrypt a ciphertext or to generate a digital signature, Alice only needs her decryption exponent d, she does not need to know the primes p and q. Nevertheless, knowing p and q can be helpful for her (e.g. to speed up decryption, see below). At any time, Alice can derive the primes from n, e and d by an efficient algorithm with very high probability (see Exercise 4).

An adversary should not have the slightest idea what Alice's primes are. Therefore, we proceed as follows to get the primes p and q. First we choose a large number x at random. If x is even, we replace x by $x + 1$ and apply a probabilistic primality test to check whether x is a prime (see Appendix A.8). If x is not a prime number, we replace x with $x + 2$, and so on until the first prime is reached. We expect to test $O(\ln(x))$ numbers for primality before reaching the first prime (see Corollary A.69). The method described does not produce primes with mathematical certainty (we use a probabilistic primality test), but it is sufficient for practical purposes. At the moment, it is suggested to take 512-bit prime numbers. No one can predict for how long such numbers will be secure, because it is difficult to predict improvements in factorization algorithms and computer technology.

The number e can also be chosen at random. Whether e is prime to $\varphi(n)$ is tested with Euclid's algorithm (Algorithm A.4). Another method for obtaining e is to choose a prime between $\max(p, q)$ and $\varphi(n)$ which guarantees that it will be relatively prime to $\varphi(n)$. We can do this in the same way as choosing p and q. The number d can be computed with the extended Euclidean algorithm (Algorithm A.5).

To choose a number at random, we may use a pseudorandom number generator. This is an algorithm that generates a sequence of digits which look like a sequence of random digits. There is a wide array of literature concerning efficient and secure generation of pseudorandom numbers (see, e.g., [MenOorVan96]). We also discuss the subject in Chapter 8.

Encryption and Decryption. We encrypt messages in $\{0, \ldots, n-1\}$, considered as elements of $\mathbb{Z}_n$.

1. The encryption function is defined by

$$E : \mathbb{Z}_n \longrightarrow \mathbb{Z}_n, \ x \longmapsto x^e.$$

2. The decryption function is of the same type, and is defined by

$$D : \mathbb{Z}_n \longrightarrow \mathbb{Z}_n, \ x \longmapsto x^d.$$

E and D are bijective maps and inverse to each other: $E \circ D = D \circ E = \mathrm{id}_{\mathbb{Z}_n}$ (see Proposition 3.3). Encryption and decryption can be implemented using an efficient algorithm (Algorithm A.26).

With the basic encryption procedure, we can encrypt bit sequences up to $k := \lfloor \log_2(n) \rfloor$ bits. If our messages are longer, we may decompose them into blocks of length k and apply the scheme described in Section 3.3.4 or a suitable mode of operation, for example the electronic codebook mode or the cipher-block chaining mode (see Section 2.2.3). The cipher feedback mode and the output feedback mode are not immediately applicable. Namely, if the initial value is not kept secret everyone can decrypt the cryptogram. See Chapter 9 for an application of the output feedback mode with RSA.

Security. An adversary knowing the factors p and q of n also knows $\varphi(n) = (p-1)(q-1)$, and then derives d from the public encryption key e using the extended Euclidean algorithm (Algorithm A.5). Thus, the security of RSA depends on the difficulty of finding the factors p and q of n, if p and q are large primes. It is widely believed that it is impossible today to factor n by an efficient algorithm if p and q are sufficiently large. This fact is known as the factoring assumption (for a precise definition, see Definition 6.9). An efficient factoring algorithm would break RSA. It is not proven whether factoring is necessary to decrypt RSA ciphertexts, but it is also believed that inverting the RSA function is intractable. This statement is made precise by the RSA assumption (see Definition 6.7).

In the construction of the RSA keys, the only inputs for the computation of the exponents e and d are $\varphi(n)$ and n. Since $\varphi(n) = (p-1)(q-1)$, we have

$$p + q = n - \varphi(n) + 1 \text{ and } p - q = \sqrt{(p+q)^2 - 4n} \text{ (if } p > q).$$

Therefore, it is easy to compute the factors of n if $\varphi(n)$ is known.

The factorization of n can be reduced to an algorithm A that computes d from n and e (see Exercise 4). The resulting factoring algorithm A' is probabilistic and of the same complexity as A. (This fact was already mentioned in [RivShaAdl78]).

It is an open question as to whether an efficient algorithm for factoring can be derived from an efficient algorithm inverting RSA, i.e., an efficient algorithm that on inputs n, e and x^e outputs x. A result of Boneh and Venkatesan (see [BonVen98]) provides evidence that, for a small encryption exponent e, inverting the RSA function might be easier than factoring n. They show that an efficient factoring algorithm A which uses, as a subroutine, an algorithm for computing e-th roots – called oracle for e-th roots – can be converted into an efficient factoring algorithm B which does not call the oracle. However, they use a restricted computation model for the algorithm A – only algebraic reductions are allowed. Their result says that factoring is easy, if an efficient algorithm like A exists in the restricted computational model. The result of Boneh and Venkatesan does not expose any weakness in the RSA cryptosystem.

The decryption exponent d should be greater than $n^{1/4}$. For $d < n^{1/4}$, a polynomial-time algorithm to compute d has been developed ([Wiener90]). The algorithm uses the continued fraction expansion of e/n.

Efficient factoring algorithms are known for special types of primes p and q. To give these algorithms no chance, we have to avoid such primes. First we require that the absolute value $|p-q|$ is large. This prevents the following attack: We have $(p+q)^2/4 - n = (p+q)^2/4 - pq = (p-q)^2/4$. If $|p-q|$ is small, then $(p-q)^2/4$ is also small and therefore $(p+q)^2/4$ is slightly larger than n. Thus $p+q/2$ is slightly larger than $\sqrt{n}$ and the following factoring method could be successful:

1. Choose successive numbers $x > \sqrt{n}$ and test whether $x^2 - n$ is a square.
2. In this case, we have $x^2 - n = y^2$. Thus $x^2 - y^2 = (x-y)(x+y) = n$, and we have found a factorization of n.

This idea for factoring numbers goes back to Fermat.

To prevent other attacks on the RSA cryptosystem, the notion of *strong primes* has been defined. A prime number is called strong if the following conditions are satisfied:

1. $p-1$ has a large prime factor, denoted by r.
2. $p+1$ has a large prime factor.
3. $r-1$ has a large prime factor.

What "large" means can be derived from the attacks to prevent (see below). Strong primes can be generated by Gordon's algorithm (see [Gordon84]). If used in conjunction with a probabilistic primality test, the running time of Gordon's algorithm is only about 20% more than the time needed to generate a prime factor of the RSA modulus in the way described above. Gordon's algorithm yields a prime with high probability. The size of the resulting prime p can be controlled to guarantee a large absolute value $|p-q|$.

Strong primes are intended to prevent the $p-1$ and the $p+1$ factoring attacks. These are efficient if $p-1$ or $p+1$ have only small prime factors (see, e.g., [Forster96]; [Riesel94]). Note that $p-1$ and $p+1$ can be expected to have a large prime factor if the prime p is chosen large and at random. Moreover, choosing strong primes does not increase the protection against factoring attacks with a modern algorithm like the number-field sieve (see [Cohen95]). Thus, the notion of strong primes has lost significance.

There is another attack which should be prevented by strong primes: decryption by iterated encryption. The idea is to repeatedly apply the encryption algorithm to the cryptogram until $c = c^{e^i}$. Then $c = \left(c^{e^{i-1}}\right)^e$, and the plaintext $m = c^{e^{i-1}}$ can be recovered. Condition 1 and 3 ensure that this attack fails, since the order of c in $\mathbb{Z}_n^*$ and the order of e in $\mathbb{Z}_{\varphi(n)}^*$ are, with high probability, very large (see Exercises 6 and 7). If p and q are chosen at random and are sufficiently large, then the probability of success of a decryption-by-iterated-encryption attack is negligible (see [MenOorVan96], p. 313). Thus, to prevent this attack, there is no compelling reason for choosing strong primes, too.

Speeding Up Encryption and Decryption. The modular exponentiation algorithm is especially efficient, if the exponent has many zeros in its binary encoding. For each zero we have one less multiplication. We can take advantage of this fact by choosing an encryption exponent e with many zeros in its binary encoding. The primes 3, 17 or $2^{16} + 1$ are good examples, with only two ones in their binary encoding.

The efficiency of decryption can be improved by use of the Chinese Remainder Theorem (Theorem A.29). The receiver of the message knows the factors p and q of the modulus n. Let ϕ be the isomorphism

$$\phi : \mathbb{Z}_n \longrightarrow \mathbb{Z}_p \times \mathbb{Z}_q, \quad [x] \longmapsto ([x \bmod p], [x \bmod q]).$$

Compute $c^d = \phi^{-1}(\phi(c^d)) = \phi^{-1}((c \bmod p)^d, (c \bmod q)^d)$. The computation of $(c \bmod p)^d$ and $(c \bmod q)^d$ is executed in $\mathbb{Z}_p$ and $\mathbb{Z}_q$, respectively. In $\mathbb{Z}_p$ and $\mathbb{Z}_q$ we have much smaller numbers than in $\mathbb{Z}_n$. Moreover, the decryption exponent d can be replaced by $d \bmod (p-1)$ and $d \bmod (q-1)$, respectively, since $(c \bmod p)^d \bmod p = (c \bmod p)^{d \bmod (p-1)} \bmod p$ and $(c \bmod q)^d \bmod q = (c \bmod q)^{d \bmod (q-1)} \bmod q$ (by Proposition A.24, also see "Computing modulo a prime" on page 303).

3.3.2 Digital Signatures

The RSA cryptosystem may also be used for digital signatures. Let (n, e) be the public key and d be the secret decryption exponent of Alice. We first discuss signing messages that are encoded by numbers $m \in \{0, \ldots, n-1\}$. As usual, we consider those numbers as the elements of $\mathbb{Z}_n$, and the computations are done in $\mathbb{Z}_n$.

Signing. If Alice wants to sign a message m, she uses her secret key and computes her *signature* $\sigma = m^d$ of m by applying her decryption algorithm. We call (m, σ) a *signed message*.

Verification. Assume that Bob received a signed message (m, σ) from Alice. To verify the signature, Bob uses the public key of Alice and computes σ^e. He accepts the signature if $\sigma^e = m$.

If Alice signed the message, we have $(m^d)^e = (E \circ D)(m) = m$ and Bob accepts (see Proposition 3.3). However, the converse is not true. It might happen that Bob accepts a signature not produced by Alice. Suppose Eve uses Alice's public key, computes m^e and says that (m^e, m) is a message signed by Alice. Everyone verifying Alice's signature gets $m^e = m^e$ and is convinced that Alice really signed the message m^e. The message m^e is not likely to be meaningful if the message belongs to some natural language. This example shows that RSA signatures can be *existentially forged*. This means that an adversary can forge a signature for some message, but not for a message of his choice.

Another attack uses the fact that the RSA encryption and decryption functions are *ring homomorphisms* (see Appendix A.3). The image of a product is the product of the images and the image of the unit element is the unit element. If Alice signed m_1 and m_2, then the signatures for $m_1 m_2$ and m_1^{-1} are $\sigma_1 \sigma_2$ and σ_1^{-1}. These signatures can easily be computed without the secret key. We will now discuss how to overcome these difficulties.

Signing with Redundancy and Hash Functions. If the messages to be signed belong to some natural language, it is very unlikely that the above attacks will succeed. The messages m^e and $m_1 m_2$ will rarely be meaningful. When embedding the messages into $\{0,1\}^*$, the message space is sparse. The probability that a randomly chosen bit string belongs to the message space is small. By adding redundancy to each message we can always guarantee that the message space is sparse, even if arbitrary bit strings are admissible messages. A possible redundancy function is

$$R : \{0,1\}^* \longrightarrow \{0,1\}^*, \ x \longmapsto x \| x.$$

This principle is also used in error-detection and error-correction codes. Doubling the message we can detect transmission errors if the first half of the transmitted message does not match the second half. The redundancy function R has the additional advantage that the composition of R with the RSA function no longer preserves products.

If the message does not need to be recovered from the signature, another approach to prevent the attacks is the use of a hash function (see Section 3.4).

3.3.3 Attacks Against RSA

We now describe attacks not primarily directed against the RSA algorithm itself, but against the environment in which the RSA cryptosystem is used.

The Common-Modulus Attack. Suppose two users Bob and Bridget of the RSA cryptosystem have the same modulus n. Let (n, e_1) be the public key of Bob and (n, e_2) be the public key of Bridget, and assume that e_1 and e_2 are relatively prime. Let $m \in \mathbb{Z}_n$ be a message sent to both Bob and Bridget and encrypted as $c_i = m^{e_i}$, $i = 1, 2$. The problem now is that the plaintext can be computed from c_1, c_2, e_1, e_2 and n. Since e_1 is prime to e_2, integers r and s with $re_1 + se_2 = 1$ can be derived by use of the extended Euclidean algorithm (see Algorithm A.5). Either r or s, say r, is negative. If $c_1 \notin \mathbb{Z}_n^*$, we can factor n by computing $\gcd(c_1, n)$, thereby breaking the cryptosystem. Otherwise we again apply the extended Euclidean algorithm and compute c_1^{-1}. We can recover the message m using $(c_1^{-1})^{-r} c_2^s = (m^{e_1})^r (m^{e_2})^s = m^{re_1 + se_2} = m$. Thus, the cryptosystem fails to protect a message m if it is sent to two users with common modulus whose encryption exponents are relatively prime.

With common moduli, secret keys can be recovered. If Bob and Bridget have the same modulus n, then Bob can determine Bridget's secret key.

Namely, either Bob already knows the prime factors of n or he can compute them from his encryption and decryption exponents, with a very high probability (see Exercise 4). Therefore, common moduli should be avoided in RSA cryptosystems – each user should have his own modulus. If the prime factors (and hence the modulus) are randomly chosen, as described above, then the probability that two users share the same modulus is negligibly small.

Low-Encryption-Exponent Attack. Suppose that the RSA cryptosystem will be used for k users, and each user has a small encryption exponent. We discuss the case of three users, Bob, Bridget and Bert, with public keys $(n_i, 3)$, $i = 1, 2, 3$. Of course, the moduli n_i and n_j must satisfy $\gcd(n_i, n_j) = 1$, for $i \neq j$, since otherwise factoring of n_i and n_j is possible by computing $\gcd(n_i, n_j)$. We assume that Alice sends the same message m to Bob, Bridget and Bert. The following attack is possible: let $c_1 := m^3 \bmod n_1, c_2 := m^3 \bmod n_2$ and $c_3 := m^3 \bmod n_3$. The inverse of the Chinese Remainder isomorphism (see Theorem A.29)

$$\phi : \mathbb{Z}_{n_1 n_2 n_3} \longrightarrow \mathbb{Z}_{n_1} \times \mathbb{Z}_{n_2} \times \mathbb{Z}_{n_3}$$

can be used to compute $m^3 \bmod n_1 n_2 n_3$. Since $m^3 < n_1 n_2 n_3$, we can get m by computing the ordinary cube root of m^3 in $\mathbb{Z}$.

Small-Message-Space Attack. If the number of all possible messages is small, and if these messages are known in advance, an adversary can encrypt all messages with the public key. He can decrypt an intercepted cryptogram by comparing it with the precomputed cryptograms.

A Chosen-Ciphertext Attack. In the first phase of a chosen-ciphertext attack against an encryption scheme (see Section 1.3), adversary Eve has access to the decryption device. She obtains the plaintexts for ciphertexts of her choosing. Then, in the second phase, she attempts to decrypt another ciphertext for which she did not request decryption in the first phase.

Basic RSA encryption does not resist the following chosen-ciphertext attack. Let (n, e) be Bob's public RSA key, and let c be any ciphertext, encrypted with Bob's public key.

To find the plaintext m of c, adversary Eve first chooses a random unit $r \in \mathbb{Z}_n^*$ and requests the decryption of the random-looking message

$$r^e c \bmod n.$$

She obtains the plaintext $\tilde{m} = rm \bmod n$. Then, in the second phase of the attack, Eve easily derives the plaintext m, because we have in $\mathbb{Z}_n$

$$r^{-1}\tilde{m} = r^{-1}rm = m.$$

The attack relies on the fact that the RSA function is a ring isomorphism.

Analogously, there is a chosen-plaintext attack against digital signatures[6] generated with basic RSA (i.e., RSA without a hash function). Eve can successfully forge Bob's signature for a message m, if she is first supplied with a valid signature for $r^e m \bmod n$, where r is randomly chosen by Eve.

A setting in which the chosen-ciphertext attack against RSA encryption may work is described in the following attack.

Attack on Encryption and Signing with RSA. The attack is possible if Bob has only one public-secret RSA key pair and uses this key pair for both encryption and digital signatures. Assume that the cryptosystem is also used for mutual authentication. On request, Bob proves his identity to Eve by signing a random number, supplied by Eve, with his secret key. Eve verifies the signature of Bob with his public key. In this situation, Eve can successfully attack Bob as follows. Suppose Eve intercepts a ciphertext c intended for Bob:

1. Eve selects $r \in \mathbb{Z}_n^*$ at random.
2. Eve computes $x = r^e c \bmod n$, where (n, e) is the public key of Bob. She sends x to Bob to get a signature x^d (d the secret key of Bob). Note that x looks like a random number to Bob.
3. Eve computes $r^{-1} x^d$, which is the plaintext of c.

Bleichenbacher's 1-Million-Chosen-Ciphertext Attack. The 1-Million-Chosen-Ciphertext Attack of Bleichenbacher ([Bleichenbacher98]) is an attack against PKCS#1 v1.5 ([RFC 2313]).[7] The widely used PKCS#1 is part of the Public-Key Cryptography Standards series PKCS. These are de facto standards that are developed and published by RSA Security ([RSALabs]) in conjunction with system developers worldwide. The RSA standard PKCS#1 defines mechanisms for encrypting and signing data using the RSA public key system. We explain Bleichenbacher's attack against encryption.

Let (n, e) be a public RSA key with encryption exponent e and modulus n. The modulus n is assumed to be a k-byte integer, i.e., $256^{k-1} < n < 256^k$. PKCS#1 defines a padding format. Messages (which are assumed to be shorter than k bytes) are padded out to obtain a formatted plaintext block m consisting of k bytes, and this plaintext block is then encrypted by using the RSA function. The ciphertext is $m^e \bmod n$, as usual. The first byte of the plaintext block m is 00, and the second byte is 02 (in hexadecimal notation). Then a padding string follows. It consists of at least 8 randomly chosen bytes, different from 00. The end of the padding block is marked by the zero byte 00. Then the original message bytes are appended. After the padding, we get

$$m = 00\|02\|\text{padding string}\|00\|\text{original message}.$$

The leading 00-byte ensures that the plaintext block, when converted to an integer, is less than the modulus.

---

[6] A detailed discussion of types of attacks against digital signature schemes is given in Section 10.1.

[7] PKCS#1 has been updated. The current version is Version 2.1 ([RFC 3447]).

We call a k-byte message m *PKCS conforming*, if it has the above format. A message $m \in \mathbb{Z}$ is PKCS conforming, if and only if

$$2B \leq m \leq 3B - 1,$$

with $B = 256^{k-2}$.

Adversary Eve wants to decrypt a ciphertext c. The attack is an adaptively-chosen-ciphertext attack (see Section 1.3). Eve chooses ciphertexts $c_1, c_2, \ldots$, different from c, and gets information about the plaintexts from a "decryption oracle" (imagine that Eve can supply ciphertexts to the decryption device and obtain some information on the decryption results). Adaptively means that Eve can choose a ciphertext, get information about the corresponding plaintext and do some analysis. Depending on the results of her analysis, she can choose a new ciphertext, and so on. With the help of the oracle, she computes the plaintext m of the ciphertext c. If Eve does not get the full plaintexts of the ciphertexts $c_1, c_2, \ldots$, as in Bleichenbacher's attack, such an attack is also called, more precisely, a *partial chosen-ciphertext attack.*

In Bleichenbacher's attack, on input of a ciphertext, the decryption oracle answers whether the corresponding plaintext is PKCS conforming or not. There were implementations of the SSL/TLS-protocol that contained such an oracle in practice (see below).

Eve wants to decrypt a ciphertext c which is the encryption of a PKCS conforming message m. She successively constructs intervals $[a_i, b_i] \subset \mathbb{Z}, i = 0, 1, 2, \ldots$, which all contain m and become shorter in each step, usually by a factor of 2. Eve finds m as soon as the interval has become sufficiently small and contains only one integer.

Eve knows that m is PKCS conforming and hence in $[2B, 3B - 1]$. Thus, she starts with the interval $[a_0, b_0] = [2B, 3B - 1]$. In each step, she chooses integers s, computes the ciphertext

$$\tilde{c} := s^e c \bmod n$$

of $sm \bmod n$ and queries the oracle with input $\tilde{c}$. The oracle outputs whether

$$sm \bmod n$$

is PKCS conforming or not. Whenever $sm \bmod n$ is PKCS conforming, Eve can narrow the interval $[a, b]$. The choice of the multipliers s depends on the output of the previous computations. So, the ciphertexts $\tilde{c}$ are chosen adaptively.

We now describe the attack in more detail.

Let $[a_0, b_0] = [2B, 3B - 1]$. We have $m \in [a_0, b_0]$, and the length of $[a_0, b_0]$ is $B - 1$.

Step 1: Eve searches for the smallest integer $s_1 > 1$, such that $s_1 m \bmod n$ is PKCS conforming. Since $2m \geq 4B$, the residue $s_1 m \bmod n$ can be PKCS conforming, only if $s_1 m \geq n + 2B$. Therefore, Eve can start her search with $s_1 \geq \lceil n + 2B/3B - 1 \rceil$.

We have $s_1 m \in [s_1 a_0, s_1 b_0]$. If $s_1 m \bmod n$ is PKCS conforming, then

$$a_0 + tn \leq s_1 m \leq b_0 + tn$$

for some $t \in \mathbb{N}$ with $s_1 a_0 \leq b_0 + tn$ and $a_0 + tn \leq s_1 b_0$.

This means that m is contained in one of the intervals

$$[a_{1,t}, b_{1,t}] := [a_0, b_0] \cap \left[a_0 + tn/s_1, b_0 + tn/s_1 \right],$$

with $\lceil s_1 a_0 - b_0/n \rceil \leq t \leq \lfloor s_1 b_0 - a_0/n \rfloor$. We call the intervals $[a_{1,t}, b_{1,t}]$ the candidate intervals of step 1. They are pairwise disjoint and have length $< B/s_1$.

Step 2: Eve searches for the smallest integer $s_2, s_2 > s_1$, such that $s_2 m \bmod n$ is PKCS conforming. With high probability (see [Bleichenbacher98]), only one of the candidate intervals $[a_{1,t}, b_{1,t}]$ of step 1 contains a message x, such that $s_2 x \bmod n$ is PKCS conforming. Eve can easily find out whether an interval $[a, b]$ contains a message x, such that $s_2 x \bmod n$ is PKCS conforming. By comparing the interval boundaries, she simply checks whether

$$[s_2 a, s_2 b] \cap [a_0 + rn, b_0 + rn] \neq \emptyset \text{ for some } r \text{ with } \lceil s_2 a/n \rceil \leq r \leq \lfloor s_2 b/n \rfloor.$$

By performing this check for all of the candidate intervals $[a_{1,t}, b_{1,t}]$ of step 1, Eve finds the candidate interval containing m. We denote this interval by $[a_1, b_1]$. With a high probability, we have $s_2(b_1 - a_1) < n - B$ and then, $[s_2 a_1, s_2 b_1]$ is sufficiently short to meet only one of the intervals $[a_0 + rn, b_0 + rn]$, $r \in \mathbb{N}$, say for $r = r_2$. Now, Eve knows that

$$m \in [a_2, b_2] := [a_1, b_1] \cap \left[a_0 + r_2 n/s_2, b_0 + r_2 n/s_2 \right].$$

The length of this interval is $< B/s_2$.

In the rare case that more than one of the candidate intervals of step 1 contains a message x, such that $s_2 x$ is PKCS conforming, or that more than one values for r_2 exist, Eve is left with more than one interval $[a_2, b_2]$. Then, she repeats step 2, starting with the candidate intervals $[a_2, b_2]$ in place of the candidate intervals $[a_{1,t}, b_{1,t}]$ (and searching for $s_2' > s_2$, such that $s_2' m \bmod n$ is PKCS conforming).

Step 3: Step 3 is repeatedly executed, until the plaintext m is determined. Eve starts with $[a_2, b_2]$ and (r_2, s_2), which she has computed in step 2. She iteratively computes pairs (r_i, s_i) and intervals $[a_i, b_i]$ of length $\leq B/s_i$, such that $m \in [a_i, b_i]$. The numbers r_i, s_i are chosen, such that

1. $s_i \approx 2 s_{i-1}$,
2. $[s_i a_{i-1}, s_i b_{i-1}] \cap [a_0 + r_i n, b_0 + r_i n] \neq \emptyset$,
3. $s_i m$ is PKCS conforming.

The number of multipliers s that Eve has to test by querying the decryption oracle is much smaller than in steps 1 and 2. She searches for s_i only in the neighborhood of $2s_{i-1}$. This kind of choosing s_i works, because, after step 2, the intervals $[a_i, b_i]$ are sufficiently small. The length of $[a_{i-1}, b_{i-1}]$ is $< B/s_{i-1}$ and $s_i \approx 2s_{i-1}$. Hence, the length of the interval $[s_i a_{i-1}, s_i b_{i-1}]$ is less than $\approx 2B$, and it therefore meets at most one of the intervals $[a_0 + rn, b_0 + rn], r \in \mathbb{N}$, say for $r = r_i$. From properties 2 and 3, we conclude that $s_i m \in [a_0 + r_i n, b_0 + r_i n]$. Eve sets

$$[a_i, b_i] := [a_{i-1}, b_{i-1}] \cap \left[a_0 + r_i n / s_i, b_0 + r_i n / s_i \right].$$

Then, $[a_i, b_i]$ contains m and its length is $< B/s_i$.

The upper bound B/s_i for the length of $[a_i, b_i]$ decreases by a factor of two in each iteration ($s_i \approx 2s_{i-1}$). Step 3 is repeated, until $[a_i, b_i]$ contains only one integer. This integer is the searched plaintext m.

The analysis in [Bleichenbacher98] shows that for a 1024-bit modulus n, the total number of ciphertexts, for which Eve queries the oracle, is typically about 2^{20}.

Chosen-ciphertext attacks were considered to be only of theoretical interest. Bleichenbacher's attack proved the contrary. It was used against a web server with the ubiquitous SSL/TLS protocol ([RFC 4346]). In the interactive key establishment phase of SSL/TLS, a secret session key is encrypted by using RSA and PKCS#1 v1.5. For a communication server it is natural to process many messages, and to report the success or failure of an operation. Some implementations of SSL/TLS reported an error to the client, when the RSA-encrypted message was not PKCS conforming. Thus, they could be used as the oracle. The adversary could anonymously attack the server, because SSL/TLS is often applied without client authentication. To prevent Bleichenbacher's attack, the implementation of SSL/TLS-servers was improved and the PKCS#1 standard was updated. Now it uses the OAEP padding scheme, which we describe below in Section 3.3.4.

Encryption schemes that are provably secure against adaptively-chosen-ciphertext attacks are studied in more detail in Section 9.5.

The predicate "PKCS conforming" reveals one bit of information about the plaintext. Bleichenbacher's attack shows that, if we were able to compute the bit "PKCS conforming" for RSA-ciphertexts, then we could easily compute complete (PKCS conforming) plaintexts from the ciphertexts. To compute the bit "PKCS conforming" from c is as difficult as to compute m from c. Such a bit is called a secure bit. The bit security of one-way functions is carefully studied in Chapter 7. There we show, for example, that the least significant bit of an RSA-encrypted message is as secure as the whole message. In particular, we develop an algorithm that inverts the RSA function given an oracle with only a small advantage on the least significant bit.

3.3.4 Probabilistic RSA Encryption

Before applying an encryption algorithm such as RSA, preprocessing of the message is necessary. The message is divided into blocks and then some padding or formatting mechanisms are performed. Such preprocessing is provided by the OAEP (optimal asymmetric encryption padding) scheme. It is not only applicable with the RSA cryptosystem; it can be used with any encryption scheme based on a bijective trapdoor function f, such as the RSA function or the modular squaring function of Rabin's cryptosystem (see Section 3.6.1).

In addition to the trapdoor function

$$f : D \longrightarrow D, D \subset \{0,1\}^n,$$

a pseudorandom bit generator

$$G : \{0,1\}^k \longrightarrow \{0,1\}^l$$

and a hash function

$$h : \{0,1\}^l \longrightarrow \{0,1\}^k$$

are used, with $n = l + k$. Given a random seed $s \in \{0,1\}^k$ as input, G generates a pseudorandom bit sequence of length l (see Chapter 8 for more details on pseudorandom bit generators). Hash functions will be discussed in Section 3.4.

Encryption. To encrypt a message $m \in \{0,1\}^l$, we proceed in three steps:

1. We choose a random bit string $r \in \{0,1\}^k$.
2. We set $x = (m \oplus G(r)) \| (r \oplus h(m \oplus G(r)))$.
   (If $x \notin D$ we return to step 1.)
3. We compute $c = f(x)$.

As always, let $\|$ denote the concatenation of strings and $\oplus$ the bitwise XOR operator.

OAEP is an embedding scheme. The message m is embedded into the input x of f such that all bits of x depend on the bits of m. The length of the message m is l. Shorter messages are padded with some additional bits to get length l. The first l bits of x, namely $m \oplus G(r)$ are obtained from m by masking with the pseudorandom bits $G(r)$. The seed r is encoded in the last k bits masked with $h(m \oplus G(r))$. The encryption depends on a randomly chosen r. Therefore, the resulting encryption scheme is not deterministic – encrypting a message m twice will produce different ciphertexts.

Decryption. To decrypt a ciphertext c, we use the function f^{-1}, the same pseudorandom random bit generator G and the same hash function h as above:

1. Compute $f^{-1}(c) = a\|b$, with $|a| = l$ and $|b| = k$.
2. Set $r = h(a) \oplus b$ and get $m = a \oplus G(r)$.

To compute the plaintext m from the ciphertext $c = f(x)$, an adversary must figure out all the bits of x from $c = f(x)$. He needs the first l bits to compute $h(a)$ and the last k bits to get r. Therefore, an adversary cannot exploit any advantage from some partial knowledge of x.

The OAEP scheme is published in [BelRog94]. OAEP has great practical importance. It has been adopted in PKCS#1 v2.0, a widely used standard which is implemented by Internet browsers and used in the secure socket layer protocol (SSL/TLS, [RFC 4346]). Using OAEP prevents Bleichenbacher's attack, which we studied in the preceding section. Furthermore, OAEP is included in electronic payment protocols to encrypt credit card numbers, and it is part of the IEEE P1363 standard.

For practical purposes, it is recommended to implement the hash function h and the random bit generator G using the secure hash algorithm SHA-1 (see Section 3.4.2 below) or some other cryptographic hash algorithm which is considered secure (for details, see [BelRog94]).

If h and G are implemented with efficient hash algorithms, the time to compute h and G is negligible compared to the time to compute f and f^{-1}. Formatting with OAEP does not increase the length of the message substantially.

Using OAEP with RSA encryption no longer preserves the multiplicative structure of numbers, and it is probabilistic. This prevents the previously discussed small-message-space attack. The low encryption exponent attack against RSA is also prevented, provided that the plaintext is individually re-encoded by OAEP for each recipient before it is encrypted with RSA.

We explained the so-called basic OAEP scheme. A slight modification of the basic scheme is the following. Let k, l be as before and let k' be another parameter, with $n = l + k + k'$. We use a pseudorandom generator

$$G : \{0,1\}^k \longrightarrow \{0,1\}^{l+k'}$$

and a cryptographic hash function

$$h : \{0,1\}^{l+k'} \longrightarrow \{0,1\}^k.$$

To encrypt a message $m \in \{0,1\}^l$, we first append k' 0-bits to m, and then we encrypt the extended message as before, i.e., we randomly choose a bit string $r \in \{0,1\}^k$ and the encryption c of m is defined by

$$c = f(((m\|0^{k'}) \oplus G(r))\|(r \oplus h((m\|0^{k'}) \oplus G(r)))).$$

Here, we denote by $0^{k'}$ the constant bit string $000\ldots0$ of length k'.

In [BelRog94], the modified scheme is proven to be secure in the random oracle model against adaptively-chosen-ciphertext attacks. The proof assumes that the hash function and the pseudorandom generator used behave like truly random functions. We describe the random oracle model in Section 3.4.5.

In [Shoup2001], it was observed that there is a gap in the security proof of OAEP. This does not imply that a particular instantiation of OAEP, such as

OAEP with RSA, is insecure. In the same paper, it is shown that OAEP with RSA is secure for an encryption exponent of 3. In [FujOkaPoiSte2001], this result is generalized to arbitrary encryption exponents. In Section 9.5.1, we describe SAEP – a simplified OAEP – and we give a security proof for SAEP in the random oracle model against adaptively-chosen-ciphertext attacks.

3.4 Cryptographic Hash Functions

Cryptographic hash functions such as SHA-1 or MD5 are widely used in cryptography. In digital signature schemes, messages are first hashed and the hash value $h(m)$ is signed in place of m. Hash values are used to check the integrity of public keys. Pseudorandom bit strings are generated by hash functions. When used with a secret key, cryptographic hash functions become *message authentication codes* (MACs), the preferred tool in protocols like SSL and IPSec to check the integrity of a message and to authenticate the sender.

A *hash function* is a function that takes as input an arbitrarily long string of bits (called a message) and outputs a bit string of a fixed length n. Mathematically, a hash function is a function

$$h : \{0,1\}^* \longrightarrow \{0,1\}^n, \ m \longmapsto h(m).$$

The length n of the output is typically between 128 and 512 bits[8]. Later, when discussing the birthday attack, we will see why the output lengths are in this range.

One basic requirement is that the hash values $h(m)$ are easy to compute, making both hardware and software implementations practical.

3.4.1 Security Requirements for Hash Functions

A classical application of cryptographic hash functions is the "encryption" of passwords. Rather than storing the cleartext of a user password *pwd* in the password file of a system, the hash value $h(pwd)$ is stored in place of the password itself. If a user enters a password, the system computes the hash value of the entered password and compares it with the stored value. This technique of non-reversible "encryption" is applied in operating systems. It prevents, for example, passwords becoming known to privileged users of the system such as administrators, provided it is not possible to compute a password *pwd* from its hash value $h(pwd)$. This leads to our first security requirement.

A cryptographic hash function must be a *one-way function*: Given a value $y \in \{0,1\}^n$, it is computationally infeasible to find an m with $h(m) = y$.

If a hash function is used in conjunction with digital signature schemes, the message is hashed first and then the hash value is signed in place of

---

[8] The output lengths of MD5 and SHA-1 are 128 and 160 bits.

the original message. Suppose Alice signs $h(m)$ for a message m. Adversary Eve should have no chance to find a message $m' \neq m$ with $h(m') = h(m)$. Otherwise, she could pretend that Alice signed m' instead of m.

Thus, the hash function must have the property that given a message m, it is computationally infeasible to obtain a second message m' with $m \neq m'$ and $h(m) = h(m')$. This property is called the *second pre-image resistance*.

When using hash functions with digital signatures, we require an even stronger property. The legal user Alice of a signature scheme with hash function h should have no chance of finding two distinct messages m and m' with $h(m) = h(m')$. If Alice finds such messages, she could sign m and say later that she has signed m' and not m.

Such a pair (m, m') of messages, with $m \neq m'$ and $h(m) = h(m')$, is called a *collision* of h. If it is computationally infeasible to find a collision (m, m') of h, then h is called *collision resistant*.

Sometimes, collision resistant hash functions are called *collision free*, but that's misleading. The function h maps an infinite number of elements to a finite number of elements. Thus, there are lots of collisions (in fact, infinitely many). Collision resistance merely states that they cannot be found.

Proposition 3.5. *A collision-resistant hash function h is second-pre-image resistant.*

Proof. An algorithm computing second pre-images can be used to compute collisions in the following way: Choose m at random. Compute a pre-image $m' \neq m$ of $h(m)$. (m, m') is a collision of h. $\square$

Proposition 3.5 says that collision resistance is the stronger property. Therefore, second pre-image resistance is sometimes also called *weak collision resistance*, and collision resistance is referred to as *strong collision resistance*.

Proposition 3.6. *A second-pre-image-resistant hash function is a one-way function.*

Proof. If h were not one-way, there would be a practical algorithm A that on input of a randomly chosen value v computes a message $\tilde{m}$ with $h(\tilde{m}) = v$, with a non-negligible probability. Given a random message m, attacker Eve could find, with a non-negligible probability, a second pre-image of $h(m)$ in the following way: She applies A to the hash value $h(m)$ and obtains $\tilde{m}$ with $h(\tilde{m}) = h(m)$. The probability that $\tilde{m} \neq m$ is high. $\square$

Our definitions and the argument in the previous proof lack some precision and are not mathematically rigorous. For example, we do not explain what "computationally infeasible" and a "non-negligible probability" mean. It is possible to give precise definitions and a rigorous proof of Proposition 3.6 (see Chapter 10, Exercise 2).

Definition 3.7. A hash function is called a *cryptographic hash function* if it is collision resistant.

Sometimes, hash functions used in cryptography are referred to as *one-way hash functions*. We have seen that there is a stronger requirement, collision resistance, and the one-way property follows from it. Therefore, we prefer to speak of collision-resistant hash functions.

3.4.2 Construction of Hash Functions

Merkle-Damgård's construction. There are no known examples of hash functions whose collision resistance can be proven without any assumptions. In Section 10.2, we give examples of (rather inefficient) hash functions that are provably collision resistant under standard assumptions in public-key cryptography, such as the factoring assumption.

Many cryptographic hash functions used in practice are obtained by the following method, known as *Merkle-Damgård's construction* or *Merkle's meta method*. The method reduces the problem of constructing a collision-resistant hash function $h : \{0,1\}^* \longrightarrow \{0,1\}^n$ to the problem of constructing a collision-resistant function

$$f : \{0,1\}^{n+r} \longrightarrow \{0,1\}^n \qquad (r \in \mathbb{N}, r > 0)$$

with finite domain $\{0,1\}^{n+r}$. Such a function f is called a *compression function*. A compression function maps messages m of a fixed length $n + r$ to messages $f(m)$ of length n. We call r the *compression rate*.

We discuss Merkle-Damgård's construction. Let $f : \{0,1\}^{n+r} \longrightarrow \{0,1\}^n$ be a compression function with compression rate r. By using f, we define a hash function

$$h : \{0,1\}^* \longrightarrow \{0,1\}^n.$$

Let $m \in \{0,1\}^*$ be a message of arbitrary length. The hash function h works iteratively. To compute the hash value $h(m)$, we start with a fixed initial n-bit hash value $v = v_0$ (the same for all m). The message m is subdivided into blocks of length r. One block after the other is taken from m, concatenated with the current value v and compressed by f to get a new v. The final v is the hash value $h(m)$.

More precisely, we pad m out, i.e., we append some bits to m, to obtain a message $\tilde{m}$, whose bit length is a multiple of r. We apply the following padding method: A single 1-bit followed by as few (possibly zero) 0-bits as necessary are appended. Every message m is padded out with such a string $100\ldots0$, even if the length of the original message m is a multiple of r. This guarantees that the padding can be removed unambiguously – the bits which are added during padding can be distinguished from the original message bits.[9]

---

[9] There are also other padding methods which may be applied. See, for example, [RFC 3369].

After the padding, we decompose

$$\tilde{m} = m_1 \| \ldots \| m_k, \ m_i \in \{0,1\}^r, \ 1 \leq i \leq k,$$

into blocks m_i of length r.

We add one more r-bit block m_{k+1} to $\tilde{m}$ and store the original length of m (i.e., the length of m before padding it out) into this block right-aligned. The remaining bits of m_{k+1} are filled with zeros:

$$\tilde{m} = m_1 \| m_2 \| \ldots \| m_k \| m_{k+1}.$$

Starting with the initial value $v_0 \in \{0,1\}^n$, we set recursively

$$v_i := f(v_{i-1} \| m_i), \ 1 \leq i \leq k+1.$$

The last value of v is taken as hash value $h(m)$:

$$h(m) := v_{k+1}.$$

The last block m_{k+1} is added to prevent certain types of collisions. It might happen that we obtain $v_i = v_0$ for some i. If we had not added m_{k+1}, then (m, m') would be a collision of h, where m' is obtained from $m_{i+1} \| \ldots \| m_k$ by removing the padding string $10 \ldots 0$ from the last block m_k. Since m and m' have different lengths, the additional length blocks differ and prevent such collisions.[10]

Proposition 3.8. *Let f be a collision-resistant compression function. The hash function h constructed by Merkle's meta method is also collision resistant.*

Proof. The proof runs by contradiction. Assume that h is not collision resistant, i.e., that we can efficiently find a collision (m, m') of h. Let $(\tilde{m}, \tilde{m}')$ be the modified messages as above. The following algorithm efficiently computes a collision of f from $(\tilde{m}, \tilde{m}')$. This contradicts our assumption that f is collision resistant.

---

[10] Sometimes, the padding and the length block are combined: the length of the original message is stored into the rightmost bits of the padding string. See, for example, SHA-1 ([RFC 3174]).

Algorithm 3.9.
collision *FindCollision*(bitString $\tilde{m}, \tilde{m}'$)
1   $\tilde{m} = m_1 \| \dots \| m_{k+1}, \tilde{m}' = m_1' \| \dots \| m_{k'+1}'$ decomposed as above
2   $v_1, \dots, v_{k+1}, v_1', \dots, v_{k'+1}'$ constructed as above
3   if $|m| \neq |m'|$
4   then return $(v_k \| m_{k+1}, v_{k'}' \| m_{k'+1}')$
5   for $i \leftarrow 1$ to k do
6   if $v_i \neq v_i'$ and $v_{i+1} = v_{i+1}'$
7   then return $(v_i \| m_{i+1}, v_i' \| m_{i+1}')$
8   for $i \leftarrow 0$ to $k - 1$ do
9   if $m_{i+1} \neq m_{i+1}'$
10   then return $(v_i \| m_{i+1}, v_i' \| m_{i+1}')$

Note that $h(m) = v_{k+1} = v_{k'+1}' = h(m')$. If $|m| \neq |m'|$ we have $m_{k+1} \neq m_{k'+1}'$, since the length of the string is encoded in the last block. Hence $v_k \| m_{k+1} \neq v_{k'}' \| m_{k'+1}'$. We obtain a collision $(v_k \| m_{k+1}, v_{k'}' \| m_{k'+1}')$, because $f(v_k \| m_{k+1}) = h(m) = h(m') = f(v_{k'}' \| m_{k'+1}')$. On the other hand, if $|m| = |m'|$, then $k = k'$, and we are looking for an index i with $v_i \neq v_i'$ and $v_{i+1} = v_{i+1}'$. $(v_i \| m_{i+1}, v_i' \| m_{i+1}')$ is then a collision of f, because $f(v_i \| m_{i+1}) = v_{i+1} = v_{i+1}' = f(v_i' \| m_{i+1}')$. If no index with the above condition exists, we have $v_i = v_i'$, $1 \leq i \leq k+1$. In this case, we search for an index i with $m_{i+1} \neq m_{i+1}'$. Such an index exists, because $m \neq m'$. $(v_i \| m_{i+1}, v_i' \| m_{i+1}')$ is then a collision of f, because $f(v_i \| m_{i+1}) = v_{i+1} = v_{i+1}' = f(v_i' \| m_{i+1}')$. $\square$

The Birthday Attack. One of the main questions when designing a hash function $h : \{0,1\}^* \longrightarrow \{0,1\}^n$ is how large to choose the length n of the hash values. A lower bound for n is obtained by analyzing the birthday attack.

The *birthday attack* is a brute-force attack against collision resistance. Adversary Eve randomly generates messages $m_1, m_2, m_3, \dots$. For each newly generated message m_i, Eve computes and stores the hash value $h(m_i)$ and compares it with the previous hash values. If $h(m_i)$ coincides with one of the previous hash values, $h(m_i) = h(m_j)$ for some $j < i$, Eve has found a collision (m_i, m_j)[11]. We show below that Eve can expect to find a collision after choosing about $2^{n/2}$ messages. Thus, it is necessary to choose n so large that it is impossible to calculate and store $2^{n/2}$ hash values. If $n = 128$ (as with MD5), about $2^{64} \approx 10^{20}$ messages have to be chosen for a successful attack.[12] Many people think that today a hash length of 128 bits is no longer large enough, and that 160 bits (as in SHA-1 and RIPEMD-160) should be the lower bound (also see Section 3.4.2 below).

---

[11] In practice, the messages are generated by a deterministic pseudorandom generator. Therefore, the messages themselves can be reconstructed and need not be stored.

[12] To store 2^{64} 16-byte hash values, you need 2^{28} TB of storage. There are memoryless variations of the birthday attack which avoid these extreme storage requirements, see [MenOorVan96].

Attacking second-pre-image resistance or the one-way property of h with brute force would mean to generate, for a given hash value $v \in \{0,1\}^n$, random messages $m_1, m_2, m_3, \ldots$ and check each time whether $h(m_i) = v$. Here we expect to find a pre-image of v after choosing 2^n messages (see Exercise 9). For $n = 128$, we need $2^{128} \approx 10^{39}$ messages. To protect against this attack, a smaller n would be sufficient.

The surprising efficiency of the birthday attack is based on the *birthday paradox*. It says that the probability of two persons in a group sharing the same birthday is greater than $1/2$, if the group is chosen at random and has more than 23 members. It is really surprising that this happens with such a small group.

Considering hash functions, the 365 days of a year correspond to the number of hash values. We assume in our discussion that the hash function $h : \{0,1\}^* \longrightarrow \{0,1\}^n$ behaves like the birthdays of people. Each of the $s = 2^n$ values has the same probability. This assumption is reasonable. It is a basic design principle that a cryptographic hash function comes close to a random function, which yields random and uniformly distributed values (see Section 3.4.4).

Evaluating h, k times with independently chosen inputs, the probability that no collisions occur is

$$p = p(s, k) = \frac{1}{s^k} \prod_{i=0}^{k-1} (s - i) = \prod_{i=1}^{k-1} \left(1 - \frac{i}{s} \right).$$

We have $1 - x \le e^{-x}$ for all real numbers x and get

$$p \le \prod_{i=1}^{k-1} e^{-i/s} = e^{-(1/s) \sum_{i=1}^{k-1} i} = e^{-k(k-1)/2s}.$$

The probability that a collision occurs is $1 - p$, and $1 - p \ge 1/2$ if $k \ge 1/2 \left(\sqrt{1 + 8 \ln 2 \cdot s} + 1 \right) \approx 1.18 \sqrt{s}$.

For $s = 365$, we get an explanation for the original birthday paradox, since $1.18 \cdot \sqrt{s} = 22.54$.

For the hash function h, we conclude that it suffices to choose about $2^{n/2}$ many messages at random to obtain a collision with probability $> 1/2$.

In a hash-then-decrypt digital signature scheme, where the hash value is signed in place of the message (see Section 3.4.5 below), the birthday attack might be practically implemented in the following way. Suppose that Eve and Bob want to sign a contract m_1. Later, Eve wants to say that Bob has signed a different contract m_2. Eve generates $O(2^{n/2})$ minor variations of m_1 and m_2. In many cases, for example, if m_1 includes a bitmap, Bob might not observe the slight modification of m_1. If the birthday attack is successful, Eve gets messages $\tilde{m}_1$ and $\tilde{m}_2$ with $h(\tilde{m}_1) = h(\tilde{m}_2)$. Eve lets Bob sign the contract $\tilde{m}_1$. Later, she can pretend that Bob signed $\tilde{m}_2$.

Compression Functions from Block Ciphers. We show in this section how to derive compression functions from a block cipher, such as DES or AES. From these compression functions, cryptographic hash functions can be obtained by using Merkle-Damgård's construction.

Symmetric block ciphers are widely used and well studied. Encryption is implemented by efficient algorithms (see Chapter 2). It seems natural to also use them for the construction of compression functions. Though no rigorous proofs exist, the hope is that a good block cipher will result in a good compression function. Let

$$E : \{0,1\}^r \times \{0,1\}^n \longrightarrow \{0,1\}^n, \ (k,x) \longmapsto E(k,x)$$

be the encryption function of a symmetric block cipher, which encrypts blocks x of bit length n with r-bit keys k.

First, we consider constructions where the bit length of the hash value is equal to the block length of the block cipher. These schemes are called *single-length MDCs*[13]. To obtain a collision-resistant compression function, the block length n of the block cipher should be at least 128 bits.

The compression function

$$f_1 : \{0,1\}^{n+r} \longrightarrow \{0,1\}^n, \ (x\|y) \longmapsto E(y,x)$$

maps bit blocks of length $n + r$ to blocks of length n. A block of length $n + r$ is split into a left block x of length n and a right block y of length r. The right block y is used as key to encrypt the left block x.

The second example of a compression function – it is the basis of the Matyas-Meyer-Oseas hash function ([MatMeyOse85]) – has been included in [ISO/IEC 10118-2]. Its compression rate is n. A block of bit length $2n$ is split into two halves, x and y, each of length n. Then x is encrypted with a key $g(y)$ which is derived from y by a function $g : \{0,1\}^n \longrightarrow \{0,1\}^r$.[14] The resulting ciphertext is bitwise XORed with x:

$$f_2 : \{0,1\}^{2n} \longrightarrow \{0,1\}^n, \ (x\|y) \longmapsto E(g(y),x) \oplus x.$$

If the block length of the block cipher is less than 128, *double-length MDCs* are used. Compression functions whose output length is twice the block length can be obtained by combining two types of the above compression functions (for details, see [MenOorVan96]).

Real Hash Functions. Most cryptographic hash functions used in practice today do not rely on other cryptographic primitives such as block ciphers. They are derived from custom-designed compression functions by applying

---

[13] The acronym MDC is explained in Section 3.4.3 below.

[14] For example, we can take $g(y) = y$, if the block length of E is equal to the key length, or, more generally, we can compute r key bits $g(y)$ from y by using a Boolean function.

Merkle-Damgård's construction. The functions are especially designed for the purpose of hashing, with performance efficiency in mind.

In [Rivest90], R. Rivest proposed MD4, which is algorithm number 4 in a family of hash algorithms. MD4 was designed for software implementation on a 32-bit processor. The MD4 algorithm is not strong enough, as early attacks showed. However, the design principles of the MD4 algorithm were subsequently used in the construction of hash functions. These functions are often called the MD4 family. The family contains the most popular hash functions in use today, such as MD5, SHA-1 and RIPEMD-160. The hash values of MD5 are 128 bits long, those of RIPEMD-160 and SHA-1 160 bits. All of these hash functions are iterative hash functions; they are constructed with Merkle-Damgård's method. The compression rate of the underlying compression functions is 512 bits.

SHA-1 is included in the Secure Hash Standard FIPS 180 of NIST ([RFC 1510]; [RFC 3174]). It is an improvement of SHA-0, which turned out to have a weakness. The standard was updated in 2002 ([FIPS 180-2]). Now it includes additional algorithms that produce 256-bit, 384-bit and 512-bit outputs.

Since no rigorous mathematical proofs for the security of these hash functions exist, there is always the chance of a surprise attack.

For example, the MD5 algorithm is very popular, but there have been very successful attacks.

In 1996, H. Dobbertin detected collisions $(v_0 \| m, v_0 \| m')$ of the underlying compression function, where v_0 is a common 128-bit string and m, m' are distinct 512-bit messages ([Dobbertin96a]; [Dobbertin96]). Dobbertin's v_0 is different from the initial value that is specified for MD5 in the Merkle-Damgård iteration. Otherwise, the collision would have immediately implied a collision of MD5 (note that the same length block is appended to both messages). Already in 1993, B. den Boer and A. Bosselaers had detected collisions of MD5's compression function. Their collisions $(v_0 \| m, v_0' \| m)$ were made of distinct initial values v_0 and the same message m. Thus, they did not fit Merkle-Damgård's method and were sometimes called pseudocollisions.

The recent attacks by the Chinese researchers X. Wang, D. Feng, X. Lai and H. Yu showed that MD5 can no longer be considered collision-resistant. In August 2004, they published collisions for the hash functions MD4, MD5, HAVAL-128, RIPEMD-128 ([WanFenLaiYu04]). V. Klima published an algorithm which works for any initial value and computes collisions of MD5 on a standard PC within a minute ([Klima06]). MD5 is really broken.

Moreover, in February 2005, X. Wang, Y. L. Yin and H. Yu cast serious doubts on the security of SHA-1 ([WanYinYu05]). They announced that they found an algorithm which computes collisions with 2^{69} hash operations. This is much less than the expected 2^{80} steps of the brute-force birthday attack. With current technology, 2^{69} steps are still on the far edge of feasibility. For example, the RC5-64 Challenge was finished in 2002. A worldwide network of

Internet users was able to figure out a 64-bit RC5 key by a brute-force search. The search took almost 5 years, and more than 300,000 users participated (see [RSALabs]; [DistributedNet]).

All of these attacks are against collision resistance, and they are relevant for digital signatures. They are not attacks against second-pre-image resistance or the one-way property. Therefore, applications like HMAC (see Section 3.4.3), whose security is based on these properties, are not yet affected.

In the future, hash functions with longer hash values, such as SHA-256 or SHA-512, will be used in place of MD5 and SHA-1.

3.4.3 Data Integrity and Message Authentication

Modification Detection Codes. Cryptographic hash functions are also known as *message digest functions*, and the hash value $h(m)$ of a message m is called the *digest* or *fingerprint* or *thumbprint* of m [15]. The hash value $h(m)$ is indeed a "fingerprint" of m. It is a very compact representation of m, and, as an immediate consequence of second-pre-image resistance, this representation is practically unique. Since it is computationally infeasible to obtain a second message m' with $m \neq m'$ and $h(m') = h(m)$, a different hash value would result, if the message m were altered in any way.

This implies that a cryptographic hash function can be used to control the integrity of a message m. If the hash value of m is stored in a secure place, a modification of m can be detected by calculating the hash value and comparing it with the stored value. Therefore, hash functions are also called *modification detection codes* (MDCs).

Let us consider an example. If you install a new root certificate in your Internet browser, you have to make sure (among other things) that the source of the certificate is the one you think it is and that your copy of the certificate was not modified. You can do this by checking the certificate's thumbprint. For this purpose, you can get the fingerprint of the root certificate from the issuing certification authority's web page or even on real paper by ordinary mail (certificates and certification authorities are discussed in Section 4.1.5).

Message Authentication Codes. A very important application of hash functions is message authentication, which means to authenticate the origin of the message. At the same time, the integrity of the message is guaranteed. If hash functions are used for message authentication, they are called *message authentication codes*, or MACs for short.

MACs are the standard symmetric technique for message authentication and integrity protection and widely used, for example, in protocols such as SSL/TLS ([RFC 4346]) and IPSec. They depend on secret keys shared between the communicating parties. In contrast to digital signatures, where

---

[15] MD5 is a "message digest function".

only one person knows the secret key and is able to generate the signature, each of the two parties can produce the valid MAC for a message.

Formally, the secret keys k are used to parameterize hash functions. Thus, MACs are families of hash functions

$$(h_k : \{0,1\}^* \longrightarrow \{0,1\}^n)_{k \in K}.$$

MACs may be derived from block ciphers or from cryptographic hash functions. We describe two methods to obtain MACs.

The standard method to convert a cryptographic hash function into a MAC is called *HMAC*. It is published in [RFC 2104] (and [FIPS 198], [ISO/IEC 9797-2]) and can be applied to a hash function h that is derived from a compression function f by using Merkle-Damgård's method. You can take as h, for example, MD5, SHA-1 or RIPEMD-160 (see Section 3.4.2).

We have to assume that the compression rate of f and the length of the hash values are multiples of 8, so we can measure them in bytes. We denote by r the compression rate of f in bytes. The secret key k can be of any length, up to r bytes. By appending zero bytes, the key k is extended to a length of r bytes (e.g., if k is of length 20 bytes and $r = 64$, then k will be appended with 44 zero bytes 0x00).

Two fixed and distinct strings ipad and opad are defined (the 'i' and 'o' are mnemonics for inner and outer):

ipad := the byte 0x36 repeated r times,

opad := the byte 0x5C repeated r times.

The keyed hash value HMAC of a message m is calculated as follows:

$$\text{HMAC}(k, m) := h((k \oplus \text{opad}) \| h((k \oplus \text{ipad}) \| m)).$$

The hash function h is applied twice in order to guarantee the security of the MAC. If we apply h only once and define $\text{HMAC}(k, m) := h((k \oplus \text{ipad}) \| m)$, an adversary Eve could take a valid MAC value, modify the message m and compute the valid MAC value of the modified message, without knowing the secret key. For example, Eve may take any message m' and compute the hash value v of m' by applying Merkle-Damgård's iteration with $\text{HMAC}(k, m) = h((k \oplus \text{ipad}) \| m)$ as initial value v_0. Before iterating, Eve appends the padding bits and the additional length block to m'. She does not store the length of m' into the length block, but the length of $\tilde{m} \| m'$, where $\tilde{m}$ is the padded message m (including the length block for m, see Section 3.4.2). Then v is the MAC of the extended message $\tilde{m} \| m'$, and Eve has computed it without knowing the secret key k. This problem is called the *length extension problem of iterated hash functions*. Applying the hash function twice prevents the length extension attack.

MACs can also be constructed from block ciphers. The most important construction is CBC-MAC. Let E be the encryption function of a block cipher, such as DES or AES. Then, with k a secret key, the MAC value for a message m is the last ciphertext block when encrypting m, with E in the Cipher-Block Chaining Mode CBC and key k (see Section 2.2.3). We need an initialization vector IV for CBC. For encryption purposes, it is important not to use the same value twice. Here, the IV is fixed and typically set to $0\ldots0$. If the block length of E is n, then m is split into blocks of length n, $m_1 \| m_2 \| \ldots \| m_l$ (pad out the last block, if necessary, for example, by appending zeros), and we compute

$$c_0 := IV$$
$$c_i := E(k, m_i \oplus c_{i-1})$$
$$\text{CBC-MAC} := c_l$$

Sometimes, the output of the CBC-MAC function is taken only to be a part of the last block. There are various standards for CBC-MAC, for example, [FIPS 113] and [ISO/IEC 9797-1]. A comprehensive discussion of hash functions and MACs can be found in [MenOorVan96].

3.4.4 Hash Functions as Random Functions

A random function would be the perfect cryptographic hash function h. Random means that for all messages m, each of the n bits of the hash value $h(m)$ is determined by tossing a coin. Such a perfect cryptographic hash function is also called a *random oracle*[16]. Unfortunately, it is obvious that a perfect random oracle can not be implemented. To determine only the hash values for all messages of fixed length l would require exponentially many $(n \cdot 2^l)$ coin tosses and storage of all the results, which is clearly impossible.

Nevertheless, it is a design goal to construct hash functions which approximate random functions. It should be computationally infeasible to distinguish the hash function from a truly random function. Recall that there is a similar design goal for symmetric encryption algorithms. The ciphertext should appear random to the attacker. That is the reason why we hoped in Section 3.4.2 that a good block cipher induces a good compression function for a good hash function.

If we assume that the designers of a hash function h have done a good job and h comes close to a random oracle, then we can use h as a generator of pseudorandom bits. Therefore, we often see popular cryptographic hash functions such as SHA-1 or MD5 as sources of pseudorandomness. For example, in the Transport Layer Security (TLS) protocol ([RFC 4346]), also known as Secure Socket Layer (SSL), client and server agree on a shared 48-bit master secret, and then they derive further key material (for example, the MAC keys, encryption keys and initialization vectors) from this master secret by

---

[16] Security proofs in cryptography sometimes rely on the assumption that the hash function involved is a random oracle. An example of such a proof is given in Section 3.4.5

using a pseudorandom function. The pseudorandom function of TLS is based on the HMAC construction, with the hash functions SHA-1 or MD5.

3.4.5 Signatures with Hash Functions

Let (n, e) be the public RSA key and d be the secret decryption exponent of Alice. In the basic RSA signature scheme (see Section 3.3.2), Alice can sign messages that are encoded by numbers $m \in \{0, \dots, n - 1\}$. To sign m, she applies the RSA decryption algorithm and obtains the signature $\sigma = m^d \bmod n$ of m.

Typically, n is a 1024–bit number. Alice can sign a bit string m that, when interpreted as a number, is less than n. This is a text string of at most 128 ASCII-characters. Most documents are much larger, and we are not able to sign them with basic RSA. This problem, which exists in all digital signatures schemes, is commonly solved by applying a collision resistant hash function h.

Message m is first hashed, and the hash value $h(m)$ is signed in place of m. Alice's RSA signature of m is

$$\sigma = h(m)^d \bmod n.$$

To verify Alice's signature σ for message m, Bob checks whether

$$\sigma^e = h(m) \bmod n.$$

This way of generating signatures is called the *hash-then-decrypt paradigm*. This term is even used for signature schemes, where the signing algorithm is not the decryption algorithm as in RSA (see, for example, ElGamal's Signature Scheme in Section 3.5.2).

Messages with the same hash value have the same signature. Collision resistance of h is essential for non-repudiation. It prevents Alice from first signing m and pretending later that she has signed a different message m' and not m. To do this, Alice would have to generate a collision (m, m'). Collision resistance also prevents that an attacker Eve takes a signed message (m, σ) of Alice, generates another message m' with the same hash value and uses σ as a (valid) signature of Alice for m'. To protect against the latter attack, second-pre-image resistance of h would be sufficient.

The hash-then-decrypt paradigm has two major advantages. Messages of any length can be signed by applying the basic signature algorithm, and the attacks, which we discussed in Section 3.3.2, are prevented. Recall that the hash function reduces a message of arbitrary length to a short digital fingerprint of less than 100 bytes.

The schemes which we discuss now implement the hash-then-decrypt paradigm.

Full-Domain-Hash RSA signatures. We apply the hash-then-decrypt paradigm in an RSA signature scheme with public key (n, e) and secret key d and use a hash function

$$h : \{0,1\}^* \longrightarrow \{0, \ldots, n-1\},$$

whose values range through the full set $\{0, \ldots, n-1\}$ rather than a smaller subset. Such a hash function h is called a *full-domain* hash function, because the image of h is the full domain $\{0, \ldots, n-1\}$ of the RSA function[17]. The signature of a message $m \in \{0,1\}^*$ is $h(m)^d \bmod n$.

The hash functions that are typically used in practical RSA schemes, like SHA, MD5 or RIPEMD, are not full-domain hash functions. They produce hash values of bit length between 128 and 512 bits, whereas the typical bit length of n is 1024 or 2048.

It can be mathematically proven that full-domain-hash RSA signatures are secure in the *random oracle model* ([BelRog93]), and we will give such a proof in this section.

For this purpose, we consider an adversary F, who attacks Bob, the legitimate owner of an RSA key pair, and tries to forge at least one signature of Bob, without knowing Bob's private key d. More precisely, F is an efficiently computable algorithm that, with some probability of success, on input of Bob's public RSA key (n, e) outputs a message m together with a valid signature σ of m.

The random oracle model. In this model, the hash function h is assumed to operate as a *random oracle*. This means that

1. the hash function h is a random function (as explained in Section 3.4.4), and
2. whenever the adversary F needs the hash value for a message m, it has to call the oracle h with m as input. Then it obtains the hash value $h(m)$ from the oracle.

Condition 2 means that F always calls h as a "black box" (for example, by calling it as a subroutine or by communicating with another computer program), whenever it needs a hash value, and this may appear as a trivial condition. But it includes, for example, that the adversary has no algorithm to compute the hash values by itself; it has no knowledge about the internal structure of h, and it is stateless with respect to hash values. It does not store and reuse any hash values from previous executions. The hash values $h(m)$ appear as truly random values to him.

We assume from now on that our full-domain hash function h is a random oracle. Given m, each element of $\mathbb{Z}_n$ has the same probability $1/n$ of being the hash value $h(m)$.

The security of RSA signatures relies, of course, on the RSA assumption, which states that the RSA function is a one-way function. Without knowing

---

[17] As often, we identify $\mathbb{Z}_n$ with $\{0, \ldots, n-1\}$.

the secret exponent d, it is infeasible to compute e-th roots modulo n, i.e., for a randomly chosen e-th power $y = x^e \bmod n$, it is impossible to compute x from y with more than a negligible probability (see Definition 6.7 for a precise statement).

Our security proof for full-domain-hash RSA signatures is a typical one. We develop an efficient algorithm A which attacks the underlying assumption – here the RSA assumption. In our example, A tries to compute the e-th root of a randomly chosen $y \in \mathbb{Z}_n$. The algorithm A calls the forger F as a subroutine. If F is successful in its forgery, then A is successful in computing the e-th root. Now, we conclude: since it is infeasible to compute e-th roots (by the RSA assumption), F can not be successful, i.e., it is impossible to forge signatures. By A, the security of the signature scheme is reduced to the security of the RSA trapdoor function. Therefore, such proofs are called *security proofs by reduction*.

The security of full-domain-hash signatures is guaranteed, even if forger F is supplied with valid signatures for messages m' of its choice. Of course, to be successful, F has to produce a valid signature for a message m which is different from the messages m'. F can request the signature for a message m' at any time during its attack, and it can choose the messages m' adaptively, i.e., F can analyze the signatures that it has previously obtained, and then choose the next message to be signed. F performs an *adaptively-chosen-message attack* (see Section 10.1 for a more detailed discussion of the various types of attacks against signature schemes).

In the real attack, the forger F interacts with Bob, the legitimate owner of the secret key, to obtain signatures, and with the random oracle to obtain hash values. Algorithm A is constructed to replace both, Bob and the random oracle h, in the attack. It "simulates" the signer Bob and h.

Since A has no access to the secret key, it has a problem to produce a valid signature, when F issues a signature request for message m'. Here, the random oracle model helps A. It is not the message that is signed, but its hash value. To check if a signature is valid, the forger F must know the hash value, and to get the hash value it has to ask the random oracle. Algorithm A answers in place of the oracle. If asked for the hash value of a message m', it selects $s \in \mathbb{Z}_n$ at random and supplies s^e as the hash value. Then, it can provide s as the valid signature of m'.

Forger F can not detect that A sends manipulated hash values. The elements s, s^e and the real hash values (generated by a random oracle) are all random and uniformly distributed elements of $\mathbb{Z}_n$. This means that forger F, when interacting with A, runs in the same probabilistic setting as in the real attack. Therefore, its probability of successfully forging a signature is the same as in the real attack against Bob.

A takes as input the public key (n, e) and a random element $y \in \mathbb{Z}_n$. Let F query the hash values of the r messages $m_1, m_2, \ldots, m_r$. The structure of A is the following.

Algorithm 3.10.

```
int A(int n, e, y)
```
1   choose $t \in \{1, \ldots, r\}$ at random and set $h_t \leftarrow y$
2   choose $s_i \in \mathbb{Z}_n$ at random and set $h_i \leftarrow s_i^e$, $i = 1, \ldots, r, i \neq t$
3   call $F(n, e)$
4   if F queries the hash value of m_i, then respond with h_i
5   if F requests the signature of m_i, $i \neq t$, then respond with s_i
6   if F requests the signature of m_t, then terminate with failure
7   if F requests the signature of $m', m' \neq m_i$ for $i = 1, \ldots, r$,
8   then respond with a random element of $\mathbb{Z}_n$
9   if F returns (m, s), return s

In step 1, A tries to guess the message $m \in \{m_1, \ldots, m_r\}$, for which F will output a forged signature. F must know the hash value of m. Otherwise, the hash value $h(m)$ of m would be randomly generated independently from F's point of view. Then, the probability that a signature s, generated by F, satisfies the verification condition $s^e \bmod n = h(m)$ is $1/n$, and hence negligibly small. Thus, m is necessarily one of the messages m_i, for which F queries the hash value.

If F requests the signature of m_i, $i \neq t$, then A responds with the valid signature s_i (line 5). If F requests the signature of $m', m' \neq m_i$ for $i = 1, \ldots, r$, and F never asks for the hash value of m', then A can respond with a random value (line 7) – F is not able to check the validity of the answer.

Suppose that A guesses the right m_t in step 1 and F forges successfully. Then, F returns a valid signature s for m_t, which is an e-th root of y, i.e., $s^e = h(m_t) = y$. In this case, A returns s. It has successfully computed an e-th root of y modulo n.

The probability that A guesses the right m_t in step 1 is $1/r$. Hence, the success probability of A is $1/r \cdot \alpha$, where α is the success probability of forger F. Assume for a moment that forger F is always successful, i.e., $\alpha = 1$. By independent repetitions of A, we then get an algorithm which successfully computes e-th roots with a probability close to 1. In general, we get an algorithm to compute e-th roots with about the same success probability α as F.

We described the notion of provable security in the random oracle model by studying full-domain hash RSA signatures. The proof says that a successful forger can not exist in the random oracle model. But since real hash functions are never perfect random oracles (see Section 3.4.4 above), our argument can never be completed to a security proof of the real signature scheme, where a real implementation of the hash function has to be used.

In Section 9.5, we will give a random-oracle proof for Boneh's SAEP encryption scheme.

Our proof requires a full-domain hash function h – it is essential that each element of $\mathbb{Z}_n$ has the same probability of being the hash value. The hash functions used in practice usually are not full-domain hash functions, as we observed above. The scheme we describe in the next section does not rely on a full-domain hash function. It provides a clever embedding of the hashed message into the domain of the signature function.

PSS. The probabilistic signature scheme (PSS) was introduced in [BelRog96]. The signature of a message depends on the message and some randomly chosen input. The resulting signature scheme is therefore probabilistic. To set up the scheme, we need the decryption function of a public-key cryptosystem like the RSA decryption function or the decryption function of Rabin's cryptosystem (see Section 3.6). More generally, it requires a trapdoor permutation

$$f : D \longrightarrow D, D \subset \{0,1\}^n,$$

a pseudorandom bit generator

$$G : \{0,1\}^l \longrightarrow \{0,1\}^k \times \{0,1\}^{n-(l+k)}, \ w \longmapsto (G_1(w), G_2(w))$$

and a hash function

$$h : \{0,1\}^* \longrightarrow \{0,1\}^l.$$

The PSS is applicable to messages of arbitrary length. The message m cannot be recovered from the signature σ.

Signing. To sign a message $m \in \{0,1\}^*$, Alice proceeds in three steps:

1. Alice chooses $r \in \{0,1\}^k$ at random and calculates $w := h(m\|r)$.
2. She computes $G(w) = (G_1(w), G_2(w))$ and $y := w\|(G_1(w) \oplus r)\|G_2(w)$. (If $y \notin D$, she returns to step 1.)
3. The signature of m is $\sigma := f^{-1}(y)$.

As usual, $\|$ denotes the concatenation of strings and $\oplus$ the bitwise XOR operator. If Alice wants to sign message m, she concatenates a random seed r to the message and applies the hash function h to $m\|r$. Then Alice applies the generator G to the hash value w. The first part $G_1(w)$ of $G(w)$ is used to mask r; the second part of $G(w)$, $G_2(w)$, is appended to $w\|G_1(w) \oplus r$ to obtain a bit string y of appropriate length. All bits of y depend on the message m. The hope is that mapping m into the domain of f by $m \longmapsto y$ behaves like a truly random function. This assumption guarantees the security of the scheme. Finally, y is decrypted with f to get the signature. The random seed r is selected independently for each message m – signing a message twice yields distinct signatures.

Verification. To verify the signature of a signed message (m, σ), we use the same trapdoor function f, the same random bit generator G and the same hash function h as above, and proceed as follows:

1. Compute $f(\sigma)$ and decompose $f(\sigma) = w\|t\|u$, where $|w| = l, |t| = k$ and $|u| = n - (k + l)$.
2. Compute $r = t \oplus G_1(w)$.

3. We accept the signature σ if $h(m\|r) = w$ and $G_2(w) = u$; otherwise we reject it.

PSS can be proven to be secure in the random oracle model under the RSA assumption. The proof assumes that the hash functions G and h are random oracles.

For practical applications of the scheme, it is recommended to implement the hash function h and the random bit generator G with the secure hash algorithm SHA-1 or some other cryptographic hash algorithm that is considered collision resistant. Typical values of the parameters n, k and l are $n = 1024$ bits and $k = l = 128$ bits.

3.5 The Discrete Logarithm

In Section 3.3 we discussed the RSA cryptosystem. The RSA function raises an element m to the e-th power. It is a bijective function and is efficient to compute. If the factorization of n is not known, there is no efficient algorithm for computing the e-th root. There are other functions in number theory that are easy to compute but hard to invert. One of the most important is exponentiation in finite fields. Let p be a prime and g be a primitive root in $\mathbb{Z}_p^*$ (see Appendix A.4). The discrete exponential function

$$\mathrm{Exp} : \mathbb{Z}_{p-1} \longrightarrow \mathbb{Z}_p^*, \ x \longmapsto g^x,$$

is a one-way function. It can be efficiently computed, for example, by the repeated squaring algorithm (Section 3.2). No efficient algorithm for computing the inverse function Log of Exp, i.e., for computing x from $y = g^x$, is known, and it is widely believed that no such algorithm exists. This assumption is called the *discrete logarithm assumption* (for a precise definition, see Definition 6.1).

3.5.1 ElGamal's Encryption

In contrast to the RSA function, Exp is a one-way function without a trapdoor. It does not have any additional information, which makes the computation of the inverse function easy. Nevertheless, Exp is the basis of ElGamal's cryptosystem ([ElGamal84]).

Key Generation. The recipient of messages, Bob, proceeds as follows:

1. He chooses a large prime p, such that $p - 1$ has a big prime factor and a primitive root $g \in \mathbb{Z}_p^*$.
2. He chooses at random an integer x in the range $0 \le x \le p - 2$. The triple (p, g, x) is the secret key of Bob.
3. He computes $y = g^x$ in $\mathbb{Z}_p$. The public key of Bob is (p, g, y), and x is kept secret.

The number $p-1$ will have a large prime factor if Bob is looking for primes p of the form $2kq + 1$, where q is a large prime. Thus, Bob first chooses a large prime q. Here he proceeds in the same way as in the RSA key generation procedure (see Section 3.3.1). Then, to get p, Bob randomly generates a k of appropriate bit length and applies a probabilistic primality test to $z = 2kq+1$. He replaces k by $k+1$ until he succeeds in finding a prime. He expects to test $O(\ln z)$ numbers for primality before reaching the first prime (see Corollary A.71). Having found a prime $p = 2kq + 1$, he randomly selects elements g in $\mathbb{Z}_p^*$ and tests whether g is a primitive root. The factorization of k is required for this test (see Algorithm A.39). Thus q must be chosen to be sufficiently large, such that k is small enough to be factored efficiently.

Bob has to avoid that all prime factors of $p-1$ are small. Otherwise, there is an efficient algorithm for the computation of discrete logarithms developed by Silver, Pohlig and Hellman (see [Koblitz94]).

Encryption and Decryption. Alice encrypts messages for Bob by using Bob's public key (p, g, y). She can encrypt elements $m \in \mathbb{Z}_p$. To encrypt a message $m \in \mathbb{Z}_p$, Alice chooses at random an integer $k, 1 \leq k \leq p - 2$. The encrypted message is the following pair (c_1, c_2) of elements in $\mathbb{Z}_p$:

$$(c_1, c_2) := (g^k, y^k m).$$

The computations are done in $\mathbb{Z}_p$. By multiplying m with y^k, Alice hides the message m behind the random element y^k.

Bob decrypts a ciphertext (c_1, c_2) by using his secret key x. Since $y^k = (g^x)^k = (g^k)^x = c_1^x$, he obtains the plaintext m by multiplying c_2 with the inverse c_1^{-x} of c_1^x:

$$c_1^{-x} c_2 = y^{-k} y^k m = m.$$

Recall that $c_1^{-x} = c_1^{p-1-x}$, because $c_1^{p-1} = [1]$ (see "Computing modulo a prime" on page 303). Therefore, Bob can decrypt the ciphertext by raising c_1 to the $(p-1-x)$-th power, $m = c_1^{p-1-x} c_2$. Note that $p-1-x$ is a positive number.

The encryption algorithm is not a deterministic algorithm. The cryptogram depends on the message, the public key and on a randomly chosen number. If the random number is chosen independently for each message, it rarely happens that two plaintexts lead to the same ciphertext.

The security of the scheme depends on the following assumption: it is impossible to compute g^{xk} (and hence $g^{-xk} = (g^{xk})^{-1}$ and m) from g^x and g^k, which is called the *Diffie-Hellman problem*. An efficient algorithm to compute discrete logarithms would solve the Diffie-Hellman problem. It is unknown whether the Diffie-Hellman problem is equivalent to computing discrete logarithms, but it is believed that no efficient algorithm exists for this problem (also see Section 4.1.2).

Like basic RSA (see Section 3.3.3), ElGamal's encryption is vulnerable to a chosen-ciphertext attack. Adversary Eve, who wants to decrypt a ciphertext

$c = (c_1, c_2)$, with $c_1 = g^k$ and $c_2 = my^k$, chooses random elements $\tilde{k}$ and $\tilde{m}$ and gets Bob to decrypt $\tilde{c} = (c_1 g^{\tilde{k}}, c_2 \tilde{m} y^{\tilde{k}})$. Bob sends $m\tilde{m}$, the plaintext of $\tilde{c} = (g^{k+\tilde{k}}, m\tilde{m} y^{k+\tilde{k}})$, to Eve. Eve simply divides by $\tilde{m}$ and obtains the plaintext m of c: $m = (m\tilde{m})\tilde{m}^{-1}$. Bob's suspicion is not aroused, because the plaintext $m\tilde{m}$ looks random to him.

3.5.2 ElGamal's Signature Scheme

Key Generation. To generate a key for signing, Alice proceeds as Bob in the key generation procedure above to obtain a public key (p, g, y) and a secret key (p, g, x) with $y = g^x$.

Signing. We assume that the message m to be signed is an element in $\mathbb{Z}_p$. In practice, a hash function h is used to map the messages into $\mathbb{Z}_p$. Then the hash value is signed. The signed message is produced by Alice using the following steps:

1. She selects a random integer k, $1 \leq k \leq p - 2$, with $\gcd(k, p - 1) = 1$.
2. She sets $r := g^k$ and $s := k^{-1}(m - rx) \bmod (p - 1)$.
3. (m, r, s) is the signed message.

Verification. Bob verifies the signed message (m, r, s) as follows:

1. He verifies whether $1 \leq r \leq p - 1$. If not, he rejects the signature.
2. He computes $v := g^m$ and $w := y^r r^s$, where y is Alice's public key.
3. The signature is accepted if $v = w$; otherwise it is rejected.

Proposition 3.11. *If Alice signed the message* (m, r, s), *we have* $v = w$.

Proof.

$$w = y^r r^s = (g^x)^r (g^k)^s = g^{rx} g^{kk^{-1}(m-rx)} = g^m = v.$$

Here, recall that exponents of g can be reduced modulo $(p - 1)$, since $g^{p-1} = [1]$ (see "Computing modulo a prime" on page 303). □

Remarks. The following observations concern the security of the system:

1. The security of the system depends on the discrete logarithm assumption. Someone who can compute discrete logarithms can get everyone's secret key and thereby break the system totally. To find an s, such that $g^m = y^r r^s$, on given inputs m and r, is equivalent to the computation of discrete logarithms.

   To forge a signature for a message m, one has to find elements r and s, such that $g^m = y^r r^s$. It is not known whether this problem is equivalent to the computation of discrete logarithms. However, it is also believed that no efficient algorithm for this problem exists.

2. If adversary Eve succeeds in getting the chosen random number k for some signed message m, she can compute $rx \equiv (m - sk) \bmod (p-1)$ and the secret key x, because with high probability $\gcd(r, p - 1) = 1$. Thus, the random number generator used to get k must be of superior quality.

3. It is absolutely necessary to choose a new random number for each message. If the same random number is used for different messages $m_1 \neq m_2$, it is possible to compute k: $s - s' \equiv (m - m')k^{-1} \bmod (p-1)$ and hence $k \equiv (s - s')^{-1}(m - m') \bmod (p - 1)$.

4. When used without a hash function, ElGamal's signature scheme is existentially forgeable; i.e., an adversary Eve can construct a message m and a valid signature (m, r, s) for m.

   This is easily done. Let b and c be numbers such that $\gcd(c, p - 1) = 1$. Set $r = g^b y^c$, $s = -rc^{-1} \bmod (p-1)$ and $m = -rbc^{-1} \bmod (p-1)$. Then (m, r, s) satisfies $g^m = y^r r^s$. Fortunately in practice, as observed above, a hash function h is applied to the original message, and it is the hash value that is signed. Thus, to forge the signature for a real message is not so easy. Adversary Eve has to find some meaningful message $\tilde{m}$ with $h(\tilde{m}) = m$. If h is a collision-resistant hash function, her probability of accomplishing this is very low.

5. D. Bleichenbacher observed in [Bleichenbacher96] that step 1 in the verification procedure is essential. Otherwise Eve would be able to sign messages of her choice, provided she knows one valid signature (m, r, s), where m is a unit in $\mathbb{Z}_{p-1}$.

   Let m' be a message of Eve's choice, $u = m'm^{-1} \bmod (p - 1)$, $s' = su \bmod (p - 1)$, $r' \in \mathbb{Z}$, such that $r' \equiv r \bmod p$ and $r' \equiv ru \bmod (p-1)$. r' is obtained by the Chinese Remainder Theorem (see Theorem A.29). Then (m', r', s') is accepted by the verification procedure.

3.5.3 Digital Signature Algorithm

In 1991 NIST proposed a digital signature standard (DSS) (see [NIST94]). DSS was intended to become a standard digital signature method for use by government and financial organizations. The DSS contains the digital signature algorithm (DSA), which is very similar to ElGamal's algorithm.

Key Generation. The keys are generated in a similar way as in ElGamal's signature scheme. As above, a prime p, an element $g \in \mathbb{Z}_p^*$ and an exponent x are chosen. x is kept secret, whereas p, g and $y = g^x$ are published. The difference is that g is not a primitive root in $\mathbb{Z}_p^*$, but an element of order q, where q is a prime divisor of $p-1$.[18] Moreover, the binary size of q is required to be 160 bits.

To generate a public and a secret key, Alice proceeds as follows:

---

[18] The order of g is the smallest $e \in \mathbb{N}$ with $g^e = [1]$. The order of a primitive root in $\mathbb{Z}_p^*$ is $p - 1$.

1. She chooses a 160-bit prime q and a prime p, such that q divides $p - 1$ (p should have the binary length $|p| = 512 + 64t, 0 \leq t \leq 8$). She can do this in a way analogous to the key generation in ElGamal's encryption scheme. First, she selects q at random, and then she looks for a prime p in $\{2kq + 1, 2(k + 1)q + 1, 2(k + 2)q + 1, \ldots\}$, with k a randomly chosen number of appropriate size.
2. To get an element g of order q, she selects elements $h \in \mathbb{Z}_p^*$ at random until $g := h^{(p-1)/q} \neq [1]$. Then g has order q, and it generates the unique cyclic group G_q of order q in $\mathbb{Z}_p^*$.[19] Note that in G_q elements are computed modulo p and exponents are computed modulo q.[20]
3. Finally, she chooses an integer x in the range $1 \leq x \leq q - 1$ at random.
4. (p, q, g, x) is the secret key, and the public key is (p, q, g, y), with $y := g^x$.

Signing. Messages m to be signed by DSA must be elements in $\mathbb{Z}_q$. In DSS, a hash function h is used to map real messages to elements of $\mathbb{Z}_q$. The signed message is produced by Alice using the following steps:

1. She selects a random integer k, $1 \leq k \leq q - 1$.
2. She sets $r := (g^k \bmod p) \bmod q$ and $s := k^{-1}(m + rx) \bmod q$. If $s = 0$, she returns to step 1, but it is extremely unlikely that this occurs.
3. (m, r, s) is the signed message.

Recall the verification condition of ElGamal's signature scheme. It says that $(m, \tilde{r}, \tilde{s})$ with $\tilde{r} = g^k \bmod p$ and $\tilde{s} = k^{-1}(m - \tilde{r}x) \bmod (p - 1)$ can be verified by

$$y^{\tilde{r}} \tilde{r}^{\tilde{s}} \equiv g^m \bmod p.[21]$$

Now suppose that, as in DSA, $\tilde{s}$ is defined by use of $(m + \tilde{r}x)$, $\tilde{s} = (m + \tilde{r}x)k^{-1} \bmod (p - 1)$, and not by use of $(m - \tilde{r}x)$, as in ElGamal's scheme. Then the equation remains valid if we replace the exponent $\tilde{r}$ of y by $-\tilde{r}$:

$$y^{-\tilde{r}} \tilde{r}^{\tilde{s}} \equiv g^m \bmod p. \tag{3.1}$$

In the DSA, g and hence $\tilde{r}$ and y are elements of order q in $\mathbb{Z}_p^*$. Thus, we can replace the exponents $\tilde{s}$ and $\tilde{r}$ in (3.1) by $\tilde{s} \bmod q = s$ and $\tilde{r} \bmod q = r$. So, we have the idea that a verification condition for (r, s) may be derived from (3.1) by reducing $\tilde{r}$ and $\tilde{s}$ modulo q. This is not so easy, because the exponent $\tilde{r}$ also appears as a base on the left-hand side of (3.1). The base cannot be reduced without destroying the equality. To overcome this difficulty, we first transform (3.1) to

$$\tilde{r}^s \equiv g^m y^r \bmod p.$$

---

[19] There is a unique subgroup G_q of order q of $\mathbb{Z}_p^*$. G_q consists of the unit element and all elements $x \in \mathbb{Z}_p^*$ of order q. It is cyclic and each member except the unit element is a generator, see Lemma A.40.

[20] See "Computing modulo a prime" on page 303.

[21] We write "mod p" to make clear that computations are done in $\mathbb{Z}_p$.

Now, the idea of DSA is to remove the exponentiation on the left-hand side. This is possible because s is a unit in $\mathbb{Z}_q^*$. For $t = s^{-1} \bmod q$, we get

$$\tilde{r} \equiv (g^m y^r)^t \bmod p.$$

Now we can reduce by modulo q on both sides and obtain the verification condition of DSA:

$$r = \tilde{r} \bmod q \equiv ((g^m y^r)^t \bmod p) \bmod q.$$

A complete proof of the verification condition is given below (Proposition 3.12). Note that the exponentiations on the right-hand side are done in $\mathbb{Z}_p$.

Verification. Bob verifies the signed message (m, r, s) as follows:

1. He verifies that $1 \leq r \leq q - 1$ and $1 \leq s \leq q - 1$; if not, then he rejects the signature.
2. He computes the inverse $t := s^{-1}$ of s modulo q and $v := ((g^m y^r)^t \bmod p) \bmod q$, where y is the public key of Alice.
3. The signature is accepted if $v = r$; otherwise it is rejected.

Proposition 3.12. *If Alice signed the message (m, r, s), we have $v = r$.*

Proof.

$$v = ((g^m y^r)^t \bmod p) \bmod q = (g^{mt} g^{rxt} \bmod p) \bmod q$$
$$= (g^{(m+rx)t \bmod q} \bmod p) \bmod q = (g^k \bmod p) \bmod q$$
$$= r.$$

Note that exponents can be reduced by modulo q, because $g^q \equiv 1 \bmod p$. $\square$

Remarks:

1. Compared with ElGamal, the DSA has the advantage that signatures are fairly short, consisting of two 160-bit numbers.
2. In DSA, most computations – in particular the exponentiations – take place in the field $\mathbb{Z}_p^*$. The security of DSA depends on the difficulty of computing discrete logarithms. So it relies on the discrete logarithm assumption. This assumption says that it is infeasible to compute the discrete logarithm x of an element $y = g^x$ randomly chosen from $\mathbb{Z}_p^*$, where p is a sufficiently large prime and g is a primitive root of $\mathbb{Z}_p^*$ (see Definition 6.1 for a precise statement). Here, as in some cryptographic protocols discussed in Chapter 4 (commitments, electronic elections and digital cash), the base g is not a primitive root (with order $p - 1$), but an element of order q, where q is a large prime divisor of $p-1$. To get the secret key x, it would suffice to find discrete logarithms for random elements $y = g^x$ from the much smaller subgroup G_q generated by g. Thus, the security of DSA (and some protocols discussed in Chapter 4) requires the (widely believed) stronger assumption that finding discrete logarithms for elements randomly chosen from the subgroup G_q is infeasible.

3. The remarks on the security of ElGamal's signature scheme also apply to DSA and DSS.
4. In the DSS, messages are first hashed before signed by DSA. The DSS suggests taking SHA-1 for the hash function.

3.6 Modular Squaring

Breaking the RSA cryptosystem might be easier than solving the factoring problem. It is widely believed to be equivalent to factoring, but no proof of this assumption exists. Rabin proposed a cryptosystem whose underlying encryption algorithm is provably as difficult to break as the factorization of large numbers (see [Rabin79]).

3.6.1 Rabin's Encryption

Rabin's system is based on the modular squaring function

$$\text{Square}: \mathbb{Z}_n \longrightarrow \mathbb{Z}_n, \ m \longmapsto m^2.$$

This is a one-way function with trapdoor, if the factoring of large numbers is assumed to be infeasible (see Section 3.2.2).

Key Generation. As in the RSA scheme, we randomly choose two large distinct primes, p and q, for Bob. The scheme works with arbitrary primes, but primes p and q, such that $p, q \equiv 3 \bmod 4$ speed up the decryption algorithm. Such primes are found as in the RSA key generation procedure. We are looking for primes p and q of the form $4k + 3$ in order to get the condition $p, q \equiv 3 \bmod 4$. Then $n = pq$ is used as the public key and p and q are used as secret key.

Encryption and Decryption. We suppose that the messages to be encrypted are elements in $\mathbb{Z}_n$. The modular squaring one-way function is used as the encryption function E:

$$\text{E}: \mathbb{Z}_n \longrightarrow \mathbb{Z}_n, \ m \longmapsto m^2.$$

To decrypt a ciphertext c, Bob has to compute the square roots of c in $\mathbb{Z}_n$. The computation of modular square roots is discussed in detail in Appendix A.7. For example, square roots modulo n can be efficiently computed, if and only if the factors of n can be efficiently computed. Bob can compute the square roots of c because he knows the secret key p and q.

Using the Chinese Remainder Theorem (Theorem A.29), he decomposes $\mathbb{Z}_n$:

$$\phi: \mathbb{Z}_n \longrightarrow \mathbb{Z}_p \times \mathbb{Z}_q, \ c \longmapsto (c \bmod p, c \bmod q).$$

Then he computes the square roots of $c \bmod p$ and the square roots of $c \bmod q$. Combining the solutions, he gets the square roots modulo n. If p divides

c (or q divides c), then the only square root modulo p (or modulo q) is 0. Otherwise, there are two distinct square roots modulo p and modulo q. Since the primes are $\equiv 3 \bmod 4$, the square roots can be computed by one modular exponentiation (see Algorithm A.61). Bob combines the square roots modulo p and modulo q by using the Chinese Remainder Theorem, and obtains four distinct square roots of c (or two roots in the rare case that p or q divides c, or the only root 0 if $c = 0$). He has to decide which of the square roots represents the plaintext. There are different approaches. If the message is written in some natural language, it should be easy to choose the right one. If the messages are unstructured, one way to solve the problem is to add additional information. The sender, Alice, might add a header to each message consisting of the Jacobian symbol $\left(\frac{m}{n}\right)$ and the sign bit b of m. The sign bit b is defined as 0 if $0 \leq m < n/2$, and 1 otherwise. Now Bob can easily determine the correct square root (see Proposition A.66).

Remark. The difficulty of computing square roots modulo n is equivalent to the difficulty of computing the factors of n (see Proposition A.64). Hence, Rabin's encryption resists ciphertext-only attacks as long as factoring is impossible. The basic scheme, however, is completely insecure against a chosen-ciphertext attack. If adversary Eve can use the decryption algorithm as a black box, she can determine the secret key using the following attack. She selects m in the range $0 < m < n$ and computes $c = m^2 \bmod n$. Decrypting c delivers y. There is a 50% chance that $m \not\equiv \pm y \bmod n$, and in this case Eve can easily compute the prime factors of n from m and y (see Lemma A.63). If the Jacobian symbol $\left(\frac{m}{n}\right)$ is added to each message m as sketched above, Eve may choose m with $\left(\frac{m}{n}\right) = -1$, but add +1 to the header. Then she will be certain to get a square root y with $m \not\equiv \pm y$. Applying some proper formatting, as in the OAEP scheme, can prevent this attack.

3.6.2 Rabin's Signature Scheme

The decryption function in Rabin's cryptosystem is only applicable to quadratic residues modulo n. Therefore, the system is not immediately applicable as a digital signature scheme. Before applying the decryption function, we usually apply a hash function to the message to be signed. Here we need a collision-resistant hash function whose values are quadratic residues modulo n. Such a hash function is obtained by the following construction. Let M be the message space and let

$$h : M \times \{0,1\}^k \longrightarrow \mathbb{Z}_n, \ (m, x) \longmapsto h(m, x)$$

be a hash function. To sign a message m, Bob generates pseudorandom bits x using a pseudorandom bit generator and computes $h(m, x)$. Knowing the factors of n, he can easily test whether $z := h(m, x) \in \mathrm{QR}_n$. z is a square if and only if $z \bmod p$ and $z \bmod q$ are squares, and $z \bmod p$ is a square in $\mathbb{Z}_p$, if and only if $z^{(p-1)/2} \equiv 1 \bmod p$ (Proposition A.52). He repeatedly chooses pseudorandom bit strings x until $h(m, x)$ is a square in $\mathbb{Z}_n$. Then he computes a square root y of $h(m, x)$ (e.g. using Proposition A.62). The signed message is defined as

$$(m, x, y).$$

A signed message (m, x, y) is verified by testing

$$h(m, x) = y^2.$$

If an adversary Eve is able to make Bob sign hash values of her choice, she can figure out Bob's secret key (see above).

Exercises

1. Set up an RSA encryption scheme by generating a pair of public and secret keys. Choose a suitable plaintext and a ciphertext. Encrypt and decrypt them.

2. Let n denote the product of two distinct primes p and q, and let $e \in \mathbb{N}$. Show that e is prime to $\varphi(n)$ if

$$\mu : \mathbb{Z}_n^* \longrightarrow \mathbb{Z}_n^*, \ x \longmapsto x^e$$

is bijective.

3. Let $\mathrm{RSA}_e : \mathbb{Z}_n^* \longrightarrow \mathbb{Z}_n^*, \ x \longmapsto x^e$. Show that

$$|\{x \in \mathbb{Z}_n^* \mid \mathrm{RSA}_e(x) = x\}| = \gcd(e - 1, p - 1) \cdot \gcd(e - 1, q - 1).$$

   Hint: Show that $|\{x \in \mathbb{Z}_p^* \mid x^k = 1\}| = \gcd(k, p - 1)$, where p is a prime, and use the Chinese Remainder Theorem (Theorem A.29).

4. Let (n, e) be the public key of an RSA cryptosystem and d be the associated decryption exponent. Construct an efficient probabilistic algorithm A which on input n, e and d computes the prime factors p and q of n with very high probability.
   Hint: Use the idea of the Miller-Rabin primality test, especially case 2 in the proof of Proposition A.78.

5. Consider RSA encryption. Discuss, in detail, the advantage you get using, for encryption, a public exponent which has only two 1 digits in its binary encoding and using, for decryption, the Chinese Remainder Theorem.

6. Let p, p', q and q' be prime numbers, with $p' \neq q'$, $p = ap' + 1, q = bq' + 1$ and $n := pq$:
   a. Show $|\{x \in \mathbb{Z}_p^* \mid p' \text{ does not divide } \mathrm{ord}(x)\}| = a$.

b. Assume that p' and q' are large (compared to a and b). Let $x \in \mathbb{Z}_n^*$ be a randomly chosen element. Show that the probability that x has large order is $\geq 1 - \left(1/p' + 1/q' - 1/p'q'\right)$. More precisely, show $|\{x \in \mathbb{Z}_n^* \mid p'q' \text{ does not divide } ord(x)\}| = a(q-1) + b(p-1) - ab$.

7. Consider RSA encryption $\text{RSA}_e : \mathbb{Z}_n^* \longrightarrow \mathbb{Z}_n^*, \ x \longmapsto x^e$:

   a. Show that $\text{RSA}_e^l = id_{\mathbb{Z}_n^*}$ for some $l \in \mathbb{N}$.

   b. Consider the decryption-by-iterated-encryption attack (see Section 3.3.1). Let p, p', p'', q, q' and q'' be prime numbers, with $p' \neq q'$, $p = ap' + 1, q = bq' + 1, \ p' = a'p'' + 1, q' = b'q'' + 1, \ n := pq$ and $n' := p'q'$. Assume that p' and q' are large (compared to a and b) and that p'' and q'' are large (compared to a' and b'). (This means that the factors of n satisfy the conditions 1 and 3 required for strong primes.)

   Show that the number of iterations necessary to decrypt a ciphertext c is $\geq p''q''$ (and thus very large) for all but an exponentially small fraction of ciphertexts. By exponentially small, we mean $\leq 2^{-|n|/k}$ (k, constant).

8. Let p be a large prime, such that $q := (p-1)/2$ is also prime. Let G_q be the subgroup of order q in $\mathbb{Z}_p^*$. Let g and h be randomly chosen generators in G_q. We assume that it is infeasible to compute discrete logarithms in G_q. Show that

$$f : \{0, \ldots, q-1\}^2 \longrightarrow G_q, \ (x, y) \longmapsto g^x h^y$$

can be used to obtain a collision-resistant compression function.

9. Let $h : \{0,1\}^* \longrightarrow \{0,1\}^n$ be a cryptographic hash function. We assume that the hash values are uniformly distributed, i.e., each value $v \in \{0,1\}^n$ has the same probability $1/2^n$. How many steps do you expect until you succeed with the brute-force attacks against the one-way property and second pre-image resistance?

10. Set up an ElGamal encryption scheme by generating a pair of public and secret keys.

   a. Choose a suitable plaintext and a ciphertext. Encrypt and decrypt them.

   b. Generate ElGamal signatures for suitable messages. Verify the signatures.

   c. Forge a signature without using the secret key.

   d. Play the role of an adversary Eve, who learns the random number k used to generate a signature, and break the system.

   e. Demonstrate that checking the condition $1 \leq r \leq p-1$ is necessary in the verification of a signature (r, s).

11. Weak generators (see [Bleichenbacher96]; [MenOorVan96]).

   Let p be a prime, $p \equiv 1 \bmod 4$, and $g \in \mathbb{Z}$, such that $g \bmod p$ is a

primitive root in $\mathbb{Z}_p^*$. Let (p, g, x) be Bob's secret key and $(p, g, y = g^x)$ be Bob's public key in an ElGamal signature scheme. We assume that: (1) $p - 1 = gt$ and (2) discrete logarithms can be efficiently computed in the subgroup H of order g in $\mathbb{Z}_p^*$ (e.g. using the Pohlig-Hellman algorithm). To sign a message m, adversary Eve does the following: (1) she sets $r = t$; (2) she computes z, such that $g^{tz} = y^t$; and (3) she sets $s = \frac{1}{2}(p - 3)(m - tz) \bmod (p - 1)$.
Show:

  a. That it is possible to compute z in step 2.

  b. That (r, s) is accepted as Bob's signature for m.

  c. How the attack can be prevented.

12. Set up a DSA signature scheme by generating a pair of public and secret keys. Generate and verify signatures for suitable messages. Take small primes p and q.

13. Set up a Rabin encryption scheme by generating a pair of public and secret keys. Choose a suitable plaintext and a ciphertext. Encrypt and decrypt them. Then play the role of an adversary Eve, who succeeds in a chosen-ciphertext attack and recovers the secret key.

4. Cryptographic Protocols

One of the major contributions of modern cryptography is the development of advanced protocols providing high-level cryptographic services, such as secure user identification, voting schemes and digital cash. Cryptographic protocols use cryptographic mechanisms – such as encryption algorithms and digital signature schemes – as basic components.

A protocol is a multi-party algorithm which requires at least two participating parties. Therefore, the algorithm is distributed and invoked in at least two different places. An algorithm that is not distributed, is not called a protocol. The parties of the algorithm must communicate with each other to complete the task. The communication is described by the messages to be exchanged between the parties. These are referred to as the communication interface. The protocol requires precise definitions of the interface and the actions to be taken by each party.

A party participating in a protocol must fulfill the syntax of the communication interface, since not following the syntax would be immediately detected by the other parties. The party can behave honestly and follow the behavior specified in the protocol. Or she can behave dishonestly, only fulfill the syntax of the communication interface and do completely different things otherwise. These points must be taken into account when designing cryptographic protocols.

4.1 Key Exchange and Entity Authentication

Public- and secret-key cryptosystems assume that the participating parties have access to keys. In practice, one can only apply these systems if the problem of distributing the keys is solved.

The security concept for keys, which we describe below, has two levels. The first level embraces long-lived, secret keys, called *master keys*. The keys of the second level are associated with a session, and are called *session keys*. A session key is only valid for the short time of the duration of a session. The master keys are usually keys of a public-key cryptosystem.

There are two main reasons for the two-level concept. The first is that symmetric key encryption is more efficient than public-key encryption. Thus, session keys are usually keys of a symmetric cryptosystem, and these keys must be exchanged in a secure way, by using other keys. The second, probably more important reason is that the two-level concept provides more security. If a session key is compromised, it affects only that session; other sessions in the past or in the future are not concerned. Given one session key, the number of ciphertexts available for cryptanalysis is limited. Session keys are generated when actually required and discarded after use; they need not be stored. Thus, there is no need to protect a large amount of stored keys.

A master key is used for the generation of session keys. Special care is taken to prevent attacks on the master key. The access to the master key is severely limited. It is possible to store the master key on protected hardware, accessible only via a secure interface. The main focus of this section is to describe how to establish a session key between two parties.

Besides key exchange, we introduce *entity authentication*. Entity authentication prevents impersonation. By entity authentication, Alice can convince her communication partner Bob that, in the current communication, her identity is as declared. This might be achieved, for example, if Alice signs a specific message. Alice proves her identity by her signature on the message. If an adversary Eve intercepts the message signed by Alice, she can use it later to authenticate herself as Alice. Such attacks are called *replay attacks*. A replay attack can be prevented if the message to be signed by Alice varies. For this purpose we introduce two methods. In the first method, Alice puts Bob's name and a time stamp into the message she signs. Bob accepts a message only if it appears the first time. The second method uses random numbers. A random number is chosen by Bob and transmitted to Alice. Then Alice puts the random number into the message, signs the message and returns it to Bob. Bob can check that the returned random number matches the random number he sent and the validity of Alice's signature. The random number is viewed as a challenge. Bob sends a challenge to Alice and Alice returns a response to Bob's challenge. We speak of *challenge-response* identification.

Some of the protocols we discuss provide both entity authentication and key exchange.

4.1.1 Kerberos

Kerberos denotes the distributed authentication service originating from MIT's project Athena. Here we use the term Kerberos in a restricted way: we define it as the underlying protocol that provides both entity authentication and key establishment, by use of symmetric cryptography and a trusted third party.

We continue with our description in [NeuTs'o94]. In that overview article, a simplified version of the basic protocol is introduced to make the basic principles clear. Kerberos is designed to authenticate clients who try to get access to servers in a network. A central role is played by a trusted server called the *Kerberos authentication server*.

The Kerberos authentication server T shares a secret key of a symmetric key encryption scheme E with each client A and each server B in the network, denoted by k_A and k_B respectively. Now, assume that client A wants to access server B. At the outset, A and B do not share any secrets. The execution of the Kerberos protocol involves A, B and T, and proceeds as follows:

Client A sends a request to the authentication server T, requesting *credentials* for server B. T responds with these credentials. The credentials consist of:

1. A *ticket* t for the server B, encrypted with B's secret key k_B.
2. A session key k, encrypted with A's key k_A.

The ticket t contains A's identity and a copy of the session key. It is intended for B. A will forward it to B. The ticket is encrypted with k_B, which is known only to B and T. Thus, it is not possible for A to modify the ticket without being detected. A creates an *authenticator* which also contains A's identity, encrypts it with the session key k (by this encryption A proves to B that she knows the session key embedded in the ticket) and transmits the ticket and the authenticator to B. B trusts the ticket (it is encrypted with k_B, hence it originates from T), decrypts it and gets k. Now B uses the session key to decrypt the authenticator. If B succeeds, he is convinced that A encrypted the authenticator, because only A and the trusted T can know k. Thus A is authenticated to B. Optionally, the session key k can also be used to authenticate B to A. Finally, k may be used to encrypt further communication between the two parties in the current session.

Kerberos protects the ticket and the authenticator against modification by encrypting it. Thus, the encryption algorithm E is assumed to have built-in data integrity mechanisms.

Protocol 4.1.
Basic Kerberos authentication protocol (simplified):

1. A chooses r_A at random[1] and sends (A, B, r_A) to T.
2. T generates a new session key k, and creates a ticket
   $t := (A, k, l)$. Here l defines a validity period (consisting of a starting and an ending time) for the ticket. T sends
   $(E_{k_A}(k, r_A, l, B), E_{k_B}(t))$ to A.
3. A recovers k, r_A, l and B, and verifies that r_A and B match those sent in step 1. Then she creates an authenticator
   $a := (A, t_A)$, where t_A is a time stamp from A's local clock, and sends $(E_k(a), E_{k_B}(t))$ to B.
4. B recovers $t = (A, k, l)$ and $a = (A, t_A)$, and checks that:
   a. The identifier A in the ticket matches the one in the authenticator.

---

[1] In this chapter all random choices are with respect to the uniform distribution. All elements have the same probability (see Appendix B.1).

    b. The time stamp t_A is fresh, i.e., within a small time interval around B's local time.

    c. The time stamp t_A is in the validity period l.

If all checks pass, B considers the authentication of A as successful.

If, in addition, B is to be authenticated to A, steps 5 and 6 are executed:

    5. B takes t_A and sends $E_k(t_A)$ to A.

    6. A recovers t_A from $E_k(t_A)$ and checks whether it matches with the t_A sent in step 3. If yes, she considers B as authenticated.

Remarks:

1. In step 1, a random number is included in the request. It is used to match the response in step 2 with the request. This ensures that the Kerberos authentication server is alive and created the response. In step 3, a time stamp is included in the request to the server. This prevents replay attacks of such requests. To avoid perfect time synchronization, a small time interval around B's local time (called a window) is used. The server accepts the request if its time stamp is in the current window and appears the first time. The use of time stamps means that the network must provide secure and synchronized clocks. Modifications of local time clocks by adversaries must be prevented to guarantee the security of the protocol.

2. The validity period of a ticket allows the reuse of the ticket in that period. Then steps 1 and 2 in the protocol can be omitted. The client can use the ticket t to repeatedly get access to the server for which the ticket was issued. Each time, she creates a new authenticator and executes steps 3 and 4 (or steps 3–6) of the protocol.

3. Kerberos is a popular authentication service. Version 5 of Kerberos (the current version) was specified in [RFC 1510]. Kerberos is based in part on Needham and Schroeder's trusted third-party authentication protocol [NeeSch78].

4. In the non-basic version of Kerberos, the authentication server is only used to get tickets for the ticket-granting server. These tickets are called ticket-granting tickets. The ticket-granting server is a specialized server, granting tickets (server tickets) for the other servers (the ticket-granting server must have access to the servers' secret keys, so usually the authentication server and the ticket granting server run on the same host).

   Client A executes steps 1 and 2 of Protocol 4.1 with the authentication server in order to obtain a ticket-granting ticket. Then A uses the ticket-granting ticket – more precisely, the session key included in the ticket granting ticket – to authenticate herself to the ticket-granting server and to get server tickets. The ticket-granting ticket can be reused during its validity period for the intended ticket-granting server. As long as the same ticket-granting ticket is used, the client's secret key k_A is not used again. This reduces the risk of exposing k_A. We get a three-level key scheme. The first embraces the long-lived secret keys of the participating clients and servers. The keys of the second level are the session keys of the ticket-granting tickets, and the keys of the third level are the session keys of the server tickets.

A ticket-granting ticket is verified by the ticket-granting server in the same way as any other ticket (see above). The ticket-granting server decrypts the ticket, extracts the session key and decrypts the authenticator with the session key. The client uses a ticket from the ticket-granting server as in the basic model to authenticate to a service in the network.

4.1.2 Diffie-Hellman Key Agreement

Diffie-Hellman key agreement (also called exponential key exchange) provided the first practical solution to the key distribution problem. It is based on public-key cryptography. W. Diffie and M.E. Hellman published their fundamental technique of key exchange together with the idea of public-key cryptography in the famous paper, "New Directions in Cryptography", in 1976 in [DifHel76]. Exponential key exchange enables two parties that have never communicated before to establish a mutual secret key by exchanging messages over a public channel. However, the scheme only resists passive adversaries.

Let p be a sufficiently large prime, such that it is intractable to compute discrete logarithms in $\mathbb{Z}_p^*$. Let g be a primitive root in $\mathbb{Z}_p^*$. p and g are publicly known. Alice and Bob can establish a secret shared key by executing the following protocol:

Protocol 4.2.
Diffie-Hellman key agreement:

1. Alice chooses a, $0 \leq a \leq p - 2$, at random, sets $c := g^a$ and sends c to Bob.
2. Bob chooses b, $0 \leq b \leq p - 2$, at random, sets $d := g^b$ and sends d to Alice.
3. Alice computes the shared key $k = d^a = (g^b)^a$.
4. Bob computes the shared key $k = c^b = (g^a)^b$.

Remarks:

1. If an attacker can compute discrete logarithms in $\mathbb{Z}_p^*$, he can compute a from c and then $k = d^a$. However, to get the secret key k, it would suffice to compute g^{ab} from g^a and g^b. This problem is called the *Diffie-Hellman problem*. The security of the protocol relies on the assumption that no efficient algorithms exist to solve this problem. This assumption is called the *Diffie-Hellman assumption*. It implies the discrete logarithm assumption. For certain primes, the Diffie-Hellman and the discrete logarithm

assumption have been proven to be equivalent ([Boer88]; [Maurer94]; [MauWol96]; [MauWol98]; [MauWol2000]).

2. Alice and Bob can use the randomly chosen element $k = g^{ab} \in \mathbb{Z}_p^*$ as their session key. The Diffie-Hellman assumption does not guarantee that individual bits or groups of bits of k cannot be efficiently derived from g^a and g^b – this would be a stronger assumption.

   It is recommended to make prime p 1024 bits long. Usually, the length of a session key in a symmetric key encryption scheme is much smaller, say 128 bits. If we take, for example, the 128 most-significant bits of k as a session key $\tilde{k}$, then $\tilde{k}$ is hard to compute from g^a and g^b. However, we do not know that all the individual bits of $\tilde{k}$ are secure (on the other hand, none of the more significant bits is known to be easy to compute). In [BonVen96] it is shown that computing the $\sqrt{|p|}$ most-significant bits of g^{ab} from g^a and g^b is as difficult as computing g^{ab} from g^a and g^b. For a 1024-bit prime p, this result implies that the 32 most-significant bits of g^{ab} are hard to compute. The problem of finding a more secure random session key can be solved by applying an appropriate hash function h to g^{ab}, and taking $\tilde{k} = h(g^{ab})$.

3. This protocol provides protection against passive adversaries. An active attacker Eve can intercept the message sent to Bob by Alice and then play Bob's role. The protocol does not provide authentication of the opposite party. Combined with authentication techniques, the Diffie-Hellman key agreement can be used in practice (see Section 4.1.4).

4.1.3 Key Exchange and Mutual Authentication

The problem of spontaneous key exchange, like Diffie-Hellman's key agreement, is the authenticity of the communication partners in an open network. Entity authentication (also called entity identification) guarantees the identity of the communicating parties in the current communication session, thereby preventing impersonation. Mutual entity authentication requires some mutual secret, which has been exchanged previously, or access to predistributed authentic material, like the public keys of a digital signature scheme.

   The protocol we describe is similar to the X.509 strong three-way authentication protocol (see [ISO/IEC 9594-8]). It provides entity authentication and key distribution, two different cryptographic mechanisms. The term "strong" distinguishes the protocol from simpler password-based schemes. To set up the scheme, a public-key encryption scheme (E, D) and a digital signature scheme $(Sign, Verify)$ are needed. Each user Alice has a key pair (e_A, d_A) for encryption and another key pair (s_A, v_A) for digital signatures. It is assumed that everyone has access to Alice's authentic public keys, e_A and v_A, for encryption and the verification of signatures.

Executing the following protocol, Alice and Bob establish a secret session key. Furthermore, the protocol guarantees the mutual authenticity of the communication parties.

Protocol 4.3.
Strong three-way authentication:

1. Alice chooses r_A at random, sets $t_1 := (B, r_A)$ (where B represents Bob's identity), $s_1 := Sign_{s_A}(t_1)$ and sends (t_1, s_1) to Bob.
2. Bob verifies Alice's signature, checks that he is the intended recipient, chooses r_B and a session key k at random, encrypts the session key with Alice's public key, $c := E_{e_A}(k)$, sets $t_2 := (A, r_A, r_B, c)$, signs t_2 to get $s_2 := Sign_{s_B}(t_2)$ and sends (t_2, s_2) to Alice.
3. Alice verifies Bobs signature, checks that she is the intended recipient and that the r_A she received matches the r_A from step 1 (this prevents replay attacks). If both verifications pass, she is convinced that her communication partner is Bob. Now Alice decrypts the session key k, sets $t_3 := (B, r_B)$, $s_3 := Sign_{s_A}(t_3)$ and sends (t_3, s_3) to Bob.
4. Bob verifies Alice's signature and checks that the r_B he received matches the r_B from step 2 (this again prevents replay attacks). If both verifications pass, Bob and Alice use k as the session key.

Remarks:

1. The protocol identifies the communication partner by checking that she possesses the secret key of the signature scheme. The check is done by the *challenge-response* principle. First the challenge, a random number, used only once, is submitted. If the partner can return a signature of this random number, he necessarily possesses the secret key, thus proving his identity. The messages exchanged (the communication tokens) are signed by the sender and contain the recipient. This guarantees that the token was constructed for the intended recipient by the sender. Three messages are exchanged in the above protocol. Therefore, it is called the *three-way* or *three-pass authentication protocol*.
   There are also *two-way authentication protocols*. To prevent replay attacks, the communication tokens must be stored. Since these communication tokens have to be deleted from time to time, they are given a time stamp and an expiration time. This requires a network with secure and synchronized clocks. A three-way protocol requires more messages to be exchanged, but avoids storing tokens and maintaining secure and synchronized clocks.
2. The session key is encrypted with a public-key cryptosystem. Suppose adversary Eve records all the data that Alice and Bob have exchanged,

hoping that Alice's secret encryption key will be compromised in the future. If Eve really gets Alice's secret key, she can decrypt the data of all sessions she recorded. A key-exchange scheme which resists this attack is said to have *forward secrecy*. The Diffie-Hellman key agreement does not encrypt a session key. Thus, the session key cannot be revealed by a compromised secret key. If we combine the Diffie-Hellman key agreement with the authentication technique of the previous protocol, as in Section 4.1.4, we achieve a key-exchange protocol with authentication and forward secrecy.

4.1.4 Station-to-Station Protocol

The station-to-station protocol, combines Diffie-Hellman key agreement with authentication. It goes back to earlier work on ISDN telephone security, as outlined by W. Diffie in [Diffie88], in which the protocol is executed between two ISDN telephones (stations). The station-to-station protocol enables two parties to establish a shared secret key k to be used in a symmetric encryption algorithm E. Additionally, it provides mutual authentication.

Let p be a prime such that it is intractable to compute discrete logarithms in $\mathbb{Z}_p^*$. Let g be a primitive root in $\mathbb{Z}_p^*$. p and g are publicly known. Further, we assume that a digital signature scheme (*Sign*, *Verify*) is available. Each user Alice has a key pair (s_A, v_A) for this signature scheme. s_A is the secret key for signing and v_A is the public key for verifying Alice's signatures. We assume that each party has access to authentic copies of the other's public key.

Alice and Bob can establish a secret shared key and authenticate each other if they execute the following protocol:

Protocol 4.4.
Station-to-station protocol:

1. Alice chooses a, $0 \le a \le p-2$, at random, sets $c := g^a$ and sends c to Bob.
2. Bob chooses b, $0 \le b \le p - 2$, at random, computes the shared secret key $k = g^{ab}$, takes his secret key s_B and signs the concatenation of g^a and g^b to get $s := Sign_{s_B}(g^a \| g^b)$. Then he sends $(g^b, E_k(s))$ to Alice.
3. Alice computes the shared key $k = g^{ab}$, decrypts $E_k(s)$ and verifies Bob's signature. If this verification succeeds, Alice is convinced that the opposite party is Bob. She takes her secret key s_A, generates the signature $s := Sign_{s_A}(g^b \| g^a)$ and sends $E_k(s)$ to Bob.
4. Bob decrypts $E_k(s)$ and verifies Alice's signature. If the verification succeeds, Bob accepts that he actually shares k with Alice.

Remarks:

1. Both Alice and Bob contribute to the random strings $g^a\|g^b$ and $g^b\|g^a$. Thus each string can serve as a challenge.
2. Encrypting the signatures with the key k guarantees that the party who signed also knows the secret key k.

4.1.5 Public-Key Management Techniques

In the public-key-based key-exchange protocols discussed in the previous sections, we assumed that each party has access to the other parties' (authentic) public keys. This requirement can be met by public-key management techniques. A trusted third party C is needed, similar to the Kerberos authentication server in the Kerberos protocol. However, in contrast to Kerberos, the authentication transactions do not include an online communication with C. C prepares information in advance, which is then available to Alice and Bob during the execution of the authentication protocol. We say that C is offline. Offline third parties reduce network traffic, which is advantageous in large networks.

Certification Authority. The offline trusted party is referred to as a *certification authority*. Her tasks are:

1. To verify the authenticity of the entity to be associated with a public key.
2. To bind a public key to a distinguished name and to register it.
3. (Optionally) to generate a party's private and public key.

Certificates play a fundamental role. They enable the storage and forwarding of public keys over insecure media, without danger of undetected manipulation. Certificates are signed by a certification authority, using a public-key signature scheme. Everyone knows the certification authority's public key. The authenticity of this key may be provided by non-cryptographic means, such as couriers or personal acquisition. Another method would be to publish the key in all newspapers. The public key of the certification authority can be used to verify certificates signed by the certification authority. Certificates prove the binding of a public key to a distinguished name. The signature of the certification authority protects the certificate against undetected manipulation. We list some of the most important data stored on a certificate:

1. A distinguished name (the real name or a pseudonym of the owner of the certificate).
2. The owner's public key.
3. The name of the certification authority.
4. A serial number identifying the certificate.
5. A validity period of the certificate.

Creation of Certificates. If Alice wants to get a certificate, she goes to a certification authority C. To prove her identity, Alice shows her passport. Now, Alice needs public- and private-key pairs for encryption and digital signatures. Alice can generate the key pair and hand over a copy of the public key to C. This alternative might be taken if Alice uses a smart card to store her secret key. Smart cards often involve key generation functionality. If the keys are generated inside the smart card, the private key never leaves it. This reduces the risk of theft. Another model is that C generates the key pair and transmits the secret key to Alice. The transmission requires a secret channel. C must of course be trustworthy, because she has the opportunity to steal the secret key. After the key generation, C puts the public key on the certificate, together with all the other necessary information, and signs the certificate with her secret key.

Storing Certificates. Alice can take her certificate and store it at home. She provides the certificate to others when needed, for example for signature verification. A better solution in an open system is to provide a certificate directory, and to store the certificates there. The certificate directory is a (distributed) database, usually maintained by the certification authority. It enables the search and retrieval of certificates.

Usage of Certificates. If Bob wants to encrypt a message for Alice or to verify a signature allegedly produced by Alice, he retrieves Alice's certificate from the certificate directory (or from Alice) and verifies the certification authority's signature. If the verification is successful, he can be sure that he really receives Alice's public key from the certificate and can use it.

For reasons of operational efficiency, multiple certification authorities are needed in large networks. If Alice and Bob belong to different certification authorities, Bob must access an authentic copy of the public key of Alice's certification authority. This is possible if Bob's certification authority C_B has issued a certificate for Alice's certification authority C_A. Then Bob retrieves the certificate for C_A, verifies it and can then trust the public key of C_A. Now he can retrieve and verify Alice's certificate.

It is not necessary that each certification authority issues a certificate for each other certification authority in the network. We may use a directed graph as a model. The vertices correspond to the certification authorities, and an edge from C_A to C_B corresponds to a certificate of C_A for C_B. Then, a directed path should connect any two certification authorities. This is the minimal requirement which guarantees that each user in the network can verify each other user's certificate.

However, the chaining of authentications may reduce the trust in the final result: the more people you have to trust, the greater the risk that you have a cheater in the group.

Revocation of Certificates. If Alice's secret key is compromised, the corresponding public key can no longer be used for encrypting messages. If the key is used in a signature scheme, Alice can no longer sign messages with this key. Moreover, it should be possible for Alice to deny all signatures produced with this key from then on. Therefore, the fact that Alice's secret key is compromised must be publicized. Of course, the certification authority will remove Alice's certificate from the certificate directory. However, certificates may have been retrieved before and may not yet have expired. It is not possible to notify all users possessing copies of Alice's certificate: they are not known to the certification authority. A solution to this problem is to maintain certificate revocation lists. A certificate revocation list is a list of entries corresponding to revoked certificates. To guarantee authenticity, the list is signed by the certification authority.

4.2 Identification Schemes

There are many situations where it is necessary to "prove" one's identity. Typical scenarios are to login to a computer, to get access to an account for electronic banking or to withdraw money from an automatic teller machine. Older methods use passwords or PINs to implement user identification. Though successfully used in certain environments, these methods also have weaknesses. For example, anyone to whom you must give your password to be verified has the ability to use that password and impersonate you. Zero-knowledge (and other) identification schemes provide a new type of user identification. It is possible for you to authenticate yourself without giving to the authenticator the ability to impersonate you. We will see that very efficient implementations of such schemes exist.

4.2.1 Interactive Proof Systems

There are two participants in an interactive proof system, the *prover* and the *verifier*. It is common to call the prover Peggy and the verifier Vic. Peggy knows some fact (e.g. a secret key sk of a public-key cryptosystem or a square root of a quadratic residue x), which we call the *prover's secret*. In an *interactive proof of knowledge*, Peggy wishes to convince Vic that she knows the prover's secret. Peggy and Vic communicate with each other through a communication channel. Peggy and Vic alternately perform *moves* consisting of:

1. Receive a message from the opposite party.
2. Perform some computation.
3. Send a message to the opposite party.

Usually, Peggy starts and Vic finishes the protocol. In the first move, Peggy does not receive a message. The interactive proof may consist of several *rounds*. This means that the protocol specifies a sequence of moves, and this sequence is repeated a specified number of times. Typically, a move consists of a challenge by Vic and a response by Peggy. Vic accepts or rejects

Peggy's proof, depending on whether Peggy successfully answers all of Vic's challenges.

Proofs in interactive proof systems are quite different from proofs in mathematics. In mathematics, the prover of some theorem can sit down and prove the statement by himself. In interactive proof systems, there are two computational tasks, namely producing a proof (Peggy's task) and verifying its validity (Vic's task). Additionally, communication between the prover and verifier is necessary.

We have the following requirements for interactive proof systems.

1. *(Knowledge) completeness.* If Peggy knows the prover's secret, then Vic will always accept Peggy's proof.
2. *(Knowledge) soundness.* If Peggy can convince Vic with reasonable probability, then she knows the prover's secret.

If the prover and the verifier of an interactive proof system follow the behavior specified in the protocol, they are called an *honest verifier* and an *honest prover*. A prover who does not know the prover's secret and tries to convince the verifier is called a *cheating or dishonest prover*. A verifier who does not follow the behavior specified in the protocol is called a *cheating or dishonest verifier*. Sometimes, the verifier can get additional information from the prover if he does not follow the protocol. Note that each prover (or verifier), whether she is honest or not, fulfills the syntax of the communication interface, because not following the syntax is immediately detected. She may only be dishonest in her private computations and the resulting data that she transmits.

Password Scheme. In a simple password scheme, Peggy uses a secret password to prove her identity. The password is the only message, and it is sent from the prover Peggy to the verifier Vic. Vic accepts Peggy's identity if the transmitted password and the stored password are equal. Here, only one message is transmitted, and obviously the scheme meets the requirements. If Peggy knows the password, Vic accepts. If a cheating prover Eve does not know the password, Vic does not accept. The problem is that everyone who observed the password during communication can use the password.

Identification Based on Public-Key Encryption. In Section 4.1, we considered an identification scheme based on a public-key cryptosystem. We recall the basic scenario. Each user Peggy has a secret key sk only known to her and a public key pk known to everyone. Suppose that everyone who can decrypt a randomly chosen encrypted message must know the secret key. This assumption should be true if the cryptosystem is secure. Hence, the secret key sk can be used to identify Peggy.

Peggy proves her identity to Vic using the following steps:

1. Vic chooses a random message m, encrypts it with the public key pk and sends the cryptogram c to Peggy.

2. Peggy decrypts c with her secret key sk and sends the result m' back to Vic.
3. Vic accepts the identity of Peggy if and only if $m = m'$.

Two messages are exchanged: it is a *two-move* protocol. The completeness of the scheme is obvious. On the other hand, a cheating prover who only knows the public key and a ciphertext should not be able to find the plaintext better than guessing at random. The probability that Vic accepts if the prover does not know the prover's secret is very small. Thus, the scheme is also sound. This reflects Vic's security requirements. Suppose that an adversary Eve observed the exchanged messages and later wants to impersonate Peggy. Vic chooses m at random and computes c. The probability of obtaining the previously observed c is very small. Thus, Eve cannot take advantage of observing the exchanged messages. At first glance, everything seems to be all right. However, there is a security problem if Vic is not honest and does not follow the protocol in step 1. If, instead of a randomly chosen encrypted message, he sends a cryptogram intended for Peggy, then he lets Peggy decrypt the cryptogram. He thereby manages to get the plaintext of a cryptogram which he could not compute by himself. This violates Peggy's security requirements.

4.2.2 Simplified Fiat-Shamir Identification Scheme

Let $n := pq$, where p and q are distinct primes. As usual, QR_n denotes the subgroup of squares in $\mathbb{Z}_n^*$ (see Definition A.48). Let $x \in QR_n$, and let y be a square root of x. The modulus n and the square x are made public, while the prime factors p, q and y are kept secret. The square root y of x is the secret of prover Peggy. Here we assume that it is intractable to compute a square root of x, without knowing the prime factors p and q. This is guaranteed by the factoring assumption (see Definition 6.9) if p and q are sufficiently large randomly chosen primes, and x is also randomly chosen. Note that the ability to compute square roots is equivalent to the ability to factorize n (Proposition A.64). We assume that Peggy chooses n and y, computes $x = y^2$ and publishes the public key (n, x) to all participants. y is Peggy's secret. Then Peggy may prove her identity by an interactive proof of knowledge by proving that she knows a square root y of x. The computations are done in $\mathbb{Z}_n^*$.

Protocol 4.5.
Fiat-Shamir identification (simplified):

1. Peggy chooses $r \in \mathbb{Z}_n^*$ at random and sets $a := r^2$. Peggy sends a to Vic.
2. Vic chooses $e \in \{0, 1\}$ at random. Vic sends e to Peggy.
3. Peggy computes $b := ry^e$ and sends b to Vic, i.e., Peggy sends r if $e = 0$, and ry if $e = 1$.
4. Vic accepts if and only if $b^2 = ax^e$.

In the protocol, three messages are exchanged – it is a *three-move* protocol:

1. The first message is a commitment by Peggy that she knows a square root of a.
2. The second message is a challenge by Vic. If Vic sends $e = 0$, then Peggy has to open the commitment and reveal r. If $e = 1$, she has to show her secret in encrypted form (by revealing ry).
3. The third message is Peggy's response to the challenge of Vic.

Completeness. If Peggy knows y, and both Peggy and Vic follow the protocol, then the response $b = ry^e$ is a square root of ax^e, and Vic will accept.

Soundness. A cheating prover Eve can convince Vic with a probability of $1/2$ if she behaves as follows:

1. Eve chooses $r \in \mathbb{Z}_n^*$ and $\tilde{e} \in \{0,1\}$ at random, and sets $a := r^2 x^{-\tilde{e}}$. Eve sends a to Vic.
2. Vic chooses $e \in \{0,1\}$ at random. Vic sends e to Eve.
3. Eve sends r to Vic.

If $e = \tilde{e}$, Vic accepts. The event $e = \tilde{e}$ occurs with a probability of $1/2$. Thus, Eve succeeds in cheating with a probability of $1/2$.

On the other hand, $1/2$ is the best probability of success that a cheating prover can reach. Namely, assume that a cheating prover Eve can convince Vic with a probability $> 1/2$. Then Eve knows an a for which she can correctly answer both challenges. This means that Eve can compute b_1 and b_2, such that

$$b_1^2 = a \text{ and } b_2^2 = ax.$$

Hence, she can compute the square root $y = b_2 b_1^{-1}$ of x. Recall that x is a random quadratic residue. Thus Eve has an algorithm A that on input $x \in \mathrm{QR}_n$ outputs a square root y of x. Then Eve can use A to factor n (see Proposition A.64). This contradicts our assumption that the factorization of n is intractable.

Security. We have to discuss the security of the scheme from the prover's and from the verifier's points of view.

The verifier accepts the proof of a cheating prover with a probability of $1/2$. The large probability of success of a cheating prover is too high in practice. It might be decreased by performing t rounds, i.e., by iterating the basic protocol t times sequentially and independently. In this way, the probability of cheating is reduced to 2^{-t}. In Section 4.2.4 we will give a generalized version of the protocol, which decreases the probability of success of a cheating prover.

Now we look at the basic protocol from an honest prover's point of view, and study Peggy's security requirements. Vic chooses his challenges from the small set $\{0,1\}$. He has no chance of producing side effects, as in the identification scheme based on public-key cryptography given above. The only information Peggy "communicates" to Vic is the fact that she knows a square root of x. The protocol has the zero-knowledge property studied in Section 4.2.3.

4.2.3 Zero-Knowledge

In the interactive proof system based on a public-key cryptosystem, which we discussed above, a dishonest verifier Vic can decrypt Peggy's cryptograms by interacting with Peggy. Since Vic is not able to decrypt them without interaction, he learns something new by interacting with Peggy. He obtains *knowledge* from Peggy. This is not desirable, because it might violate Peggy's security requirements as our example shows. It is desirable that interactive proof systems are designed so that no knowledge is transferred from the prover to the verifier. Such proof systems are called zero-knowledge. Informally, an interactive proof system is called zero-knowledge if whatever the verifier can efficiently compute after interacting with the prover, can be efficiently simulated without interaction. Below we define the zero-knowledge property more formally.

We denote the algorithm that the honest prover Peggy executes by P, the algorithm of an honest verifier by V and the algorithm of a general (possibly dishonest) verifier by V^*. The interactive proof system (including the interaction between P and V) is denoted by (P, V). Peggy knows a secret about some object x (e.g. as in the Fiat-Shamir example in Protocol 4.5, the root of a square x). This object x is the common input to P and V.

Each algorithm is assumed to have polynomial running time. It may be partly controlled by random events, i.e., it has access to a source of random bits and thus can make random choices. Such algorithms are called *probabilistic algorithms*. We study this notion in detail in Chapter 5.

Let x be the common input of (P, V). Suppose, the interactive proof takes n moves. A message is sent in each move. For simplicity, we assume that the prover starts with the first move. We denote by m_i the message sent in the i-th move. The messages $m_1, m_3, \ldots$ are sent from the prover to the verifier and the messages $m_2, m_4, \ldots$ are sent from the verifier to the prover. The *transcript* of the joint computation of P and V^* on input x is defined by

$$tr_{P, V^*}(x) := (m_1, \ldots, m_n),$$

where $tr_{P, V^*}(x)$ is called an accepting transcript if V^* accepts after the last move. Note that the transcript $tr_{P, V^*}(x)$ depends on the random bits that the algorithms P and V^* choose. Thus, it is not determined by the input x.

Definition 4.6. An interactive proof system (P, V) is *(perfect) zero-knowledge* if there is a probabilistic *simulator* $S(V^*, x)$, running in expected polynomial time, which for every verifier V^* (honest or not) outputs on input x an accepting transcript t of P and V^*, such that these simulated transcripts are distributed in the same way as if they were generated by the honest prover P and V^*.

Remark. The definition of zero-knowledge includes all verifiers (also the dishonest ones). Hence, zero-knowledge is a property of the prover P. It captures the prover's security requirements against attempts to gain "knowledge" by interacting with him.

To understand the definition, we have to clarify what a simulator is. A simulator S is an algorithm which, given some verifier V^*, honest or not, generates valid accepting transcripts for (P, V^*), without communicating with the real prover P. In particular, S does not have any access to computations that rely on the prover's secret. Trying to produce an accepting transcript, S plays the role of P in the protocol and communicates with V^*. Thus, he obtains outgoing messages of V^* which are compliant with the protocol. His task is to fill into the transcript the messages going out from P. Since P computes these messages by use of her secret and S does not know this secret, S applies his own strategy to generate the messages. Necessarily, his probability of obtaining a valid transcript in this way is significantly less than 1. Otherwise, with high probability, S could falsely convince V^* that he knows the secret, and the proof system is not sound. Thus, not every attempt of S to produce an accepting transcript is successful; he fails in many cases. Nevertheless, by repeating his attempts sufficiently often, the simulator is able to generate a valid accepting transcript. It is required that the expectation value of the running time which S needs to get an accepting transcript is bounded by a polynomial in the binary length $|x|$ of the common input x.[2]

To be zero-knowledge, the ability to produce accepting transcripts by a simulation is not sufficient. The generation of transcripts, real or simulated, includes random choices. Thus, we have a probability distribution on the set of accepting transcripts. The last condition in the definition means that the probability distribution of the transcripts that are generated by the simulator S and V^* is the same as if they were generated by the honest prover P and V^*. Otherwise, the distribution of transcripts might contain information about the secret and thus reveal some of P's knowledge.

In the following, we will illustrate the notion of zero-knowledge and the simulation of a prover by the simplified version of the Fiat-Shamir identification (Protocol 4.5).

Proposition 4.7. *The simplified version of the Fiat-Shamir identification scheme is zero-knowledge.*

Proof. The set of accepting transcripts is

$$\mathcal{T}(x) := \{(a, e, b) \in \mathrm{QR}_n \times \{0, 1\} \times \mathbb{Z}_n^* \mid b^2 = ax^e\}.$$

Let V^* be a general (honest or cheating) verifier. Then, a simulator S with the desired properties is given by the following algorithm.

---

[2] In other words, S is a Las Vegas algorithm (see Section 5.2).

Algorithm 4.8.
    transcript S(algorithm V^*, int x)
    1 repeat
    2   select $\tilde{e} \in \{0, 1\}$ and $\tilde{b} \in \mathbb{Z}_n^*$ uniformly at random
    3      $\tilde{a} \leftarrow \tilde{b}^2 x^{-\tilde{e}}$
    4      $e \leftarrow V^*(\tilde{a})$
    5 until $e = \tilde{e}$
    6 return $(\tilde{a}, \tilde{e}, \tilde{b})$

The simulator S uses the verifier V^* as a subroutine to get the challenge e. S tries to guess e in advance. If S succeeded in guessing e, he can produce a valid transcript $(\tilde{a}, \tilde{e}, \tilde{b})$. S cannot produce e by himself, because V^* is an arbitrary verifier. Therefore, V^* possibly does not generate the challenges e randomly, as it is specified in (P, V), and S must call V^* to get e. Independent of the strategy that V^* uses to output e, the guess $\tilde{e}$ coincides with e with a probability of $1/2$. Namely, if V^* outputs 0 with a probability of p and 1 with a probability of $1 - p$, the probability that $e = 0$ and $\tilde{e} = 0$ is $p/2$, and the probability that $e = 1$ and $\tilde{e} = 1$ is $(1 - p)/2$. Hence, the probability that one of both events occurs is $1/2$.

The expectation is that S will produce a result after two iterations of the while loop (see Lemma B.12). An element $(\tilde{a}, \tilde{e}, \tilde{b}) \in T$ returned by S cannot be distinguished from an element (a, e, b) produced by (P, V^*):

1. a and $\tilde{a}$ are random quadratic residues in QR_n.
2. e and $\tilde{e}$ are distributed according to V^*.
3. b and $\tilde{b}$ are random square roots.

This concludes the proof of the proposition.   □

4.2.4 Fiat-Shamir Identification Scheme

As in the simplified version of the Fiat-Shamir identification scheme, let $n := pq$, where p and q are distinct primes. Again, computations are performed in $\mathbb{Z}_n$, and we assume that it is intractable to compute square roots of randomly chosen elements in QR_n, unless the factorization of n is known (see Section 4.2.2). In the simplified version of the Fiat-Shamir identification scheme, the verifier accepts the proof of a cheating prover with a probability of $1/2$. To reduce this probability of success, now the prover's secret is a vector $y := (y_1, \ldots, y_t)$ of randomly chosen square roots. The modulus n and the vector $x := (y_1^2, \ldots, y_t^2)$ are made public. As above, we assume that Peggy chooses n and y, computes x and publishes the public key (n, x) to all participants. Peggy's secret is y.

Protocol 4.9.

Fiat-Shamir Identification:

Repeat the following k times:
1. Peggy chooses $r \in \mathbb{Z}_n^*$ at random and sets $a := r^2$. Peggy sends a to Vic.
2. Vic chooses $e := (e_1, \ldots, e_t) \in \{0,1\}^t$ at random. Vic sends e to Peggy.
3. Peggy computes $b := r \prod_{i=1}^t y_i^{e_i}$. Peggy sends b to Vic.
4. Vic rejects if $b^2 \neq a \prod_{i=1}^t x_i^{e_i}$, and stops.

Security. The Fiat-Shamir identification scheme extends the simplified scheme in two aspects. First, a challenge $e \in \{0,1\}$ in the basic scheme is replaced by a challenge $e \in \{0,1\}^t$. Then the basic scheme is iterated k times. A cheating prover Eve can convince Vic if she guesses Vic's challenge e correctly for each iteration; i.e., if she manages to select the right element from $\{0,1\}^{kt}$. Her probability of accomplishing this is 2^{-kt}. It can be shown that for $t = O(\log_2(|n|))$ and $k = O(|n|^l)$, the interactive proof system is still zero-knowledge. Observe here that the expected running time of a simulator that is constructed in a similar way as in the proof of Proposition 4.7 is no longer polynomial if t or k are too large.

Completeness. If the legitimate prover Peggy and Vic follow the protocol, then Vic will accept.

Soundness. Suppose a cheating prover Eve can convince (the honest verifier) Vic with a probability $> 2^{-kt}$, where the probability is taken over all the challenges e. Then Eve knows a vector $A = (a^1, \ldots, a^k)$ of commitments a^j (one for each iteration j, $1 \leq j \leq k$) for which she can correctly answer two distinct challenges $E = (e^1, \ldots, e^k)$ and $F = (f^1, \ldots, f^k)$, $E \neq F$, of Vic. There is an iteration j, such that $e^j \neq f^j$, and Eve can answer both challenges $e := e^j$ and $f := f^j$ for the commitment $a = a^j$. This means that Eve can compute b_1 and b_2, such that

$$b_1^2 = a \prod_{i=1}^t x_i^{e_i} \text{ and } b_2^2 = a \prod_{i=1}^t x_i^{f_i}.$$

As in Section 4.2.2, this implies that Eve can compute the square root $y = b_2 b_1^{-1}$ of the random square $x = \prod_{i=1}^t x_i^{f_i - e_i}$. This contradicts our assumption that computing square roots is intractable without knowing the prime factors p and q of n.

Remark. The number ν of exchanged bits is $k(2|n| + t)$, the average number μ of multiplications is $k(t + 2)$ and the key size s (equal to the size of the prover's secret) is $t|n|$. Choosing k and t appropriately, different values for the three numbers can be achieved. All choices of k and t, with kt constant, lead to the same level of security 2^{-kt}. Keeping the product kt constant and increasing t, while decreasing k, yields smaller values for ν and μ. However, note that the scheme is proven to be zero-knowledge only for small values of t.

4.2.5 Fiat-Shamir Signature Scheme

[FiaSha86] gives a standard method for converting an interactive identification scheme into a digital signature scheme. Digital signatures are produced by the signer without interaction. Thus, the communication between the prover and the verifier has to be eliminated. The basic idea is to take the challenge bits, which in the identification scheme are generated by the verifier, from the message to be signed. It must be guaranteed that the signer makes his commitment before he extracts the challenge bits from the message. This is achieved by the clever use of a publicly known, collision-resistant hash function (see Section 3.4):

$$h : \{0,1\}^* \longrightarrow \{0,1\}^{kt}.$$

As an example, we convert the Fiat-Shamir identification scheme (Section 4.2.4) into a signature scheme.

Signing. To sign a message $m \in \{0,1\}^*$, Peggy proceeds in three steps:

1. Peggy chooses $(r_1, \ldots, r_k) \in \mathbb{Z}_n^{*\,k}$ at random and sets $a_j := r_j^2, 1 \le j \le k$.
2. She computes $h(m\|a_1\|\ldots\|a_k)$ and writes these bits into a matrix, column by column:

$$e := (e_{i,j})_{\substack{1 \le i \le t \\ 1 \le j \le k}}.$$

3. She computes

$$b_j := r_j \prod_{i=1}^{t} y_i^{e_{ij}}, \quad 1 \le j \le k,$$

and sets $b = (b_1, \ldots, b_k)$. The signature of m is $s = (b, e)$.

Verification. To verify the signature $s = (b, e)$ of the signed message (m, s), we compute

$$c_j := b_j^2 \prod_{i=1}^{t} x_i^{-e_{ij}}, \quad 1 \le j \le k,$$

and accept if $e = h(m\|c_1\|\ldots\|c_k)$.

Here, the collision resistance of the hash function is needed. The verifier does not get the original values a_j – from step 1 of the protocol – to test $a_j = b_j^2 \prod_{i=1}^{t} x_i^{-e_{ij}}, 1 \le j \le k$.

Remark. The key size is $t|n|$ and the signature size is $k(t + |n|)$. The scheme is proven to be secure in the random oracle model, i.e., under the assumption that the hash function h is a truly random function (see Section 10.1 for a detailed discussion of what the security of a signature scheme means).

4.3 Commitment Schemes

Commitment schemes are of great importance in the construction of cryptographic protocols for practical applications (see Section 4.4.6), as well as for
protocols in theoretical computer science. They are used, for example, to construct zero-knowledge proofs for all languages in $\mathcal{NP}$ (see [GolMicWid86]).
This result is extended to the larger class of all languages in $\mathcal{IP}$, which is the
class of languages that have interactive proofs (see [BeGrGwHåKiMiRo88]).
    Commitment schemes enable a party to commit to a value while keeping
it secret. Later, the committer provides additional information to open the
commitment. It is guaranteed that after committing to a value, this value
cannot be changed. No other value can be revealed in the opening step: if
you have committed to 0, you cannot open 1 instead of 0, and vice versa. For
simplicity (and without loss of generality), we only consider the values 0 and
1 in our discussion.
    The following example demonstrates how to use commitment schemes.
Suppose Alice and Bob are getting divorced. They have decided how to split
their common possessions. Only one problem remains: who should get the
car? They want to decide the question by a coin toss. This is difficult, because Alice and Bob are in different places and can only talk to each other by
telephone. They do not trust each other to report the correct outcome of a
coin toss. This example is attributable to M. Blum. He introduced the problem of tossing a fair coin by telephone and solved it using a bit-commitment
protocol (see [Blum82]).

Protocol 4.10.
Coin tossing by telephone:

1. Alice tosses a coin, commits to the outcome b_A (heads = 0, tails
   = 1) and sends the commitment to Bob.
2. Bob also tosses a coin and sends the result b_B to Alice.
3. Alice opens her commitment by sending the additional information to Bob.

Now, both parties can compute the outcome $b_A \oplus b_B$ of the joint coin toss by
telephone. If at least one of the two parties follows the protocol, i.e., sends
the result of a true coin toss, the outcome is indeed a truly random bit.

   In a commitment scheme, there are two participants, the committer (also
called the sender) and the receiver. The commitment scheme defines two
steps:

1. *Commit.* The sender sends the bit b he wants to commit to, in encrypted
   form, to the receiver.
2. *Reveal* or *open.* The sender sends additional information to the receiver,
   enabling the receiver to recover b.

There are three requirements:

1. *Hiding property.* In the commit step, the receiver does not learn anything about the committed value. He cannot distinguish a commitment to 0 from a commitment to 1.
2. *Binding property.* The sender cannot change the committed value after the commit step. This requirement has to be satisfied, even if the sender tries to cheat.
3. *Viability.* If both the sender and the receiver follow the protocol, the receiver will always recover the committed value.

4.3.1 A Commitment Scheme Based on Quadratic Residues

The commitment scheme based on quadratic residues enables Alice to commit to a single bit. Let QR_n be the subgroup of squares in $\mathbb{Z}_n^*$ (see Definition A.48), and let $J_n^{+1} := \{x \in \mathbb{Z}_n^* \mid \left(\frac{x}{n}\right) = 1\}$ be the units modulo n with Jacobi symbol 1 (see Definition A.55). Let $QNR_n^{+1} := J_n^{+1} \setminus QR_n$ be the non-squares in J_n^{+1}.

Protocol 4.11.

QRCommitment:

1. *System setup.* Alice chooses distinct large prime numbers p and q, and $v \in QNR_n^{+1}$, $n := pq$.
2. *Commit.* To commit to a bit b, Alice chooses $r \in \mathbb{Z}_n^*$ at random, sets $c := r^2 v^b$ and sends n, c and v to Bob.
3. *Reveal.* Alice sends p, q, r and b to Bob. Bob can verify that p and q are primes, $n = pq$, $r \in \mathbb{Z}_n^*$, $v \notin QR_n$ and $c = r^2 v^b$.

Remarks:

1. There is an efficient deterministic algorithm which computes the Jacobi symbol $\left(\frac{x}{n}\right)$ of x modulo n, without knowing the prime factors p and q of n (Algorithm A.59). Thus, it is easy to determine whether a given $x \in \mathbb{Z}_n^*$ is in J_n^{+1}. However, if the factors of n are kept secret, no efficient algorithm is known that can decide whether a randomly chosen element in J_n^{+1} is a square, and it is assumed that no efficient algorithm exists for this question of quadratic residuosity (a precise definition of this quadratic residuosity assumption is given in Definition 6.11). On the other hand, if p and q are known it is easy to check whether $v \in J_n^{+1}$ is a square. Namely, v is a square if and only if $v \bmod p$ and $v \bmod q$ are squares, and this in turn is true if and only if the Legendre symbols $\left(\frac{v}{p}\right) = v^{(p-1)/2} \bmod p$ and $\left(\frac{v}{q}\right) = v^{(q-1)/2} \bmod q$ are equal to 1 (see Proposition A.52).
2. If Bob could distinguish a commitment to 0 from a commitment to 1, he could decide whether a randomly chosen element in J_n^{+1} is a square. This

contradicts the quadratic residuosity assumption stated in the preceding remark.

3. The value c is a square if and only if v^b is a square, i.e., if and only if $b = 0$. Since c is either a square or a non-square, Alice cannot change her commitment after the commit step.

4. Bob needs p and q to check that v is not a square. By not revealing the primes p and q, Alice could use them for several commitments. Then, however, she has to prove that v is not a square. She could do this by an interactive zero-knowledge proof (see Exercise 3).

5. Bob can use Alice's commitment c and commit to the same bit b as Alice, without knowing b. He chooses $\tilde{r} \in \mathbb{Z}_n^*$ at random and sets $\tilde{c} = c\tilde{r}^2$. Bob can open his commitment after Alice has opened her commitment. If commitments are used as subprotocols, it might cause a problem if Bob blindly commits to the same bit as Alice. We can prevent this by asking Bob to open his commitment before Alice does.

4.3.2 A Commitment Scheme Based on Discrete Logarithms

The commitment scheme based on discrete logarithms enables Alice to commit to a message $m \in \{0, \ldots, q-1\}$.

Protocol 4.12.

LogCommitment:

1. *System setup.* Bob randomly chooses large prime numbers p and q such that q divides $p - 1$. Then he randomly chooses g and v from the subgroup G_q of order q in $\mathbb{Z}_p^*$, $g, v \neq [1]$.[3] Bob sends p, q, g and v to Alice.

2. *Commit.* Alice verifies that p and q are primes, that q divides $p - 1$ and that g and v are elements of order q. To commit to $m \in \{0, \ldots, q-1\}$, she chooses $r \in \{0, \ldots, q-1\}$ at random, sets $c := g^r v^m$ and sends c to Bob.

3. *Reveal.* Alice sends r and m to Bob. Bob verifies that $c = g^r v^m$.

Remarks:

1. Bob can generate p, q, g and v as in the DSA (see Section 3.5.3).

2. If Alice committed to m and could open her commitment as $\tilde{m}$, $\tilde{m} \neq m$, then $g^r v^m = g^{\tilde{r}} v^{\tilde{m}}$ and $\log_g(v) = (\tilde{m} - m)^{-1}(r - \tilde{r})$.[4] Thus, Alice could compute $\log_g(v)$ of a randomly chosen element $v \in G_q$, contradicting the assumption that discrete logarithms of elements in G_q are infeasible

---

[3] There is a unique subgroup of order q in $\mathbb{Z}_p^*$. It is cyclic and each element $x \in \mathbb{Z}_p^*$ of order q is a generator (see Lemma A.40).

[4] Note that we compute in $G_q \subset \mathbb{Z}_p^*$. Hence, computations like $g^r v^m$ are done modulo p, and since the elements in G_q have order q, exponents and logarithms are computed modulo q (see "Computing modulo a prime" on page 303).

to compute (see the remarks on the security of the DSA at the end of Section 3.5.3, and Proposition 4.21).

3. g and v are generators of G_q. g^r is a uniformly chosen random element in G_q, perfectly hiding v^m and m in $g^r v^m$, as in the encryption with a one-time pad (see Section 2.1).

4. Bob has no advantage if he knows $\log_g(v)$, for example by choosing $v = g^s$ with a random s.

In the commitment scheme based on quadratic residues, the hiding property depends on the infeasibility of computing the quadratic residues property. The binding property is unconditional. The commitment scheme based on discrete logarithms has an unconditional hiding property, whereas the binding property depends on the difficulty of computing discrete logarithms.

If the binding or the hiding property depends on the complexity of a computational problem, we speak of *computational hiding* or *computational binding*. If the binding or the hiding property does not depend on the complexity of a computational problem, we speak of *unconditional hiding* or *unconditional binding*. The definitions for the hiding and binding properties, given above, are somewhat informal. To define precisely what "cannot distinguish" and "cannot change" means, probabilistic algorithms have to be used, which we introduce in Chapter 5.

It would be desirable to have a commitment scheme which features unconditional hiding and binding. However, as the following considerations show this is impossible. Suppose a deterministic algorithm

$$C : \{0,1\}^n \times \{0,1\} \longrightarrow \{0,1\}^s$$

defines a scheme with both unconditionally hiding and binding. Then when Alice sends a commitment $c = C(r, b)$ to Bob, there must exist an $\tilde{r}$ such that $c = C(\tilde{r}, 1 - b)$. Otherwise, Bob could compute (r, b) if he has unrestricted computational power, violating the unconditional hiding property. However, if Alice also has unrestricted computing power, then she can also find $(\tilde{r}, 1-b)$ and open the commitment as $1 - b$, thus violating the unconditional binding property.

4.3.3 Homomorphic Commitments

Let $\mathrm{Com}(r, m) := g^r v^m$ denote the commitment to m in the commitment scheme based on discrete logarithms. Let $r_1, r_2, m_1, m_2 \in \{0, \ldots, q-1\}$. Then

$$\mathrm{Com}(r_1, m_1) \cdot \mathrm{Com}(r_2, m_2) = \mathrm{Com}(r_1 + r_2, m_1 + m_2).$$

Commitment schemes satisfying such a property are called *homomorphic commitment schemes*.

Homomorphic commitment schemes have an interesting application in distributed computation: a sum of numbers can be computed without revealing the single numbers. This feature can be used in electronic voting schemes (see Section 4.4). We give an example using the Com function just introduced. Assume there are n voters $V_1, \ldots, V_n$. For simplicity, we assume that only "yes-no" votes are possible. A trusted center T is needed to compute the outcome of the election. The center T is assumed to be honest. If T was dishonest, it could determine each voter's vote. Let E_T and D_T be ElGamal encryption and decryption functions for the trusted center T. To vote on a subject, each voter V_i chooses $m_i \in \{0,1\}$ as his vote, a random $r_i \in \{0, \ldots, q-1\}$ and computes $c_i := \mathrm{Com}(r_i, m_i)$. Then he broadcasts c_i to the public and sends $E_T(g^{r_i})$ to the trusted center T. T computes

$$D_T(\prod_{i=1}^{n} E_T(g^{r_i})) = \prod_{i=1}^{n} g^{r_i} = g^r,$$

where $r = \sum_{i=1}^{n} r_i$, and broadcasts g^r.

Now, everyone can compute the result s of the vote from the publicly known c_i, $i = 1, \ldots, n$, and g^r:

$$v^s = g^{-r} \prod_{i=1}^{n} c_i,$$

with $s := \sum_{i=1}^{n} m_i$. s can be derived from v^s by computing $v, v^2, \ldots$ and comparing with v^s in each step, because the number of voters is not too large. If the trusted center is honest, the factor g^{r_i} – which hides V_i's vote – is never computed. Although an unconditional hiding commitment is used, the hiding property is only computational because g^{r_i} is encrypted with a cryptosystem that provides at most computational security.

4.4 Electronic Elections

In an *electronic voting scheme* there are two distinct types of participants: the voters casting the votes and the voting authority (for short, the authority) that collects the votes and computes the final tally.

Usually the following properties are required: (1) *universal verifiability* ensures that the correctness of the election, especially the correct computation of the tally, can be checked by everyone; (2) *privacy* ensures that the secrecy of an individual vote is maintained; and (3) *robustness* ensures that the scheme works even in the presence of a coalition of parties with faulty behavior. Naturally, only authorized voters should be allowed to cast their votes.

There seems to be a conflict between the last requirement and privacy. The scheme we describe resolves this conflict by establishing a group of authorities and a secret sharing scheme. It guarantees privacy, even if some of the authorities collaborate.

4.4.1 Secret Sharing

The idea of secret sharing is to start with a secret s and divide it into n pieces called shares. These shares are distributed among n users in a secure way. A coalition of some of the users is able to reconstruct the secret. The secret could, for example, be the password to open a safe shared by five people, with any three of them being able to reconstruct the password and open the safe.

In a (t, n)-threshold scheme $(t \leq n)$, a trusted center computes the shares s_j of a secret s and distributes them among n users. Each t of the n users are able to recover s from their shares. It is impossible to recover the secret from $t - 1$ or fewer shares.

Shamir's Threshold Scheme. Shamir's threshold scheme is based on the following properties of polynomials over a finite field k. For simplicity, we take $k = \mathbb{Z}_p$, where p is a prime.

Proposition 4.13. Let $f(X) = \sum_{i=0}^{t-1} a_i X^i \in \mathbb{Z}_p[X]$ be a polynomial of degree $t-1$, and let $P := \{(x_i, f(x_i)) \mid x_i \in \mathbb{Z}_p, i = 1, \ldots, t, x_i \neq x_j, i \neq j\}$. For $Q \subseteq P$, let $\mathcal{P}_Q := \{g \in \mathbb{Z}_p[X] \mid \deg(g) = t - 1, g(x) = y \text{ for all } (x, y) \in Q\}$.

1. $\mathcal{P}_P = \{f(X)\}$, i.e., f is the only polynomial of degree $t - 1$, whose graph contains all t points in P.
2. If $Q \subset P$ is a proper subset and $x \neq 0$ for all $(x, y) \in Q$, then each $a \in \mathbb{Z}_p$ appears with the same frequency as the constant coefficient of a polynomial in $\mathcal{P}_Q$.

Proof. To find all polynomials $g(X) = \sum_{i=0}^{t-1} b_i X^i \in \mathbb{Z}_p[X]$ of degree $t - 1$ through m given points $(x_i, y_i), 1 \leq i \leq m$, you have to solve the following linear equations:

$$\begin{pmatrix} 1 & x_1 & \ldots & x_1^{t-1} \\ & \vdots & & \\ 1 & x_m & \ldots & x_m^{t-1} \end{pmatrix} \begin{pmatrix} b_0 \\ \vdots \\ b_{t-1} \end{pmatrix} = \begin{pmatrix} y_1 \\ \vdots \\ y_m \end{pmatrix}$$

If $m = t$, then the above matrix (called A) is a Vandermonde matrix and its determinant

$$\det A = \prod_{1 \leq i < j \leq t} (x_j - x_i) \neq 0,$$

if $x_i \neq x_j$ for $i \neq j$. Hence, the system of linear equations has exactly one solution and part 1 of Proposition 4.13 follows.

Now let $Q \subset P$ be a proper subset. Without loss of generality, Q consists of the points $(x_1, y_1), \ldots, (x_m, y_m)$, $1 \leq m \leq t - 1$. We consider the following system of linear equations:

$$\begin{pmatrix} 1 & 0 & \dots & 0 \\ 1 & x_1 & \dots & x_1^{t-1} \\ & \vdots & & \\ 1 & x_m & \dots & x_m^{t-1} \end{pmatrix} \begin{pmatrix} b_0 \\ b_1 \\ \vdots \\ b_{t-1} \end{pmatrix} = \begin{pmatrix} a \\ y_1 \\ \vdots \\ y_m \end{pmatrix} \qquad (4.1)$$

The matrix of the system consists of rows of a Vandermonde matrix (note all $x_i \neq 0$ by assumption). Thus, the rows are linearly independent and the system (4.1) has solutions for all $a \in \mathbb{Z}_p$. The matrix has rank $m+1$ independent of a. Hence, the number of solutions is independent from a, and we see that each $a \in \mathbb{Z}_p$ appears as the constant coefficient of a polynomial in $\mathcal{P}_Q$ with the same frequency. $\qquad \square$

Corollary 4.14. Let $f(X) = \sum_{i=0}^{t-1} a_i X^i \in \mathbb{Z}_p[X]$ be a polynomial of degree $t-1$, and let $P = \{(x_i, f(x_i)) \mid i = 1, \dots, t, x_i \neq x_j, i \neq j\}$. Then

$$f(X) = \sum_{i=1}^{t} f(x_i) \prod_{1 \leq j \leq t, j \neq i} (X - x_j)(x_i - x_j)^{-1}. \qquad (4.2)$$

This formula is called the Lagrange interpolation formula.

Proof. The right-hand side of (4.2) is a polynomial g of degree $t-1$. If we substitute X by x_i in g, we get $g(x_i) = f(x_i)$. Since the polynomial $f(X)$ is uniquely defined by P, the equality holds. $\qquad \square$

Now we describe Shamir's (t, n)-threshold scheme. A trusted center T distributes n shares of a secret $s \in \mathbb{Z}$ among n users $P_1, \dots, P_n$. To set up the scheme, the trusted center T proceeds as follows:

1. T chooses a prime $p > \max(s, n)$ and sets $a_0 := s$.
2. T selects $a_1, \dots, a_{t-1} \in \{0, \dots, p-1\}$ independently and at random, and gets the polynomial $f(X) = \sum_{i=0}^{t-1} a_i X^i$.
3. T computes $s_i := f(i)$, $i = 1, \dots, n$ (we use the values $i = 1, \dots, n$ for simplicity; any n pairwise distinct values $x_i \in \{1, \dots, p-1\}$ could also be used) and transfers (i, s_i) to the user P_i in a secure way.

Any group of t or more users can compute the secret. Let $J \subset \{1, \dots, n\}$, $|J| = t$. From Corollary 4.14 we get

$$s = a_0 = f(0) = \sum_{i \in J} f(i) \prod_{j \in J, j \neq i} j(j - i)^{-1} = \sum_{i \in J} s_i \prod_{j \in J, j \neq i} j(j - i)^{-1}.$$

If only $t-1$ or fewer shares are available, then each $a \in \mathbb{Z}_p$ is equally likely to be the secret (by Proposition 4.13). Thus, knowing only $t-1$ or fewer shares provides no advantage over knowing none of them.

Remarks:

1. Suppose that each $a \in \mathbb{Z}_p$ is equally likely as the secret for someone knowing only $t-1$ or fewer shares, as in Shamir's scheme. Then the (t, n)-threshold scheme is called *perfect*: the scheme provides perfect secrecy in the information-theoretic sense (see Section 9.1). The security does not rely on the assumed hardness of a computational problem.
2. Shamir's threshold scheme is easily extendable for new users. New shares may be computed and distributed without affecting existing shares.
3. It is possible to implement varying levels of control. One user can hold one or more shares.

4.4.2 A Multi-Authority Election Scheme

For simplicity, we only consider election schemes for yes-no votes. The voters want to get a majority decision on some subject. In the scheme that we describe, each voter selects his choice (yes or no), encrypts it with a homomorphic encryption algorithm and signs the cryptogram. The signature shows that the vote is from an authorized voter. The votes are collected in a single place, the bulletin board. After all voters have posted their votes, an authority can compute the tally without decrypting the single votes. This feature depends on the fact that the encryption algorithm used is a homomorphism. It guarantees the secrecy of the votes. But if the authority was dishonest, she could decrypt the single votes. To reduce this risk, the authority consists of several parties and the decryption key is shared among these parties, by use of a Shamir (t, n)-threshold scheme. Then at least t of the n parties must be dishonest to reveal a vote. First, we assume in our discussion that a trusted center T sets up the scheme. However, the trusted center is not really needed. In Section 4.4.6 we show that it is possible to set up the scheme by a communication protocol which is executed by the parties constituting the authority.

The election scheme that we describe was introduced in [CraGenSch97]. The time and communication complexity of the scheme is remarkably low. A voter simply posts a single encrypted message, together with a compact proof that it contains a valid vote.

The Communication Model. The members of the voting scheme communicate through a *bulletin board*. The bulletin board is best viewed as publicly accessible memory. Each member has a designated section of the memory to post messages. No member can erase any information from the bulletin board. The complete bulletin board can be read by all members (including passive observers). We assume that a public-key infrastructure for digital signatures is used to guarantee the origin of posted messages.

Setting up the Scheme. To set up the scheme, we assume for now that there is a trusted center T. The trusted center T chooses primes p and q, such that q is a large divisor of $p - 1$, and an element $g \in \mathbb{Z}_p^*$ of order q, as in the key generation procedure of the DSA (see Section 3.5.3). g generates the subgroup G_q of order q of $\mathbb{Z}_p^*$.[5]

Further, we assume that T chooses a secret key $s \in \{0, \ldots, q - 1\}$ at random and publishes the public key $h := g^s$ to be used for ElGamal encryption with respect to the base g. A message $m \in G_q$ is encrypted as $(c_1, c_2) = (g^\alpha, h^\alpha m)$, where α is a randomly chosen element in $\{0, \ldots, q - 1\}$. Using the secret key s the plaintext m can be recovered as $m = c_2 c_1^{-s}$ (see Section 3.5.1). The encryption is homomorphic: if (c_1, c_2) and (c_1', c_2') are encryptions of m and m', then

$$(c_1, c_2) \cdot (c_1', c_2') = (c_1 c_1', c_2 c_2') = (g^\alpha g^{\alpha'}, h^\alpha m h^{\alpha'} m') = (g^{\alpha + \alpha'}, h^{\alpha + \alpha'} m m')$$

is an encryption of mm'.

Let $A_1, \ldots, A_n$ be the authorities and $V_1, \ldots, V_m$ be the voters in the election scheme. The trusted center T chooses a Shamir (t, n)-threshold scheme. The secret encryption key s is shared among the n authorities. A_j keeps her share (j, s_j) secret. The values $h_j := g^{s_j}$ are published on the bulletin board by the trusted center.

Decryption. Everyone can decrypt a cryptogram $c := (c_1, c_2) := (g^\alpha, h^\alpha m)$ with some help of the authorities, but without reconstructing the secret key s. Namely, the following steps are executed:

1. Each authority A_j posts $w_j := c_1^{s_j}$ to the bulletin board. Here, we assume that the authority A_j is honest and follows the protocol. In Section 4.4.3 we will see how to check that she really does (by a proof of knowledge).
2. Let J be the index set of a subset of t honest authorities. Then, everyone can recover $m = c_2 c_1^{-s}$ as soon as all A_j, $j \in J$, have finished step 1:

$$c_1^s = c_1^{\sum_{j \in J} s_j \lambda_{j,J}} = \prod_{j \in J} (c_1^{s_j})^{\lambda_{j,J}} = \prod_{j \in J} w_j^{\lambda_{j,J}},$$

   where

$$\lambda_{j,J} = \prod_{l \in J \setminus \{j\}} l(l - j)^{-1}.$$

Vote Casting. Each voter V_i selects his vote $v_i \in \{-1, 1\}$, encodes v_i as g^{v_i} and encrypts g^{v_i} by the ElGamal encryption:

$$c_i := (c_{i,1}, c_{i,2}) := (g^{\alpha_i}, h^{\alpha_i} g^{v_i}).$$

He then signs it to guarantee the origin of the message and posts it to the bulletin board. Here we assume that V_i follows the protocol and correctly forms c_i. He has to perform a proof of knowledge which shows that he really does (see Section 4.4.3); otherwise his vote is invalid.

---

[5] There is a unique subgroup of order q of $\mathbb{Z}_p^*$. It is cyclic and each element $x \in \mathbb{Z}_p^*$ of order q is a generator (see Lemma A.40 and "Computing modulo a prime" on page 303).

Tally Computing. Assume that m votes were cast:

1. Everyone can compute

$$c = (c_1, c_2) = \left(\prod_{i=1}^{m} c_{i,1}, \prod_{i=1}^{m} c_{i,2} \right).$$

Note that $c = (c_1, c_2)$ is the encryption of g^d, where d is the difference between the number of yes votes and no votes, since the encryption is homomorphic.

2. The decryption protocol from above is executed to get g^d. After sufficiently many authorities A_j have posted $w_j = c_1^{s_j}$ to the bulletin board, everyone can compute g^d.

3. Now d can be found by computing the sequence $g^{-m}, g^{-m+1}, \ldots$, and comparing with g^d in each step.

Remarks:

1. Everyone can check whether a voter or an authority was honest (see Section 4.4.3), and discard invalid votes. If he finds a subset of t honest authorities, he can compute the tally. This implies universal verifiability.

2. No coalition of $t - 1$ or fewer authorities can recover the secret key. This guarantees the robustness of the scheme.

3. Privacy depends on the security of the underlying ElGamal encryption scheme and, hence, on the assumed difficulty of the Diffie-Hellman problem. The scheme provides only computational privacy. A similar scheme is introduced in [CraFraSchYun96] which even provides perfect privacy (in the information-theoretic sense). This is achieved by using a commitment scheme with information-theoretic secure hiding to encrypt the votes.

4. The following remarks concern the communication complexity of the scheme:

   a. Each voter only has to send one message together with a compact proof that the message contains a valid vote (see below). His activities are independent of the number n of authorities.

   b. Each authority has to read m messages from the bulletin board, verify m interactive proofs of knowledge and post one message to the bulletin board.

   c. To compute the tally, you have to read t messages from the bulletin board and to verify t interactive proofs of knowledge.

5. It is possible to prepare an election beforehand. The voter V_i chooses $v_i \in \{-1, 1\}$ at random. The voting protocol is executed with the random v_i values. Later, the voter decides the alternative to choose. He selects $\tilde{v}_i \in \{-1, 1\}$, such that $\tilde{v}_i v_i$ is his vote, and posts $\tilde{v}_i$ to the bulletin board. The tally is computed with $\tilde{c}_i = (c_{i,1}^{\tilde{v}_i}, c_{i,2}^{\tilde{v}_i})$, $i = 1, \ldots, m$.

4.4.3 Proofs of Knowledge

Authority's Proof. In the decryption protocol above, each authority A_j has to prove that she really posts $w_j = c_1^{s_j}$, where s_j is her share of the secret key s. Recall that $h_j = g^{s_j}$ is published on the bulletin board. The authority has to prove that w_j and h_j have the same logarithm with respect to the bases c_1 and g and that she knows this logarithm. We simplify the notation and describe an interactive proof of knowledge of the common logarithm x of $y_1 = g_1^x$ and $y_2 = g_2^x$, where x is a random element from $\{0, \ldots, q-1\}$. As usual, we call the prover Peggy and the verifier Vic.

Of course, in our voting scheme it is desirable for practical reasons that the authority proves, in a non-interactive way, to be honest. However, it is easy to convert the interactive proof into a non-interactive one (see Section 4.4.4). Thus, we first give the interactive version of the proof.

Protocol 4.15.

$ProofLogEq(g_1, y_1, g_2, y_2)$:

1. Peggy chooses $r \in \{0, \ldots, q-1\}$ at random and sets $a := (a_1, a_2) = (g_1^r, g_2^r)$. Peggy sends a to Vic.
2. Vic chooses $c \in \{0, \ldots, q-1\}$ uniformly at random and sends c to Peggy.
3. Peggy computes $b := r - cx$ and sends b to Vic.
4. Vic accepts if and only if $a_1 = g_1^b y_1^c$ and $a_2 = g_2^b y_2^c$.

The protocol is a three-move protocol. It is very similar to the protocol used in the simplified Fiat-Shamir identification scheme (see Section 4.2.2):

1. The first message is a commitment by Peggy. She commits that two numbers have the same logarithm with respect to the different bases g_1 and g_2.
2. The second message c is a challenge by Vic.
3. The third message is Peggy's response. If $c = 0$, Peggy has to open the commitment (reveal r). If $c \neq 0$, Peggy has to show her secret in encrypted form (reveal $r - cx$).

Completeness. If Peggy knows a common logarithm for y_1 and y_2, and both Peggy and Vic follow the protocol, then $a_1 = g_1^b y_1^c$ and $a_2 = g_2^b y_2^c$, and Vic will accept.

Soundness. A cheating prover Eve can convince Vic with a probability of $1/q$ in the following way:

1. Eve chooses $r, \tilde{c} \in \{0, \ldots, q-1\}$ at random, sets $a := (g_1^r y_1^{\tilde{c}}, g_2^r y_2^{\tilde{c}})$ and sends a to Vic.
2. Vic chooses $c \in \{0, \ldots, q-1\}$ at random and sends c to Eve.
3. Eve sends r to Vic.

Vic accepts if and only if $c = \tilde{c}$. The event $c = \tilde{c}$ occurs with a probability of $1/q$. Thus, Eve succeeds in cheating with a probability of $1/q$.

If Eve can convince Vic with a probability greater than $1/q$ (the probability is taken over the challenges c), she has to answer at least two challenges correctly, for a given commitment a.

Suppose Eve knows an $a = (a_1, a_2)$ for which she can answer two distinct challenges c and $\tilde{c}$. This means that Eve can compute b and $\tilde{b}$, such that

$$a_1 = g_1^b y_1^c, \quad a_2 = g_2^b y_2^c,$$
$$a_1 = g_1^{\tilde{b}} y_1^{\tilde{c}}, \quad a_2 = g_2^{\tilde{b}} y_2^{\tilde{c}}.$$

Then she can compute

$$g_1^{\tilde{b}-b} = y_1^{c-\tilde{c}} \text{ and } g_2^{\tilde{b}-b} = y_2^{c-\tilde{c}}$$

and gets

$$(\tilde{b}-b)(c-\tilde{c})^{-1} = \log_{g_1}(y_1) \text{ and } (\tilde{b}-b)(c-\tilde{c})^{-1} = \log_{g_2}(y_2).$$

Thus, she can compute the secret x. This contradicts the assumption that it is infeasible to compute x from g^x (for randomly chosen p, g and x). We see that the probability of success of a cheating prover is bounded by $1/q$.

Honest Verifier Zero-Knowledge. The above protocol is not known to be zero-knowledge. However, it is zero-knowledge if the verifier is an honest one. An interactive proof system (P, V) is called *honest-verifier zero-knowledge* if Definition 4.6 holds for the honest verifier V, but not necessarily for an arbitrary Verifier V^*; i.e., there is a simulator S that produces correctly distributed accepting transcripts for executions of the protocol with (P, V).

To simulate an interaction with the honest verifier is quite simple. The key point is that the honest verifier V chooses the challenge $c \in \{0, \ldots, q-1\}$ independently from (a_1, a_2), uniformly and at random, and this can also be done by S.

Algorithm 4.16.
```
int S(int g₁, g₂, y₁, y₂)
```
1   select $\tilde{b} \in \{0, \ldots, q-1\}$ uniformly at random
2   select $\tilde{c} \in \{0, \ldots, q-1\}$ uniformly at random (this is V's task)
3   $\tilde{a}_1 \leftarrow g_1^{\tilde{b}} y_1^{\tilde{c}}, \tilde{a}_2 \leftarrow g_2^{\tilde{b}} y_2^{\tilde{c}}$
4   return $(\tilde{a}_1, \tilde{a}_2, \tilde{c}, \tilde{b})$

The transcript $(\tilde{a}_1, \tilde{a}_2, \tilde{c}, \tilde{b})$ returned by S is an accepting transcript and not distinguishable from a transcript (a_1, a_2, c, b) produced by (P, V):

1. a_1, a_2 and $\tilde{a}_1, \tilde{a}_2$ are randomly chosen elements in $\mathbb{Z}_q$.
2. c and $\tilde{c}$ are randomly chosen elements in $\{0, \ldots, q-1\}$.
3. b and $\tilde{b}$ are randomly chosen elements in $\{0, \ldots, q-1\}$.

Voter's Proof. In the vote-casting protocol, each voter has to prove that he really encrypted a vote $g^v \in \{g, g^{-1}\}$; i.e., he has to prove that $c = (c_1, c_2) = (g^\alpha, h^\alpha m)$ and $m \in \{g, g^{-1}\}$. For this purpose, he performs a proof of knowledge. He proves that he knows α for either $c_1 = g^\alpha$ and $c_2 g^{-1} = h^\alpha$, or for $c_1 = g^\alpha$ and $c_2 g = h^\alpha$. Each of the two alternatives could be proven as in the authority's proof. Here, however, the prover's task is more difficult. The proof must not reveal which of the two alternatives is proven. An interactive three-move proof that convinces the verifier without revealing anything about the prover's choice is the subject of Exercise 9.

4.4.4 Non-Interactive Proofs of Knowledge

It is easy to convert an interactive three-move proof into a non-interactive one using the standard method of Fiat-Shamir, which we demonstrated in Section 4.2.5. Let $h : \{0,1\}^* \longrightarrow \mathbb{Z}_q$ be a collision-resistant hash function. We get a non-interactive proof

$$(c, b) = \text{ProofLogEq}_h(g_1, y_1, g_2, y_2),$$

for $y_1 = g_1^x$ and $y_2 = g_2^x$ in the following way. The prover Peggy chooses $r \in \{0, \ldots, q-1\}$ at random and sets $a := (a_1, a_2) = (g_1^r, g_2^r)$. Then she computes the challenge $c := h(g_1 \| y_1 \| g_2 \| y_2 \| a_1 \| a_2)$ and sets $b := r - cx$. The verification condition is

$$c = h(g_1 \| y_1 \| g_2 \| y_2 \| g_1^b y_1^c \| g_2^b y_2^c).$$

The verifier needs not know a to compute the verification condition. If we trust the collision resistance of h, we can conclude that $u = v$ from $h(u) = h(v)$.

If we convert our proofs of knowledge into non-interactive proofs, honest-verifier zero-knowledge is sufficient, because here, the verifier is always honest. In our election protocol, each authority and each voter completes his message with a non-interactive proof which convinces everyone that he followed the protocol.

4.4.5 Extension to Multi-Way Elections

We describe how to extend the scheme if a choice between several, say $l > 2$, options should be possible.

To encode the votes $v_1, \ldots, v_l$, we independently choose l generators g_j, $j = 1, \ldots, l$, of G_q, and encode v_j by g_j. Voter V_i encrypts his vote v_j as

$$c_i = (c_{i,1}, c_{i,2}) = (g^\alpha, h^\alpha g_j).$$

Each voter shows – by an interactive proof of knowledge – that he knows α with

$$c_{i,1} = g^\alpha \text{ and } g_j^{-1}c_{i,2} = h^\alpha,$$

for exactly one j. We refer the interested reader to [CraGenSch97] for some hints about the proof technique used.

The problem of computing the final tally turns out to be more complicated. The result of the tally computation is

$$c_2 c_1^{-s} = g_1^{d_1} g_2^{d_2} \cdots g_l^{d_l},$$

where d_j is the number of votes for v_j. The exponents d_j are uniquely determined by $c_2 c_1^{-s}$ in the following sense: computing a different solution $(\tilde{d}_1, \ldots, \tilde{d}_l)$ would contradict the discrete logarithm assumption, because the generators were chosen independently (see Proposition 4.21 in Section 4.5.3). Again as above, $d = (d_1, \ldots, d_l)$ can be found for small values of l and d_i by searching.

4.4.6 Eliminating the Trusted Center

The trusted center sets up an ElGamal cryptosystem, generates a secret key, publishes the corresponding public key and shares the secret key among n users using a (t, n)-threshold scheme. To eliminate the trusted center, all these activities must be performed by the group of users (in our case, the authorities). For the communication between the participants, we assume below that a bulletin board, mutually secret communication channels and a commitment scheme exist.

Setting Up an ElGamal Cryptosystem. We need large primes p, q, such that q is a divisor of $p - 1$, and a generator g of the subgroup G_q of order q of $\mathbb{Z}_p^*$. They are generated jointly by the group of users (in our case the group of authorities). This can be achieved if each party runs the same generation algorithm (see Section 3.5.1). The random input to the generation algorithm, must be generated jointly. To do this, the users $P_1, \ldots, P_n$ execute the following protocol:

1. Each user P_i chooses a string r_i of random bits of sufficient length, computes a commitment $c_i = C(r_i)$ and posts the result to the bulletin board.
2. After all users have posted their commitments, each user opens his commitment.
3. They take $r = \oplus_{i=1}^n r_i$ as the random input.

Publish the Public Key. To generate and distribute the public key the users $P_1, \ldots, P_n$ execute the following protocol:

1. Each user P_i chooses $x_i \in \{0, \ldots, q - 1\}$ at random, computes $h_i := g^{x_i}$ and a commitment $c_i := C(h_i)$ for h_i. Then he posts c_i to the bulletin board.

2. After all users have posted their commitments, each user opens his commitment.
3. Everyone can compute the public key $h := \prod_{i=1}^{n} h_i$.

Share the Secret Key. The corresponding secret key

$$x := \sum_{i=1}^{n} x_i$$

must be shared. The basic idea is that each user constructs a Shamir (t, n)-threshold scheme to share his part x_i of the secret key. These schemes are combined to get a scheme for sharing x.

Let $f_i(X) \in \mathbb{Z}_p[X]$ be the polynomial of degree $t - 1$ used for sharing x_i, and let

$$f(X) = \sum_{i=1}^{n} f_i(X).$$

Then $f(0) = x$ and f can be used for a (t, n)-threshold scheme, provided $\deg(f) = t - 1$. The shares $f(j)$ can be computed from the shares $f_i(j)$:

$$f(j) = \sum_{i=1}^{n} f_i(j).$$

The group of users executes the following protocol to set up the threshold scheme:

1. The group chooses a prime $p > \max(x, n)$.
2. Each user P_i randomly chooses $f_{i,j} \in \{0, \ldots, p - 1\}$, $j = 1, \ldots, t - 1$, and sets $f_{i,0} = x_i$ and $f_i(X) = \sum_{j=0}^{t-1} f_{i,j} X^j$. He posts $F_{i,j} := g^{f_{i,j}}$, $j = 1, \ldots, t - 1$, to the bulletin board. Note that $F_{i,0} = h_i$ is P_i's piece of the public key and hence is already known.
3. After all users have posted their encrypted coefficients, each user tests whether $\sum_{i=1}^{n} f_i(X)$ has degree $t - 1$, by checking

$$\prod_{i=1}^{n} F_{i,t-1} \neq [1].$$

In the rare case that the test fails, they return to step 2. If $f(X)$ passes the test, the degree of $f(X)$ is $t - 1$.
4. Each user P_i distributes the shares $s_{i,l} = f_i(l)$, $l = 1, \ldots, n$, of his piece x_i of the secret key to the other users over secure communication channels.
5. Each user P_i verifies for $l = 1, \ldots, n$ that the share $s_{l,i}$ received from P_l is consistent with the previously posted committed coefficients of P_l's polynomial:

$$g^{s_{l,i}} = \prod_{j=0}^{t-1} (F_{l,j})^{i^j}.$$

If the test fails, he stops the protocol by broadcasting the message "failure,$(l, i), s_{l,i}$".

6. Finally, P_i computes his share s_i of x:

$$s_i = \sum_{l=1}^{n} s_{l,i},$$

signs the public key h and posts his signature to the bulletin board.

After all members have signed h, the group will use h as their public key. If all participants followed the protocol, the corresponding secret key is shared among the group. The protocol given here is described in [Pedersen91]. It only works if all parties are honest. [GenJarKraRab99] introduces an improved protocol which works if a majority of the participants is honest.

4.5 Digital Cash

The growth of electronic commerce in the Internet requires *digital cash*. Today, credit cards are used to pay on the Internet. Transactions are online; i.e., all participants – the customer, the shop and the bank – are involved at the same time (for simplicity, we assume only one bank). This requires that the bank is available even during peak traffic time, which makes the scheme very costly. Exposing the credit card number to the vendor provides him with the ability to impersonate the customer in future purchases. The bank can easily observe who pays which amount to whom and when, so the customer cannot pay anonymously, as she can with ordinary money.

A payment with ordinary money requires three different steps. First, the customer fetches some money from the bank, and his account is debited. Then he can pay anonymously in a shop. Later, the vendor can bring the money to the bank, and his account is credited.

Ordinary money has an acceptable level of security and functions well for its intended task. Its security is based on a complicated and secret manufacturing process. However, it is not as secure in the same mathematical sense as some of the proposed digital cash schemes.

Digital cash schemes are modeled on ordinary money. They involve three interacting parties: the bank, the customer and the shop. The customer and the shop have accounts with the bank. A digital cash system transfers money in a secure way from the customer's account to the shop's account. In the following, the money is called an *electronic coin*, or *coin* for short.

As with ordinary money, paying with digital cash requires three steps:

1. The customer fetches the coin from the bank: customer and bank execute the withdrawal protocol.
2. The customer pays the vendor: customer and vendor execute the payment protocol.

3. The vendor deposits the coin on his account: vendor and bank execute
   the deposit protocol.

In an offline system, each step occurs in a separate transaction, whereas in
an online system, steps 2 and 3 take place in a single transaction involving
all three parties.

The bank, the shop and the customer have different security requirements:

1. The bank is assured that only a previously withdrawn coin can be deposited. It must be impossible to deposit a coin twice without being
   detected.
2. The customer is assured that the shop will accept previously withdrawn
   coins and that he can pay anonymously.
3. In an offline system, the vendor is assured that the bank will accept a
   payment he has received from the customer.

It is not easy to enable anonymous payments, for which it must be impossible for the bank to trace a coin, i.e., to link a coin from a withdrawal
with the corresponding coin in the deposit step. This requirement protects
an honest customer's privacy, but it also enables the misuse by criminals.

Thus, to make anonymous payment systems practical, they must implement a mechanism for tracing a coin. It must be possible to revoke the customer's anonymity under certain well-defined conditions. Such systems are
sometimes called *fair payment systems*.

In the scheme we describe, anonymity may be revoked by a trusted third
party called the trusted center. During the withdrawal protocol, the customer
has to provide data which enables the trusted center to trace the coin. The
trusted center is only needed if someone asks to revoke the anonymity of
a customer. The trusted center is not involved if an account is opened or
a coin is withdrawn, paid or deposited. Using the secret sharing techniques
from Section 4.4.1, it is easy to distribute the ability to revoke a customer's
anonymity among a group of trusted parties.

A customer's anonymity is achieved by using blind signatures (see Section
4.5.1). The customer has to construct a message of a special form, and then
he hides the content. The bank signs the hidden message, without seeing its
content. Older protocols use the "cut and choose method" to ensure that the
customer formed the message correctly: the customer constructs, for example,
1000 messages and sends them to the bank. The bank selects one of the 1000
messages to sign it. The customer has to open the remaining 999 messages.
If all these messages are formed correctly, then, with high probability, the
message selected and signed by the bank is also correct. In the system that
we describe, the customer proves that he constructed the message m correctly.
Therefore, only one message is constructed and sent to the bank. This makes
digital cash efficient.

Ordinary money is transferable: the shop does not have to return a coin
to the bank after receiving it. He can transfer the money to a third person.

The digital cash system we introduce does not have this feature, but there are electronic cash systems which do implement transferable electronic coins (e.g. [OkaOht92]; [ChaPed92]).

4.5.1 Blindly Issued Proofs

The payment system which we describe provides customer anonymity by using a blind digital signature.[6] Such a signature enables the signer (the bank) to sign a message without seeing its content (the content is hidden). Later, when the message and the signature are revealed, the signer is not able to link the signature with the corresponding signing transaction. The bank's signature can be verified by everyone, just like an ordinary signature.

The Basic Signature Scheme. Let p and q be large primes, such that q divides $p - 1$. Let G_q be the subgroup of order q in $\mathbb{Z}_p^*$, g a generator of G_q[7] and $h : \{0,1\}^* \longrightarrow \mathbb{Z}_q$ be a collision-resistant hash function. The signer's secret key is a randomly chosen $x \in \{0,\ldots,q-1\}$, and the public key is (p,q,g,y), where $y = g^x$.

The basic protocol in this section is *Schnorr's identification protocol*. It is an interactive proof of knowledge. The prover Peggy proves to the verifier Vic that she knows x, the discrete logarithm of y.

Protocol 4.17.

ProofLog(g,y):

1. Peggy randomly chooses $r \in \{0,\ldots,q-1\}$, computes $a := g^r$ and sends it to Vic.
2. Vic chooses $c \in \{0,\ldots,q-1\}$ at random and sends it to Peggy.
3. Peggy computes $b := r - cx$ and sends it to Vic.
4. Vic accepts the proof if $a = g^b y^c$; otherwise he rejects it.

To achieve a signature scheme, the protocol is converted into a non-interactive proof of knowledge ProofLog_h using the same method as shown in Sections 4.4.4 and 4.2.5: the challenge c is computed by means of the collision-resistant hash function h.

A signature $\sigma(m)$ of a message m consists of a non-interactive proof that the signer (prover) knows the secret key x. The proof depends on the message m, because when computing the challenge c, the message m serves as an additional input to h:

$$\sigma(m) = (c, b) = \text{ProofLog}_h(m, g, y),$$

---

[6] Blind signatures were introduced by D. Chaum in [Chaum82] to enable untraceable electronic cash.

[7] As stated before, there is a unique subgroup of order q of $\mathbb{Z}_p^*$. It is cyclic and each element $x \in \mathbb{Z}_p^*$ of order q is a generator (see Lemma A.40 and "Computing modulo a prime" on page 303).

where $c := h(m\|a), a := g^r, r \in \{0, \ldots, q-1\}$ chosen at random, and $b := r - cx$. The signature is verified by checking the condition

$$c = h(m\|g^b y^c).$$

Here, the collision resistance of the hash function is needed. The verifier does not get a to test $a = g^b y^c$. If m is the empty string, we omit m and write

$$(c, b) = \text{ProofLog}_h(g, y).$$

In this way, we attain a non-interactive proof that the prover knows $x = \log_g(y)$.

Remarks:

1. As in the commitment scheme in Section 4.3.2 and the election protocol in Section 4.4, the security relies on the assumption that discrete logarithms of elements in G_q are infeasible to compute (also see the remarks on the security of the DSA at the end of Section 3.5.3).
2. If the signer uses the same r (i.e., he uses the same commitment $a = g^r$) to sign two different messages m_1 and m_2, then the secret key x can easily be computed:[8]
   Let $\sigma(m_i) := (c_i, b_i), i = 1, 2$. We have $g^r = g^{b_1} y^{c_1} = g^{b_2} y^{c_2}$ and derive $x = (b_1 - b_2)(c_2 - c_1)^{-1}$. Note that $c_1 \neq c_2$ for $m_1 \neq m_2$, since h is collision resistant.

The Blind Signature Scheme. The basic signature scheme can be transformed into a blind signature scheme. To understand the ideas, we first recall our application scenario. The customer (Vic) would like to submit a coin to the shop (Alice). The coin is signed by the bank (Peggy). Alice must be able to verify Peggy's signature. Later, when Alice brings the coin to the bank, Peggy should not be able to recognize that she signed the coin for Vic. Therefore, Peggy has to sign the coin blindly. Vic obtains the blind signature for a message m from Peggy by executing the interactive protocol ProofLog with Peggy. In step 2 of the protocol, he deviates a little from the original protocol: as in the non-interactive version ProofLog_h, the challenge is not chosen randomly, but computed by means of the hash function h (with m as part of the input). We denote the transcript of this interaction by $(\bar{a}, \bar{c}, \bar{b})$; i.e., $\bar{a}$ is Peggy's commitment in step 1, $\bar{c}$ is Vic's challenge in step 2 and $\bar{b}$ is sent from Peggy to Vic in step 3. $(\bar{c}, \bar{b})$ is a valid signature with $\bar{a} = g^{\bar{b}} y^{\bar{c}}$.

Now Peggy may store Vic's identity and the transcript $\bar{\tau} = (\bar{a}, \bar{c}, \bar{b})$, but later she should not be able to recognize the signature $\sigma(m)$ of m. Therefore, Vic must transform Peggy's signature $(\bar{c}, \bar{b})$ into another valid signature (c, b) of Peggy that Peggy is not able to link with the original transcript $(\bar{a}, \bar{c}, \bar{b})$.

---

[8] The elements in G_q have order q, hence all exponents and logarithms are computed modulo q. See "Computing modulo a prime" on page 303.

The idea is that Vic transforms the given transcript $\overline{\tau}$ into another accepting transcript $\tau = (a, c, b)$ of ProofLog by the following transformation:

$$\beta_{(u,v,w)} : G_q \times \mathbb{Z}_q^2 \longrightarrow G_q \times \mathbb{Z}_q^2, \ (\overline{a}, \overline{c}, \overline{b}) \longmapsto (a, c, b), \text{ where}$$

$$a := \overline{a}^u g^v y^w,$$
$$c := u\overline{c} + w,$$
$$b := u\overline{b} + v.$$

We have
$$a = \overline{a}^u g^v y^w = (g^{\overline{b}} y^{\overline{c}})^u g^v y^w = g^{u\overline{b}+v} y^{u\overline{c}+w} = g^b y^c.$$

Thus, τ is indeed an accepting transcript of ProofLog. If Vic chooses $u, v, w \in \{0, \ldots, q\}$ at random, then, $\overline{\tau}$ and τ are independent, and Peggy cannot get any information about $\overline{\tau}$ by observing the transcript τ. Namely, given $\overline{\tau} = (\overline{a}, \overline{c}, \overline{b})$, each (a, c, b) occurs exactly q times among the $\beta_{(u,v,w)}(\overline{a}, \overline{c}, \overline{b})$, where $(u, v, w) \in \mathbb{Z}_q^3$. Thus, the probability that $\tau = (a, b, c)$ is Vic's transformation of $(\overline{a}, \overline{c}, \overline{b})$ is

$$\frac{|\{(u, v, w) \in \mathbb{Z}_q^3 \mid \beta_{(u,v,w)}(\overline{a}, \overline{c}, \overline{b}) = (a, c, b)\}|}{|\mathbb{Z}_q^3|} = q^{-2},$$

and this is the same as the probability of $\tau = (a, b, c)$ if we randomly (and uniformly) select a transcript from the set $\mathcal{T}$ of all accepting transcripts of ProofLog. We see that Peggy's signature is really blind – she has no better chance than guessing at random to link her signature (c, b) with the original transcript $(\overline{a}, \overline{c}, \overline{b})$. Peggy receives no information about the transcript τ from knowing $\overline{\tau}$ (in the information-theoretic sense, see Appendix B.4).

On the other hand, the following arguments give evidence that for Vic, the only way to transform $(\overline{a}, \overline{c}, \overline{b})$ into another accepting, randomly looking transcript (a, b, c) is to randomly choose (u, v, w) and to apply $\beta_{(u,v,w)}$. First, we observe that Vic has to set $a = \overline{a}^u g^v y^w$ (for some u, v, w). Namely, assume that Vic sets $a = \overline{a}^u g^v y^w g'$ with some randomly chosen g'. Then he gets

$$a = \overline{a}^u g^v y^w g' = \left(g^{\overline{b}} y^{\overline{c}}\right)^u g^v y^w g' = g^{u\overline{b}+v} y^{u\overline{c}+w} g',$$

and, since (a, c, b) is an accepting transcript, it follows that

$$g^{u\overline{b}+v} y^{u\overline{c}+w} g' = g^b y^c$$

or

$$g^{b-(u\overline{b}+v)+x(c-(u\overline{c}+w))} = g'.$$

This equation shows that Peggy and Vic together could compute $\log_g(g')$. This contradicts the discrete logarithm assumption, because g' was chosen at random.

If (a, b, c) is an accepting transcript, we get from $a = \bar{a}^u g^v y^w$ that

$$g^b y^c = a = \bar{a}^u g^v y^w = \left(g^{\bar{b}} y^{\bar{c}}\right)^u g^v y^w = g^{u\bar{b}+v} y^{u\bar{c}+w},$$

and conclude that

$$g^{(u\bar{b}+v)-b} = y^{c-(u\bar{c}+w)}.$$

This implies that

$$b = u\bar{b} + v \text{ and } c = u\bar{c} + w,$$

because otherwise Vic could compute Peggy's secret key x as $(u\bar{b} + v - b)(c - u\bar{c} - w)^{-1}$.

Our considerations lead to *Schnorr's blind signature* scheme. In this scheme, the verifier Vic gets a blind signature for m from the prover Peggy by executing the following protocol.

Protocol 4.18.

BlindLogSig$_h$ (m):

1. Peggy randomly chooses $\bar{r} \in \{0, \ldots, q-1\}$, computes $\bar{a} := g^{\bar{r}}$ and sends it to Vic.
2. Vic chooses $u, v, w \in \{0, \ldots, q-1\}$, $u \neq 0$, at random and computes $a = \bar{a}^u g^v y^w$, $c := h(m\|a)$ and $\bar{c} := (c-w)u^{-1}$. Vic sends $\bar{c}$ to Peggy.
3. Peggy computes $\bar{b} := \bar{r} - \bar{c}x$ and sends it to Vic.
4. Vic verifies whether $\bar{a} = g^{\bar{b}} y^{\bar{c}}$, computes $b := u\bar{b} + v$ and gets the signature $\sigma(m) := (c, b)$ of m.

The verification condition for a signature (c, b) is $c = h(m\|g^b y^c)$.

A dishonest Vic may try to use a blind signature (c, b) for more than one message m. This is prevented by the collision resistance of the hash function h. In Section 4.5.2 we will use the blind signatures to form digital cash. A coin is simply the blind signature (c, b) issued by the bank for a specific message m. Thus, there is another basic security requirement for blind signatures: Vic should have no chance of deriving more than one (c, b) from one transcript $(\bar{a}, \bar{c}, \bar{b})$, i.e., from one execution of the protocol. Otherwise, in our digital cash example, Vic could derive more than one coin from the one coin issued by the bank. The Schnorr blind signature scheme seems to fulfill this security requirement. Namely, since h is collision resistant, Vic can work with at most one a. Moreover, Vic can know at most one triplet (u, v, w), with $a = \bar{a}^u g^v y^w$. This follows from Proposition 4.21 below, because $\bar{a}, g$ and y are chosen independently. Finally, the transformations $\beta_{(u,v,w)}$ are the only ones Vic can apply, as we saw above. However, we only gave convincing evidence for the latter statement, not a rigorous mathematical proof.

There is no mathematical security proof for either Schnorr's blind signature scheme or the underlying Schnorr identification scheme. A modification of the Schnorr identification scheme, the Okamoto-Schnorr identification scheme, is proven to be secure under the discrete logarithm assumption ([Okamoto92]). [PoiSte2000] gives a security proof for the Okamoto-Schnorr blind signature scheme, derived from the Okamoto-Schnorr identification scheme. It is shown that no one can derive more than l signed messages after receiving l blind signatures from the signer. The proof is in the so-called random oracle model: the hash function is assumed to behave like a truly random function. Below we use Schnorr's blind signature scheme, because it is a bit easier.

If we use $\mathrm{BlindLogSig}_h$ as a subprotocol, we simply write

$$(c, b) = \mathrm{BlindLogSig}_h(m).$$

The Blindly Issued Proof that Two Logarithms are Equal. As before, let p and q be large primes, such that q divides $p - 1$, and let G_q be the subgroup of order q in $\mathbb{Z}_p^*$, g a generator of G_q and $h : \{0,1\}^* \longrightarrow \mathbb{Z}_q$ be a collision-resistant hash function. Peggy's secret is a randomly chosen $x \in \{0, \ldots, q-1\}$, whereas $y = g^x$ is publicly known.

In Section 4.4.3 we introduced an interactive proof $\mathrm{ProofLogEq}(g, y, \tilde{g}, \tilde{y})$ that two logarithms are equal. Given a $\tilde{y}$ with $\tilde{y} = \tilde{g}^x$, Peggy can prove to the verifier Vic that she knows the common logarithm x of y and $\tilde{y}$ (with respect to the bases g and $\tilde{g}$). This proof can be transformed into a blindly issued proof. Here it is the goal of Vic to obtain, by interacting with Peggy, a z with $z = m^x$ for a given message $m \in G_q$, together with a proof (c, b) of this fact which he may then present to somebody else (recall that x is Peggy's secret and is not revealed). Now, Peggy should issue this proof blindly, i.e., she should not see m and z during the interaction, and later she should not be able to link (m, z, c, b) with this issuing transaction. We proceed in a similar way as in the blind signature scheme. Vic must not give m to Peggy, so he sends a cleverly transformed $\overline{m}$. Peggy computes $\overline{z} = \overline{m}^x$, and then both execute the ProofLogEq protocol (Section 4.4.3), with the analogous modification as above: Vic does not choose his challenge randomly, but computes it by means of the hash function h (with m as part of the input). This interaction results in a transcript $\overline{\tau} = (\overline{m}, \overline{z}, \overline{a}_1, \overline{a}_2, \overline{c}, \overline{b})$. Finally, Vic transforms $\overline{\tau}$ into an accepting transcript $\tau = (m, z, a_1, a_2, c, b)$ of the non-interactive version of ProofLogEq (see Section 4.4.4). Since τ appears completely random to Peggy, she cannot link τ to the original $\overline{\tau}$. Vic uses the following transformation:

$$a_1 := \overline{a}_1^u g^v y^w,$$
$$a_2 := (\overline{a}_2^u \overline{m}^v \overline{z}^w)^s (\overline{a}_1^u g^v y^w)^t,$$
$$c := u\overline{c} + w,$$
$$b := u\overline{b} + v,$$

$$m := \overline{m}^s g^t,$$
$$z := \overline{z}^s y^t.$$

As above, a straightforward computation shows that $a_1 = g^b y^c$ and $a_2 = m^b z^c$. Thus, the transcript τ is indeed an accepting one. Analogous arguments as those given for the blind signature scheme show that the given transformation is the only way for Vic to obtain an accepting transcript, and that $\overline{\tau}$ and τ are independent if Vic chooses $u, v, w, s, t \in \{0, \ldots, q-1\}$ at random. Hence, the proof is really blind – Peggy gets no information about the transcript τ from knowing $\overline{\tau}$.

Our considerations lead to the following blind signature scheme. The signature is issued blindly by Peggy. If Vic wants to get a signature for m, he transforms m to $\overline{m}$ and sends it to Peggy. Peggy computes $\overline{z} = \overline{m}^x$ and executes the ProofLogEq protocol with Vic to prove that $\log_{\overline{m}}(\overline{z}) = \log_g(y)$. Vic derives z and a proof that $\log_m(z) = \log_g(y)$ from $\overline{z}$ and the proof that $\log_{\overline{m}}(\overline{z}) = \log_g(y)$. The signature of m consists of $z = m^x$ and the proof that $\log_m(z) = \log_g(y)$.

Protocol 4.19.

BlindLogEqSig$_h$ (g, y, m):

1. Vic chooses $s, t \in \{0, \ldots, q-1\}$, $s \neq 0$, at random, computes $\overline{m} := m^{1/s} g^{-t/s}$ and sends $\overline{m}$ to Peggy.[9]
2. Peggy randomly chooses $\overline{r} \in \{0, \ldots, q-1\}$ and computes $\overline{z} := \overline{m}^x$ and $\overline{a} := (\overline{a}_1, \overline{a}_2) = (g^{\overline{r}}, \overline{m}^{\overline{r}})$. Peggy sends $(\overline{z}, \overline{a})$ to Vic.
3. Vic chooses $u, v, w \in \{0, \ldots, q-1\}$, $u \neq 0$, at random, and computes $a_1 := \overline{a}_1^u g^v y^w$, $a_2 := (\overline{a}_2^u \overline{m}^v \overline{z}^w)^s (\overline{a}_1^u g^v y^w)^t$ and $z := \overline{z}^s y^t$. Then Vic computes $c := h(m \| z \| a_1 \| a_2)$ and $\overline{c} := (c-w)u^{-1}$, and sends $\overline{c}$ to Peggy.
4. Peggy computes $\overline{b} := \overline{r} - \overline{c}x$ and sends it to Vic.
5. Vic verifies whether $\overline{a}_1 = g^{\overline{b}} y^{\overline{c}}$ and $\overline{a}_2 = \overline{m}^{\overline{b}} \overline{z}^{\overline{c}}$, computes $b := u\overline{b} + v$ and receives (z, c, b) as the final result.

If Vic presents the proof to Alice, then Alice may verify the proof by checking the verification condition $c = h(m \| z \| g^b y^c \| m^b z^c)$.

Below we will use BlindLogEqSig$_h$ as a subprotocol. Then we simply write

$$(z, c, b) = \text{BlindLogEqSig}_h(g, y, m).$$

Remarks:

1. Again the collision resistance of h implies that Vic can use the proof (z, c, b) for only one message m. As before, in the blind signature protocol *BlindLogSig*, we see that Vic cannot derive two different signatures

---

[9] As is common practice, we denote the s-th root $x^{s^{-1}}$ of an element x of order q by $x^{1/s}$. Here, s^{-1} is the inverse of s modulo q. See "Computing modulo a prime" on page 303.

(z, c, b) and $(\tilde{z}, \tilde{c}, \tilde{b})$ from one execution of the protocol, i.e., from one transcript $(\overline{m}, \overline{z}, \overline{a}_1, \overline{a}_2, \overline{c}, \overline{b})$.

2. The protocols BlindLogEqSig_h and BlindLogSig_h may be merged to yield not only a signature of m but also a signature of an additionally given message M. Namely, if Vic computes $c = h(M \| m \| z \| a_1 \| a_2)$ in step 3, then (c, b) is also a blind signature of M, formed in the same way as the signatures of BlindLogSig_h. We denote this merged protocol by

$$(z, c, b) = \text{BlindLogEqSig}_h(M, g, y, m),$$

and call it the proof BlindLogEqSig_h dependent on the message M. It simultaneously gives signatures of M and m (consisting of z and a proof that $\log_g(y) = \log_m(z)$).

4.5.2 A Fair Electronic Cash System

The payment scheme we describe is published in [CamMauSta96]. A coin is a bit string that is (blindly) signed by the bank. For simplicity, we restrict the scheme to a single denomination of coins: the extension to multiple denominations is straightforward, with the bank using a separate key pair for each denomination. As discussed in the introduction of Section 4.5, in a fair electronic cash system the tracing of a coin and the revoking of the customer's anonymity must be possible under certain well-defined conditions; for example to track kidnappers who obtain a ransom as electronic cash. Here, anonymity may be revoked by a trusted third party, the trusted center.

System Setup. As before, let p and q be large primes such that q divides $p - 1$. Let G_q be the subgroup of order q in $\mathbb{Z}_p^*$. Let g, g_1 and g_2 be randomly and independently chosen generators of G_q. The security of the scheme requires that the discrete logarithm of none of these elements with respect to another of these elements is known. Since g, g_1 and g_2 are chosen randomly and independently, this is true with a very high probability. Let $h : \{0, 1\}^* \longrightarrow \mathbb{Z}_q$ be a collision-resistant hash function.

1. The bank chooses a secret key $x \in \{1, \ldots, q-1\}$ at random and publishes $y = g^x$.
2. The trusted center T chooses a secret key $x_T \in \{1, \ldots, q-1\}$ at random and publishes $y_T = g_2^{x_T}$.
3. The customer – we call her Alice – has the secret key $x_C \in \{1, \ldots, q-1\}$ (randomly selected) and the public key $y_C = g_1^{x_C}$.
4. The shop's secret key x_S is randomly selected in $\{1, \ldots, q-1\}$ and $y_S = g_1^{x_S}$ is the public key.

By exploiting data observed by the bank in the withdrawal protocol, the trusted center can provide information which enables the recognition of a coin withdrawn by Alice in the deposit step. This trace mechanism is called *coin tracing*. Moreover, from data observed by the bank in the deposit protocol, the trusted center can compute information which enables the identification of the customer. This tracing mechanism is called *owner tracing*.

Opening an Account. When the customer Alice opens an account, she proves her identity to the bank. She can do this by executing the protocol $\text{ProofLog}(g_1, y_C)$ with the bank. The bank then opens an account and stores y_C in Alice's entry in the account database.

The Online Electronic Cash System. We first discuss the online system. The trusted center is involved in every withdrawal transaction, and the bank is involved in every payment.

The Withdrawal Protocol. As before, Alice has to authenticate herself to the bank. She can do this by proving that she knows $x_C = \log_{g_1}(y_C)$. To get a coin, Alice executes the *withdrawal* protocol with the bank. It is a modification of the BlindLogSig_h protocol given in Section 4.5.1. Essentially, a coin is a signature of the empty string blindly issued by the bank.

Protocol 4.20.
Withdrawal:

1. The bank randomly chooses $\bar{r} \in \{0, \ldots, q-1\}$ and computes $\bar{a} = g^{\bar{r}}$. The bank sends $\bar{a}$ to Alice.
2. Alice chooses $u, v, w \in \{0, \ldots, q-1\}$, $u \neq 0$, at random, and computes $a := \bar{a}^u g^v y^w$ and $c := h(a), \bar{c} := (c-w)u^{-1}$. Alice sends (u, v, w) encrypted with the trusted center's public key and $\bar{c}$ to the bank.
3. The bank sends $\bar{a}$ and $\bar{c}$ and the encrypted (u, v, w) to the trusted center T.
4. T checks whether $u\bar{c} + w = h(\bar{a}^u g^v y^w)$, and sends the result to the bank.
5. If the result is correct, the bank computes $\bar{b} := \bar{r} - \bar{c}x$ and sends it to Alice. Alice's account is debited.
6. Alice verifies whether $\bar{a} = g^{\bar{b}} y^{\bar{c}}$, computes $b := u\bar{b} + v$ and gets as the coin the signature $\sigma := (c, b)$ of the empty message.

Payment and Deposit. In online payment, the shop must be connected to the bank when the customer spends the coin. Payment and deposit form one transaction. Alice spends a coin by sending it to a shop. The shop verifies the coin, i.e., it verifies the signature by checking $c = h(g^b y^c)$. If the coin is valid, it passes the coin to the bank. The bank also verifies the coin and then compares it with all previously spent coins (which are stored in the database). If the coin is new, the bank accepts it and inserts it into the database. The shop's account is credited.

Coin and Owner Tracing. The trusted center can link $(\bar{a}, \bar{c}, \bar{b})$ and (c, b), which enables coin and owner tracing.

Customer Anonymity. The anonymity of the customer relies on the fact that the used signature scheme is blind and on the security of the encryption scheme used to encrypt the blinding factors u, v and w.

The Offline Electronic Cash System. In the offline system, the trusted center is not involved in the withdrawal transaction, and the bank is not involved in the payment protocol. To achieve an offline trusted center, the immediate check that is performed by the trusted center in the withdrawal protocol above is replaced by a proof that Alice correctly provides the information necessary for tracing the coin. This proof can be checked by the bank. To obtain such a proof, the BlindLogSig$_h$-protocol is replaced by the BlindLogEqSig$_h$ protocol (see Section 4.5.1).

Essentially, a coin consists of a pair (m, z) with $m = g_1 g_2^s$, where s is chosen at random and $z = m^x$, and a proof of this fact which is issued blindly by the bank. For this purpose the BlindLogEqSig$_h$ protocol is executed: the bank sees $(\overline{m} = m^{1/s}, \overline{z} = z^{1/s})$. We saw above that the general blinding transformation in the BlindLogEqSig$_h$ protocol is $\overline{m} = m^{1/s} g^{-t}$. Here Alice chooses $t = 0$, i.e., $m = \overline{m}^s$; otherwise Alice could not perform ProofLog$_h$ (M, mg_1^{-1}, g_2) as required in the payment protocol (see below). The blinding exponent s is encrypted by $d = y_T^s = g_2^{x_T s}$, where y_T is the trusted center's public key. This enables the trusted center to revoke anonymity later. It can get m by decrypting d (note that $m = g_1 g_2^s = g_1 d^{1/x_T}$), and the coin can be traced by linking m and $\overline{m}$.

The Withdrawal Protocol. As before, first Alice has to authenticate herself to the bank. The withdrawal protocol in the offline system consists of two steps. In the first of these steps, Alice generates an encryption of the message m to enable anonymity revocation by the trusted center. In the withdrawal step, Alice executes a message-dependent proof BlindLogEqSig$_h$ with the bank to obtain a (blind) signature on her coin:

1. *Enable coin and owner tracing.* Coin tracing means that, starting from the information gathered during the withdrawal the bank recognizes a specific coin in the deposit protocol. Owner tracing identifies the withdrawer of a coin, starting from the deposited coin. Tracings require the cooperation of the trusted center. To enable coin and owner tracing, Alice encrypts the message m with the trusted center's public key y_T and proves that she correctly performs this encryption:

   a. Alice chooses $s \in \{1, \ldots, q-1\}$ at random, then computes

   $$m = g_1 g_2^s, d = y_T^s, \overline{m} = m^{1/s} = g_1^{1/s} g_2 \text{ and}$$
   $$(c, b) = \text{ProofLogEq}_h\left(\overline{m} g_2^{-1}, g_1, y_T, d\right).$$

   By the proof, Alice shows that $\log_{\overline{m} g_2^{-1}}(g_1) = \log_{y_T}(d)$ and that she knows this logarithm. Alice sends $\left(c, b, \overline{m} g_2^{-1}, g_1, y_T, d\right)$ to the bank.

    b. The bank verifies the proof. If the verification condition holds, then the bank stores d in Alice's entry in the withdrawal database for a possible later anonymity revocation.

2. *The withdrawal of a coin.* Alice chooses $r \in \{0, \ldots, q-1\}$ at random, computes the so-called coin number $c\# = g^r$ and executes

$$(z, c_1, b_1) = \text{BlindLogEqSig}_h(c\#, g, y, m)$$

with the bank. Here, in the first step of BlindLogEqSig_h, Alice takes the s from the coin and owner tracing step 1 and $t = 0$. Thus, she sends the $\overline{m} = m^{1/s} = g_1^{1/s} g_2$ from step 1 to the bank. The variant of BlindLogEqSig_h is used that, in addition to the signature of m, gives a signature of $c\#$ by the bank (see the remarks after Protocol 4.19). The coin number $c\#$ is needed in the payment below. The bank debits Alice's account. The coin consists of $(c_1, b_1, c\#, g, y, m, z)$ and some additional information Alice has to supply when spending it (see the payment protocol below).

Payment. In an offline system, double spending cannot be prevented. It can only be detected by the bank in the deposit protocol. An additional mechanism in the protocols is necessary to decide whether the customer or the shop doubled the coin. If the coin was doubled by the shop, the bank refuses to credit the shop's account. If the customer doubled the coin, her anonymity should be revoked without the help of the trusted center. This can be achieved by the payment protocol. For this purpose, Alice, when paying with the coin, has to sign the message $M = y_S \| time \| (c_1, b_1)$, where y_S is the public key of the shop, *time* is the time of the payment and (c_1, b_1) is the bank's blind signature on the coin from above. Alice has to sign by means of the basic signature scheme. There she has to use $s = \log_{g_2}(mg_1^{-1})$ as her secret and the coin number $c\# = g^r$, which she computed in the withdrawal protocol, as her commitment a.

$$\sigma(M) = (c_2, b_2) = \text{ProofLog}_h\left(M, mg_1^{-1}, g_2\right)$$

Now, if Alice spent the same coin $(c_1, b_1, c\#, g, y, m, z)$ twice, she would provide two signatures $\sigma(M)$ and $\sigma(M')$ of different messages M and M' (at least the times differ!). Both signatures are computed with the same commitment $a = c\#$. Then, it is easy to identify Alice (see below). The coin submitted to the shop is defined by:

$$coin = \left((c_1, b_1, c\#, g, y, m, z), \left(c_2, b_2, M, g_2, mg_1^{-1}\right)\right).$$

The shop verifies the coin, i.e., it verifies:

1. The correct form of M.
2. Whether $c_2 = h(M \| c\#)$.
3. The proof $(z, c_1, b_1) = \text{BlindLogEqSig}_h(c\#, g, y, m)$.

4. The proof $(c_2, b_2) = \text{ProofLog}_h\left(M, {mg_1}^{-1}, g_2\right)$, by testing

$$c_2 = h\left(M \| \left({mg_1}^{-1}\right)^{b_2} g_2^{c_2}\right).$$

Since h is collision resistant and the shop checks $c_2 = h(M \| c\#)$, Alice necessarily has to use the coin number $c\#$ in the second proof. The shop accepts if the coin passes the verification.

Deposit. The shop sends the coin to the bank. The bank verifies the coin and searches in the database for an identical coin. If she finds an identical coin, she refuses to credit the shop's account. If she finds a coin with identical first and different second component, the bank revokes the customer's anonymity (see below).

Coin and Owner Tracing.

1. *Coin tracing.* If the bank provides the trusted center T with $d = y_T^s$ produced by Alice in the withdrawal protocol, T computes m:

$$g_1 d^{1/x_T} = g_1 g_2^s = m.$$

   This value can be used to recognize the coin in the deposit protocol.
2. *Owner tracing.* If the bank provides the trusted center T with the m of a spent coin, T computes d:

$$\left({mg_1}^{-1}\right)^{x_T} = (g_2^s)^{x_T} = y_T^s = d.$$

   This value can be used for searching in the withdrawal database.

Security.

1. *Double spending.* If Alice spends a coin twice (at different shops, or at the same shop but at different times), she produces signatures of two different messages. Both signatures are computed with the same commitment $c\#$ (the coin number). This reveals the signer's secret, which is the blinding exponent s in our case (see the remarks after Protocol 4.17). Knowing the blinding exponent s, the bank can derive $\overline{m} = m^{1/s}$. Searching in the withdrawal database yields Alice's identity. If the shop doubled the coin, the bank detects an identical coin in the database. Then she refuses to credit the shop's account.
2. *Customer's anonymity.* The anonymity of the customer is ensured. Namely, BlindLogEqSig_h is a perfectly blind signature scheme, as we observed above. Moreover, since ProofLogEq is honest-verifier zero knowledge (see Section 4.4.3), the bank also cannot get any information about the blinding exponent s from the ProofLogEq_h in the withdrawal protocol. Note that any information about s could enable the bank to link $\overline{m}$ with m, and hence the withdrawal of the coin with the deposit of the coin. The bank could also establish this link, if she were able to determine (without computing the logarithms) that

$$\log_{g_2}\left(mg_1{}^{-1}\right) = \log_{\overline{m}g_2{}^{-1}}(g_1). \tag{4.3}$$

Then she could link the proofs $\text{ProofLogEq}_h\left(\overline{m}g_2^{-1}, g_1, y_T, d\right)$ (from the withdrawal transaction) and $\text{ProofLog}_h\left(M, g_2, mg_1^{-1}\right)$ (from the deposit transaction). However, to find out (4.3) contradicts the decision Diffie-Hellman assumption (see Section 4.5.3).

3. *Security for the trusted center.* The trusted center makes coin and owner tracing possible. The tracings require that Alice correctly forms $m = g_1 g_2^s$ and $\overline{m} = m^{1/s} = g_1^{1/s} g_2$, and that it is really y_T^s which she sends as d to the bank (in the withdrawal transaction).

Now in the withdrawal protocol, Alice proves that $\overline{m} = g_1^{1/s} g_2$ and that $d = y_T^s$. In the payment protocol, Alice proves that $m = g_1 g_2^{\tilde{s}}$ with $\tilde{s}$ known to her. It is not clear a priori that $\tilde{s} = s$ or, equivalently, that $\overline{m} = m^{1/s}$. However, as we observed before in the blind signature scheme BlindLogEqSig_h, the only way to transform $\overline{m}$ into m is to choose σ and τ at random and to compute $m = \overline{m}^\sigma g^\tau$. From this we conclude that indeed $\tilde{s} = s$ and $m = \overline{m}^s$:

$$g_1 g_2^{\tilde{s}} = m = \overline{m}^\sigma g^\tau = (g_1^{1/s} g_2)^\sigma g^\tau = g_1^{\sigma/s} g_2^\sigma g^\tau.$$

Hence, $\tau = 0$ and $\sigma = \tilde{s} = s$ (by Proposition 4.21, below).

4.5.3 Underlying Problems

The Representation Problem. Let p and q be large primes such that q divides $p - 1$. Let G_q be the subgroup of order q in $\mathbb{Z}_p^*$.

Let $r \geq 2$ and let $g_1, \ldots, g_r$ be pairwise distinct generators of G_q.[10] Then $g = (g_1, \ldots, g_r) \in G_q^r$ is called a *generator of length* r. Let $y \in G_q$. $a = (a_1, \ldots, a_r) \in \mathbb{Z}_q^r$ is a *representation* of y (with respect to g) if

$$y = \prod_{i=1}^r g_i^{a_i}.$$

To represent y, the elements $a_1, \cdots, a_{r-1}$ can be chosen arbitrarily; a_r is then uniquely determined. Therefore, each $y \in G_q$ has q^{r-1} different representations. Given y, the probability that a randomly chosen $a \in \{0, \ldots, q\}^r$ is a representation of y is only $1/q$.

Proposition 4.21. *Assume that it is infeasible to compute discrete logarithms in G_q. Then no polynomial algorithm can exist which, on input of a randomly chosen generator of length $r \geq 2$, outputs $y \in G_q$ and two different representations of y.*

---

[10] Note that every element of G_q except $[1]$ is a generator of G_q (see Lemma A.40).

Proof. Assume that such an algorithm exists. On input of a randomly chosen generator, it outputs $y \in G_q$ and two different representations $a = (a_1, \ldots, a_r)$ and $b = (b_1, \ldots, b_r)$ of y. Then, $a - b$ is a non-trivial representation of [1]. Thus, we have a polynomial algorithm A which on input of a randomly chosen generator outputs a non-trivial representation of [1]. We may use A to define an algorithm B that on input of $g \in G_q, g \neq [1]$, and $z \in G_q$, computes the discrete logarithm of z with respect to g.

Algorithm 4.22.

   int $B(\text{int } g, z)$

  1   repeat

  2   select $i \in \{1, \ldots, r\}$ and

  3       $u_j \in \{1, \ldots, q - 1\}, 1 \leq j \leq r$, uniformly at random

  4       $g_i \leftarrow z^{u_i}, g_j \leftarrow g^{u_j}, 1 \leq j \neq i \leq r$

  5       $(a_1, \ldots, a_r) \leftarrow A(g_1, \ldots, g_r)$

  6   until $a_i u_i \not\equiv 0 \bmod q$

  7   return $-(a_i u_i)^{-1} \left(\sum_{j \neq i} a_j u_j \right) \bmod q$

We have

$$z^{-a_i u_i} = \prod_{j \neq i} g^{a_j u_j}.$$

Hence, the returned value is indeed the logarithm of z. Since at least one a_j returned by A is $\neq 0$ modulo q, the probability that $a_i \neq 0$ modulo q is $1/r$. Hence, we expect that the repeat until loop will terminate after r iterations. If r is bounded by a polynomial in the binary length $|p|$ of p, the expected running time of B is polynomial in $|p|$. $\qquad\qquad\square$

Remark. Assume there is a polynomial algorithm which, when given as input a generator of length $r \geq 2$, outputs $y \in G_q$ and two different representations of y – not with certainty, but at least with some non-negligible probability. Then, this algorithm can be used to compute discrete logarithms in G_q with an overwhelmingly high probability (see Exercise 4 in Chapter 6).

The Decision Diffie-Hellman Problem. Let p and q be large primes, such that q divides $p - 1$. Let G_q be the subgroup of order q in $\mathbb{Z}_p^*$. Let $g \in G_q$ and $a, b \in \{0, \ldots, q-1\}$ be randomly chosen. Then, the Diffie-Hellman assumption (Section 4.1.2) says that it is impossible to compute g^{ab} from g^a and g^b.

Let $g_1 = g^a, g_2 = g^b$ and g_3 be given. The *decision Diffie-Hellman problem* is to decide if

$$g_3 = g^{ab}.$$

This is equivalent to deciding whether

$$\log_g(g_3) = \log_g(g_1) \log_g(g_2), \text{ or}$$
$$\log_{g_2}(g_3) = \log_g(g_1).$$

The *decision Diffie-Hellman assumption* says that no efficient algorithm exists to solve the decision Diffie-Hellman problem if a, b and g_3 (g_1, g_2 and g_3, respectively) are chosen at random (and independently). The decision Diffie-Hellman problem is random self-reducible (see the remark on page 154). If you can solve it with an efficient probabilistic algorithm A, then you can also solve it, if $g \in G_q$ is any element of G_q and only g_1, g_2, g_3 are chosen randomly. Namely, let $g \in G_q$, then (g, g_1, g_2, g_3) has the Diffie-Hellman property, if and only if $(g^s, g_1^s, g_2^s, g_3^s)$ has the Diffie-Hellman property, with s randomly chosen in $\mathbb{Z}_q^*$.

The representation problem and the decision Diffie-Hellman problem are studied, for example, in [Brands93].

Exercises

1. Let p be a sufficiently large prime such that it is intractable to compute discrete logarithms in $\mathbb{Z}_p^*$. Let g be a primitive root in $\mathbb{Z}_p^*$. p and g are publicly known. Alice has a secret key x_A and a public key $y_A := g^{x_A}$. Bob has a secret key x_B and a public key $y_B := g^{x_B}$. Alice and Bob establish a secret shared key by executing the following protocol (see [MatTakIma86]):

   Protocol 4.23.
   A variant of the Diffie-Hellman key agreement protocol:
   1. Alice chooses at random a, $0 \le a \le p - 2$, sets $c := g^a$ and sends c to Bob.
   2. Bob chooses at random b, $0 \le b \le p - 2$, sets $d := g^b$ and sends d to Alice.
   3. Alice computes the shared key $k = d^{x_A} y_B{}^a = g^{bx_A + ax_B}$.
   4. Bob computes the shared key $k = c^{x_B} y_A{}^b = g^{ax_B + bx_A}$.

   Does the protocol provide entity authentication? Discuss the security of the protocol.

2. Let $n := pq$, where p and q are distinct primes and $x_1, x_2 \in \mathbb{Z}_n^*$. Assume that at least one of x_1 and x_2 is in QR_n. Peggy wants to prove to Vic that she knows a square root of x_i for at least one $i \in \{1, 2\}$ without revealing i. Modify Protocol 4.5 to get an interactive zero-knowledge proof of knowledge.

3. Besides interactive proofs of knowledge, there are interactive proofs for proving the membership in a language. The completeness and soundness conditions for such proofs are slightly different. Let (P, V) be an interactive proof system. P and V are probabilistic algorithms, but only V is assumed to have polynomial running time. By P^* we denote a general (possibly dishonest) prover. Let $\mathcal{L} \subseteq \{0, 1\}^*$ ($\mathcal{L}$ is called a language).

Bit strings $x \in \{0,1\}^*$ are supplied to (P,V) as common input. (P,V) is called an *interactive proof system for the language $\mathcal{L}$* if the following conditions are satisfied:

a. *Completeness.* If $x \in \mathcal{L}$, then the probability that the verifier V accepts, if interacting with the honest prover P, is $\geq 3/4$.

b. *Soundness.* If $x \notin \mathcal{L}$, then the probability that the verifier V accepts, if interacting with any prover P^*, is $\leq 1/2$.

Such an interactive proof system (P,V) is *(perfect) zero-knowledge* if there is a probabilistic *simulator* $S(V^*,x)$, running in expected polynomial time, such that for every verifier V^* (honest or not) and for every $x \in \mathcal{L}$ the distributions of the random variables $S(V^*,x)$ and $(P,V^*)(x)$ are equal.

The class of languages that have interactive proof systems is denoted by $\mathcal{IP}$. It generalizes the complexity class $\mathcal{BPP}$ (see Exercise 3 in Chapter 5).

As in Section 4.3.1, let $n := pq$, with p and q distinct primes, and let $J_n^{+1} := \{x \in \mathbb{Z}_n^* \mid \left(\frac{x}{n}\right) = 1\}$ be the units with Jacobi symbol 1. Let $\mathrm{QNR}_n^{+1} := J_n^{+1} \setminus \mathrm{QR}_n$ be the quadratic non-residues in J_n^{+1}.

The following protocol is an interactive proof system for the language QNR_n^{+1} (see [GolMicRac89]). The common input x is assumed to be in J_n^{+1} (whether or not $x \in \mathbb{Z}_n$ is in J_n^{+1} can be efficiently determined using a deterministic algorithm; see Algorithm A.59).

Protocol 4.24.

Quadratic non-residuosity:

Let $x \in J_n^{+1}$.

1. Vic chooses $r \in \mathbb{Z}_n^*$ and $\sigma \in \{0,1\}$ uniformly at random and sends $a = r^2 x^\sigma$ to Peggy.

2. Peggy computes $\tau := \begin{cases} 0 \text{ if } a \in \mathrm{QR}_n \\ 1 \text{ if } a \notin \mathrm{QR}_n \end{cases}$ and sends τ to Vic.

   (Note that it is not assumed that Peggy can solve this in polynomial time. Thus, she can find out whether $a \in \mathrm{QR}_n$, for example, by an exhaustive search.)

3. Vic accepts if and only if $\sigma = \tau$.

Show:

a. If $x \in \mathrm{QNR}_n^{+1}$ and both follow the protocol, Vic will always accept.

b. If $x \notin \mathrm{QNR}_n^{+1}$, then Vic accepts with probability $\leq 1/2$.

c. Show that the protocol is not zero-knowledge (under the quadratic residuosity assumption; see Remark 1 in Section 4.3.1).

d. The protocol is honest-verifier zero-knowledge.

e. Modify the protocol to get a zero-knowledge proof for quadratic non-residuosity.

4. We consider the identification scheme based on public-key encryption introduced in Section 4.2.1. In this scheme a dishonest verifier can obtain knowledge from the prover. Improve the scheme.

5. We modify the commitment scheme based on quadratic residues.

   Protocol 4.25.
   QRCommitment:

   1. *System setup.* Alice chooses distinct large prime numbers $p, q \equiv 3 \bmod 4$ and sets $n := pq$. (Note $-1 \in J_n^{+1} \setminus QR_n$, see Proposition A.53.)
   2. *Commit to $b \in \{0, 1\}$.* Alice chooses $r \in \mathbb{Z}_n^*$ at random, sets $c := (-1)^b r^2$ and sends c to Bob.
   3. *Reveal.* Alice sends p, q, r and b to Bob. Bob can verify that p and q are primes $\equiv 3 \bmod 4$, $r \in \mathbb{Z}_n^*$, and $c := (-1)^b r^2$.

   Show:
   a. If c is a commitment to b, then $-c$ is a commitment to $1 - b$.
   b. If c_i is a commitment to b_i, $i = 1, 2$, then $c_1 c_2$ is a commitment to $b_1 \oplus b_2$.
   c. Show how Alice can prove to Bob that two commitments c_1 and c_2 commit to equal or distinct values, without opening them.

6. Let $P = \{P_i \mid i = 1, \ldots, 6\}$. Set up a secret sharing system, such that exactly the groups $\{P_1, P_2\}, \{Q \subset P \mid |Q| \geq 3, P_1 \in Q\}$ and $\{Q \subset P \mid |Q| \geq 4, P_2 \in Q\}$ are able to reconstruct the secret.

7. Let $P = \{P_1, P_2, P_3, P_4\}$. Is it possible to set up a secret sharing system by use of Shamir's threshold scheme, such that the members of a group $Q \subset P$ are able to reconstruct the secret if and only if $\{P_1, P_2\} \subset Q$ or $\{P_3, P_4\} \subset Q$?

8. In the voting scheme of Section 4.4, it is necessary that each authority and each voter proves that he really follows the protocol. Explain why.

9. Let p and q be large primes such that q divides $p - 1$. Let G be the subgroup of order q in $\mathbb{Z}_p^*$, $g, h, y_i, z_i \in G$, $i = 1, 2$. Peggy wants to prove to Vic that she knows an x, such that $y_i := g^x$ and $z_i := h^x$ for at least one $i \in \{1, 2\}$, without revealing i. Modify Protocol 4.15 to get an interactive proof of knowledge. Show how the interactive proof can be converted into a non-interactive one.

10. We consider the problem of *vote duplication*. This means that a voter can duplicate the vote of another voter who has previously posted his vote. He can do this without knowing the content of the other voter's ballot. Discuss this problem for the voting scheme of Section 4.4.

11. **Blind RSA signatures.** Construct a blind signature scheme based on the fact that the RSA function is a homomorphism.

12. **Nyberg-Rueppel Signatures.** Let p and q be large primes such that q divides $p - 1$. Let G be the subgroup of order q in $\mathbb{Z}_p^*$, and let g be a generator of G. The secret key of the signer is a randomly chosen $x \in \mathbb{Z}_q$, the public key is $y := g^x$.

   Signing. We assume that the message m to be signed is an element in $\mathbb{Z}_p^*$. The signed message is produced using the following steps:
   1. Select a random integer k, $1 \leq k \leq q - 1$.
   2. Set $r := mg^k$ and $s := xr + k \bmod q$.
   3. (m, r, s) is the signed message.

   Verification. If $1 \leq r \leq p - 1$, $1 \leq s \leq q - 1$ and $m = ry^r g^{-s}$, accept the signature. If not, reject it.

   a. Show that the verification condition holds for a signed message.
   b. Show that it is easy to produce forged signatures.
   c. How can you prevent this attack?
   d. Show that the condition $1 \leq r \leq p - 1$ has to be checked to detect forged signatures, even if the scheme is modified as in item c.

13. **Blind Nyberg-Rueppel Signatures** (see also [CamPivSta94]). In the situation of Exercise 12, Bob gets a blind signature for a message $m \in \{1, \ldots, q - 1\}$ from Alice by executing the following protocol:

   Protocol 4.26.
   BlindNybergRueppelSig(m):
   1. Alice chooses $\tilde{k}$ at random, $1 \leq \tilde{k} \leq q - 1$, and sets $\tilde{a} := g^{\tilde{k}}$. Alice sends $\tilde{a}$ to Bob.
   2. Bob chooses α, β uniformly at random with $1 \leq \alpha \leq q - 1$ and $0 \leq \beta \leq q - 1$, sets $\tilde{m} := m\tilde{a}^{\alpha-1}g^\beta\alpha^{-1}$, and sends $\tilde{m}$ to Alice.
   3. Alice computes $\tilde{r} := \tilde{m}g^{\tilde{k}}$, $\tilde{s} := \tilde{r}x + \tilde{k} \bmod q$, and sends $\tilde{r}$ and $\tilde{s}$ to Bob.
   4. Bob checks whether $(\tilde{m}, \tilde{r}, \tilde{s})$ is a valid signed message. If it is, then he sets $r := \tilde{r}\alpha$ and $s := \tilde{s}\alpha + \beta$.

   Show that (m, r, s) is a signed message and that the protocol is really blind.

14. **Proof of Knowledge of a Representation** (see [Okamoto92]). Let p and q be large primes, such that q divides $p - 1$. Let G be the subgroup of order q in $\mathbb{Z}_p^*$, and g_1 and g_2 be independently chosen generators. The secret is a randomly chosen $(x_1, x_2) \in \{0, \ldots, q-1\}^2$, and the public key is (p, q, g_1, g_2, y), where $y := g_1^{x_1} g_2^{x_2}$ of G.
   How can Peggy convince Vic by an interactive proof of knowledge that she knows (x_1, x_2), which is a representation of y with respect to (g_1, g_2)?

15. Convert the interactive proof of Exercise 14 into a blind signature scheme.

5. Probabilistic Algorithms

Probabilistic algorithms are important in cryptography. On the one hand, the algorithms used in encryption and digital signature schemes often include random choices (as in Vernam's one-time pad or the DSA) and therefore are probabilistic. On the other hand, when studying the security of cryptographic schemes, adversaries are usually modeled as probabilistic algorithms. The subsequent chapters, which deal with provable security properties, require a thorough understanding of this notion. Therefore, we clarify what is meant precisely by a probabilistic algorithm, and discuss the underlying probabilistic model.

The output y of a *deterministic algorithm* A is completely determined by its input x. In a deterministic way, y is computed from x by a sequence of steps decided in advance by the programmer. A behaves like a mathematical mapping: applying A to the same input x several times always yields the same output y. Therefore, we may use the mathematical notation of a mapping, $A : X \longrightarrow Y$, for a deterministic algorithm A, with inputs from X and outputs in Y. There are various equivalent formal models for such algorithms. A popular one is the description of algorithms by Turing machines (see, for example, [HopUll79]). Turing machines are state machines, and deterministic algorithms are modeled by Turing machines with deterministic behavior: the state transitions are completely determined by the input.

A probabilistic algorithm A is an algorithm whose behavior is partly controlled by random events. The computation of the output y on input x depends on the outcome of a finite number of random experiments. In particular, applying A to the same input x twice may yield two different outputs.

5.1 Coin-Tossing Algorithms

Probabilistic algorithms are able to toss coins. The control flow depends on the outcome of the coin tosses. Therefore, probabilistic algorithms exhibit random behavior.

Definition 5.1. Given an input x, a *probabilistic (or randomized) algorithm* A may toss a coin a finite number of times during its computation of the output y, and the next step may depend on the results of the preceding coin tosses. The number of coin tosses may depend on the outcome of the previous ones, but it is bounded by some constant t_x for a given input x. The coin tosses are independent and the coin is a fair one, i.e., each side appears with probability $1/2$.

Examples. The encryption algorithms in Vernam's one-time pad (Section 2.1), OAEP (Section 3.3.4) and ElGamal's scheme (Section 3.5) include random choices, and thus are probabilistic, as well as the signing algorithms in PSS (Section 3.4.5), ElGamal's scheme (Section 3.5.2) and the DSA (Section 3.5.3). Other examples of probabilistic algorithms are the algorithm for computing square roots in $\mathbb{Z}_p^*$ (see Algorithm A.61) and the probabilistic primality tests discussed in Appendix A.8. Many examples of probabilistic algorithms in various areas of application can be found, for example, in [MotRag95].

Remarks and Notations:

1. A formal definition of probabilistic algorithms can be given by the notion of probabilistic Turing machines ([LeeMooShaSha55]; [Rabin63]; [Santos69]; [Gill77]; [BalDiaGab95]).[1] In a probabilistic Turing machine, the state transitions are determined by the input and the outcome of coin tosses. Probabilistic Turing machines should not be confused with non-deterministic machines. A non-deterministic Turing machine is "able to simply guess the solution to the given problem" and thus, in general, is not something that can be implemented in practice. A probabilistic machine (or algorithm) is able to find the solution by use of its coin tosses, with some probability. Thus, it is something that can be implemented in practice.

   Of course, we have to assume (and will assume in the following) that a random source of independent fair coin tosses is available. To implement such a source, the inherent randomness in physical phenomena can be exploited (see [MenOorVan96] and [Schneier96] for examples of sources which might be used in a computer).

   To derive perfectly random bits from a natural source is a non-trivial task. The output bits may be biased (i.e., the probability that 1 is emitted is different from $1/2$) or correlated (the probability of 1 depends on the previously emitted bits). The outcomes of physical processes are often affected by previous outcomes and the circumstances that led to these outcomes. If the bits are independent, the problem of biased bits can be easily solved using the following method proposed by John von Neumann ([von Neumann63]): break the sequence of bits into pairs, discard pairs 00 and 11, and interpret 01 as 0 and 10 as 1 (the pairs 01 and 10 have the same probability). Handling a correlated bit source is more difficult. However, there are effective means of generating truly random sequences

---

[1] All algorithms are assumed to have a finite description (as a Turing machine) which is independent of the size of the input. We do not consider non-uniform algorithms in this book.

of bits from a biased and correlated source. For example, Blum developed a method for a source which produces bits according to a known Markov chain ([Blum84]). Vazirani ([Vazirani85]) shows how almost independent, unbiased bits can be derived from two independent "slightly-random" sources. For a discussion of slightly random sources and their use in randomized algorithms, see [Papadimitriou94], for example.

2. The output y of a probabilistic algorithm A depends on the input x and on the binary string r, which describes the outcome of the coin tosses. Usually, the coin tosses are considered as internal operations of the probabilistic algorithm. A second way to view a probabilistic algorithm A is to consider the outcome of the coin tosses as an additional input, which is supplied by an external coin-tossing device. In this view, the model of a probabilistic algorithm is a deterministic machine. We call the corresponding deterministic algorithm A_D the *deterministic extension* of A. It takes as inputs the original input x and the outcome r of the coin tosses.

3. Given x, the output $A(x)$ of a probabilistic algorithm A is not a single constant value, but a random variable. "A outputs y on input x" is a random event, and by $\mathrm{prob}(A(x) = y)$ we mean the probability of this event. More precisely, we have

$$\mathrm{prob}(A(x) = y) := \mathrm{prob}(\{r \mid A_D(x,r) = y\}).^2$$

Here a question arises: what probability distribution of the coin tosses is meant? The question is easily answered if, as in our definition of probabilistic algorithms, the number of coin tosses is bounded by some constant t_x for a given x. In this case, adding some dummy coin tosses, if necessary, we may assume that the number of coin tosses is exactly t_x. Then the possible outcomes r of the coin tosses are the binary strings of length t_x, and since the coin tosses are independent, we have the uniform distribution of $\{0,1\}^{t_x}$. The probability of an outcome r is $1/2^{t_x}$, and hence

$$\mathrm{prob}(A(x) = y) = \frac{|\{r \mid A_D(x,r) = y\}|}{2^{t_x}}.$$

It is sufficient for all our purposes to consider only probabilistic algorithms with a bounded number of coin tosses, for a given x. In most parts of this book we consider algorithms whose running time is bounded by a function $f(|x|)$, where $|x|$ is the size of the input x. For these algorithms, the assumption is obviously true.

4. Given x, the probabilities $\mathrm{prob}(A(x) = y), y \in Y$, define a probability distribution on the range Y. We denote it by $p_{A(x)}$. The random variable $A(x)$ samples Y according to the distribution $p_{A(x)}$.

---

[2] If the probability distribution is determined by the context, we often do not specify the distribution explicitly and simply write $\mathrm{prob}(e)$ for the probability of an element or event e (see Appendix B.1).

5. The setting where a probabilistic algorithm A is executed may include further random events. Now, tossing a fair coin in A is assumed to be an independent random experiment. Therefore, the outcome of the coin tosses of A on input x is independent of all further random events in the given setting. In the following items, we apply this basic assumption.

6. Suppose that the input $x \in X$ of a probabilistic algorithm A is randomly generated. This means that a probability distribution p_X is given for the domain X (e.g. the uniform distribution). We may consider the random experiment "Randomly choose $x \in X$ according to p_X and compute $y = A(x)$". If the outputs of A are in Y, then the experiment is modeled by a joint probability space (XY, p_{XY}). The coin tosses of $A(x)$ are independent of the random choice of x. Thus, the probability that $x \in X$ is chosen and that $y = A(x)$ is

$$\text{prob}(x, A(x) = y) = p_{XY}(x, y) = p_X(x) \cdot \text{prob}(A(x) = y).$$

The probability $\text{prob}(A(x) = y)$ is the conditional probability $\text{prob}(y \,|\, x)$ of the outcome y, assuming the input x.

7. Each execution of A is a new independent random experiment: the coin tosses during one execution of A are independent of the coin tosses in other executions of A. In particular, when executing A twice, with inputs x and x', we have

$$\text{prob}(A(x) = y, A(x') = y') = \text{prob}(A(x) = y) \cdot \text{prob}(A(x') = y').$$

If probabilistic algorithms A and B are applied to inputs x and x', then the coin tosses of $A(x)$ and $B(x')$ are independent, unless $B(x')$ is called as a subroutine of $A(x)$ (such that the coin tosses of $B(x')$ are contained in the coin tosses of $A(x)$), or vice versa:

$$\text{prob}(A(x) = y, B(x') = y') = \text{prob}(A(x) = y) \cdot \text{prob}(B(x') = y').$$

8. Let A be a probabilistic algorithm with inputs from X and outputs in Y. Let $h : X \longrightarrow Z$ be a map yielding some property $h(x)$ for the elements in X (e.g. the least-significant bit of x). Let B be a probabilistic algorithm which on input $y \in Y$ outputs $B(y) \in Z$. Assume that a probability distribution p_X is given on X. B might be an algorithm trying to invert A or at least trying to determine the property $h(x)$ from $y := A(x)$. We are interested in the random experiment "Randomly choose x, compute $y = A(x)$ and $B(y)$, and check whether $B(y) = h(x)$". The random choice of x, the coin tosses of $A(x)$ and the coin tosses of $B(y)$ are independent random experiments. Thus, the probability that $x \in X$ is chosen, and that $A(x) = y$ and $B(y)$ correctly computes $h(x)$ is

$$\begin{aligned}
\text{prob}(x, &A(x) = y, B(y) = h(x)) \\
&= \text{prob}(x) \cdot \text{prob}(A(x) = y) \cdot \text{prob}(B(y) = h(x)).
\end{aligned}$$

9. Let p_X be a probability distribution on the domain X of a probabilistic algorithm A with outputs in Y. Randomly selecting an $x \in X$ and computing $A(x)$ is described by the joint probability space XY (see above). We can project to Y, $(x, y) \mapsto y$, and calculate the probability distribution p_Y on Y:

$$p_Y(y) := \sum_{x \in X} p_{XY}(x, y) = \sum_{x \in X} p_X(x) \cdot \text{prob}(A(x) = y).$$

We call p_Y the image of p_X under A. p_Y is also the image of the joint distribution of X and the coin tosses r under the deterministic extension A_D of A:

$$p_Y(y) = \text{prob}(\{(x, r) \mid A_D(x, r) = y\})$$
$$= \sum_{x \in X, r \in \{0,1\}^{t_x} : A_D(x,r) = y} p_X(x) \cdot \text{prob}(r).$$

As in the deterministic case (see Appendix B.1, p. 330), we sometimes denote the image distribution by $\{A(x) : x \leftarrow X\}$.

Let A be a probabilistic algorithm with inputs from X and outputs in Y. Then A (as a Turing machine) has a finite binary description. In particular, we can assume that both the domain X and the range Y are subsets of $\{0, 1\}^*$. The time and space complexity of an algorithm A (corresponding to the running time and memory requirements) are measured as functions of the binary length $|x|$ of the input x.

Definition 5.2. A *probabilistic polynomial algorithm* is a probabilistic algorithm A, such that the running time of $A(x)$ is bounded by $P(|x|)$, where $P \in \mathbb{Z}[X]$ is a polynomial (the same for all inputs x). The running time is measured as the number of steps in our model of algorithms, i.e., the number of steps of the probabilistic Turing machine. Tossing a coin is one step in this model.

Remark. The randomness in probabilistic algorithms is caused by random events in a very specific probability space, namely $\{0, 1\}^{t_x}$ with the uniform distribution, and at first glance this might look restrictive. Actually, it is a very general model.

For example, suppose you want to control an algorithm A on input x by r_x random events with probabilities $p_{x,1}, \ldots, p_{x,r_x}$ (the deterministic extension A_D takes as inputs x and one of the events).[3] Assume that always one of the events occurs (i.e., $\sum_{i=1}^{r_x} p_{x,i} = 1$) and that the probabilities $p_{x,i}$ have a finite binary representation $p_{x,i} = \sum_{j=1}^{t_x} a_{x,i,j} \cdot 2^{-j}$ ($a_{x,i,j} \in \{0, 1\}$). Further,

---

[3] For example, think of an algorithm A that attacks an encryption scheme. If all possible plaintexts and their probability distribution are known, then A might be based on the random choice of correctly distributed plaintexts.

assume that r_x, t_x and the probabilities $p_{x,i}$ are computable by deterministic (polynomial) algorithms with input x. The last assumption is satisfied, for example, if the events and probabilities are the same for all x.

Then the random behavior of A can be implemented by coin tosses, i.e., A can be implemented as a probabilistic (polynomial) algorithm in the sense of Definitions 5.1 and 5.2. Namely, let S be the coin-tossing algorithm which on input x:

(1) t_x times tosses the coin and obtains a binary number $b := b_{t_x-1} \ldots b_1 b_0$, with $0 \leq b < 2^{t_x}$, and

(2) returns $S(x) := i$, if $2^{t_x} \sum_{j=1}^{i-1} p_{x,j} \leq b < 2^{t_x} \sum_{j=1}^{i} p_{x,j}$.

The outputs of $S(x)$ are in $\{1, \ldots, r_x\}$ and $\mathrm{prob}(S(x) = i) = p_{x,i}$, for $1 \leq i \leq r_x$. The probabilistic (polynomial) algorithm S can be used to produce the random inputs for A_D.

5.2 Monte Carlo and Las Vegas Algorithms

The running time of a probabilistic polynomial algorithm A is required to be bounded by a polynomial P, for all inputs x. Assume A tries to compute the solution to a problem. Due to its random behavior, A might not reach its goal with certainty, but only with some probability. Therefore, the output might not be correct in some cases. Such algorithms are also called Monte Carlo algorithms if their probability of success is not too low. They are distinguished from Las Vegas algorithms.

$time_A(x)$ denotes the running time of A on input x, i.e., the number of steps A needs to generate the output $A(x)$ for the input x. As before, $|x|$ denotes the binary length of x.

Definition 5.3. Let $\mathcal{P}$ be a computational problem.

1. A *Monte Carlo algorithm* A for $\mathcal{P}$ is a probabilistic algorithm A, whose running time is bounded by a polynomial Q and which yields a correct answer to $\mathcal{P}$ with a probability of at least $2/3$:

$$time_A(x) \leq Q(|x|) \text{ and } \mathrm{prob}(A(x) \text{ is a correct answer to } \mathcal{P}) > \frac{2}{3},$$

for all instances x of $\mathcal{P}$.

2. A probabilistic algorithm A for $\mathcal{P}$ is called a *Las Vegas algorithm* if its output is always a correct answer to $\mathcal{P}$, and if the expected value for the running time is bounded by a polynomial Q:

$$E(time_A(x)) = \sum_{t=1}^{\infty} t \cdot \mathrm{prob}(time_A(x) = t) < Q(|x|),$$

for all instances x of $\mathcal{P}$.

Remarks:

1. The probabilities are computed assuming a fixed input x; they are only taken over the coin tosses during the computation of $A(x)$. The distribution of the inputs x is not considered. For example,

$$\mathrm{prob}(A(x) \text{ is a correct answer to } \mathcal{P}) = \sum_{y \in Y_x} \mathrm{prob}(A(x) = y),$$

with the sum taken over the set Y_x of correct answers for input x.
2. The running time of a Monte Carlo algorithm is bounded by one polynomial Q, for all inputs. A Monte Carlo algorithm may sometimes fail to produce a correct answer to the given problem. However, the probability of such a failure is bounded. In contrast, a Las Vegas algorithm always gives a correct answer to the given problem. The running time may vary substantially and is not necessarily bounded by a single polynomial. However, the expected value of the running time is polynomial.

Examples:

1. Typical examples of Monte Carlo algorithms are the probabilistic primality tests discussed in Appendix A.8. They check whether an integer is prime or not.
2. Algorithm A.61, which computes square roots modulo a prime p, is a Las Vegas algorithm.
3. To prove the zero-knowledge property of an interactive proof system, it is necessary to simulate the prover by a Las Vegas algorithm (see Section 4.2.3).

A Las Vegas algorithm may be turned into a Monte Carlo algorithm simply by stopping it after a suitable polynomial number of steps. To state this fact more precisely, we use the notion of positive polynomials.

Definition 5.4. A polynomial $P(X) = \sum_{i=0}^{n} a_i X^i \in \mathbb{Z}[X]$ in one variable X, with integer coefficients $a_i, 0 \leq i \leq n$, is called a *positive polynomial* if $P(x) > 0$ for $x > 0$, i.e., P has positive values for positive inputs.

Examples. The polynomials X^n are positive. More generally, each polynomial whose non-zero coefficients are positive is a positive polynomial.

Proposition 5.5. *Let $A(x)$ be a Las Vegas algorithm for a problem $\mathcal{P}$ with an expected running time $\leq Q(|x|)$ (Q a polynomial). Let P be a positive polynomial. Let $\tilde{A}$ be the algorithm obtained by stopping $A(x)$ after at most $P(|x|)$ steps. Then $\tilde{A}$ is a Monte Carlo algorithm for $\mathcal{P}$, which gives a correct answer to $\mathcal{P}$ with probability $\geq 1 - Q(|x|)/P(|x|)$.*

Proof. We have

$$P(|x|) \cdot \mathrm{prob}(time_A(x) \geq P(|x|)) \leq \sum_{t=P(|x|)}^{\infty} t \cdot \mathrm{prob}(time_A(x) = t)$$

$$\leq E(time_A(x)) \leq Q(|x|).$$

Thus, $\tilde{A}$ gives a correct answer to $\mathcal{P}$ with probability $\geq 1 - Q(|x|)/P(|x|)$. $\square$

Remark. $\tilde{A}(x)$ might return "I don't know" if $A(x)$ did not yet terminate after $P(|x|)$ steps and is stopped. Then the answer of $\tilde{A}$ is never false, though it is not always a solution to $\mathcal{P}$.

The choice of the bound $2/3$ in our definition of Monte Carlo algorithms is somewhat arbitrary. Other bounds could be used equally well, as Proposition 5.6 below shows. We can increase the probability of success of an algorithm $A(x)$ by repeatedly applying the original algorithm and by making a majority decision. This works if the probability that A produces a correct result exceeds $1/2$ by a non-negligible amount.

Proposition 5.6. *Let P and Q be positive polynomials. Let A be a probabilistic polynomial algorithm which computes a function $f : X \longrightarrow Y$, with*

$$\mathrm{prob}(A(x) = f(x)) \geq \frac{1}{2} + \frac{1}{P(|x|)} \text{ for all } x \in X.$$

Then, by repeating the computation $A(x)$ and returning the most frequent result, we get a probabilistic polynomial algorithm $\tilde{A}$, such that

$$\mathrm{prob}(\tilde{A}(x) = f(x)) > 1 - \frac{1}{Q(|x|)} \text{ for all } x \in X.$$

Proof. Consider the algorithms A_t $(t \in \mathbb{N})$, defined on input $x \in X$ as follows:

1. Execute $A(x)$ t times, and get the set $Y_x := \{y_1, \ldots, y_t\}$ of outputs.
2. Select an $i \in \{1, \ldots, t\}$, with $|\{y \in Y_x \mid y = y_i\}|$ maximal.
3. Set $A_t(x) := y_i$.

We expect that more than half of the results of A coincide with $f(x)$, and hence $A_t(x) = f(x)$ with high probability. More precisely, define the binary random variables $S_j, 1 \leq j \leq t$:

$$S_j := \begin{cases} 1 \text{ if } y_j = f(x), \\ 0 \text{ otherwise.} \end{cases}$$

The expected values $E(S_j)$ are equal to

$$E(S_j) = \mathrm{prob}(A(x) = f(x)) \geq \frac{1}{2} + \frac{1}{P(|x|)},$$

and we conclude, by Corollary B.18, that

$$\text{prob}(A_t(x) = f(x)) \geq \text{prob}\left(\sum_{j=1}^{t} S_j > \frac{t}{2}\right) \geq 1 - \frac{P(|x|)^2}{4t}.$$

For $t > \frac{1}{4} \cdot P(|x|)^2 \cdot Q(|x|)$, the probability of success of A_t is $> 1 - \frac{1}{Q(|x|)}$.
□

Remark. By using the Chernoff bound from probability theory, the preceding proof can be modified to get a polynomial algorithm $\tilde{A}$ whose probability of success is exponentially close to 1 (see Exercise 5). We do not need this stronger result. If a polynomial algorithm which checks the correctness of a solution is available, we can do even better.

Proposition 5.7. *Let P, Q and R be positive polynomials. Let A be a probabilistic polynomial algorithm computing solutions to a given problem $\mathcal{P}$ with*

$$\text{prob}(A(x) \text{ is a correct answer to } \mathcal{P}) \geq \frac{1}{P(|x|)}, \text{ for all inputs } x.$$

Let D be a deterministic polynomial algorithm, which checks whether a given answer to $\mathcal{P}$ is correct. Then, by repeating the computation $A(x)$ and by checking the results with D, we get a probabilistic polynomial algorithm $\tilde{A}$ for $\mathcal{P}$, such that

$$\text{prob}(\tilde{A}(x) \text{ is a correct answer to } \mathcal{P}) > 1 - 2^{-Q(|x|)}, \text{ for all inputs } x.$$

Proof. Consider the algorithms A_t ($t \in \mathbb{N}$), defined on input x as follows:

1. Repeat the following at most t times:
   a. Execute $A(x)$, and get the answer y.
   b. Apply D to check whether y is a correct answer.
   c. If D says "correct", stop the iteration.
2. Return y.

The executions of $A(x)$ are t independent repetitions of the same experiment. Hence, the probability that all t executions of A yield an incorrect answer is $< (1 - \frac{1}{P(|x|)})^t$, and we obtain

$$\left(1 - \frac{1}{P(|x|)}\right)^t = \left(\left(1 - \frac{1}{P(|x|)}\right)^{P(|x|)}\right)^{t/P(|x|)} < \left(e^{-1}\right)^{t/P(|x|)}$$

$$= e^{-t/P(|x|)} \leq e^{-\ln(2)Q(|x|)} = 2^{-Q(|x|)},$$

for $t \geq \ln(2)P(|x|)Q(|x|)$.
□

Remark. The iterations A_t in the preceding proof may be used to construct a Las Vegas algorithm that always gives a correct answer to the given problem (see Exercise 1).

Exercises

1. Let P be a positive polynomial. Let A be a probabilistic polynomial algorithm which on input $x \in X$ computes solutions to a given problem $\mathcal{P}$ with

$$\operatorname{prob}(A(x) \text{ is a correct answer to } \mathcal{P}) \geq \frac{1}{P(|x|)} \text{ for all } x \in X.$$

Assume that there is a deterministic algorithm D, which checks in polynomial time whether a given solution y to $\mathcal{P}$ on input x is correct.
Construct a Las Vegas algorithm that always gives the correct answer and whose expected running time is $\leq P(|x|)(R(|x|) + S(|x| + R(|x|)))$, where R and S are polynomial bounds for the running times of A and D (use Lemma B.12).

2. Let $\mathcal{L} \subset \{0,1\}^*$ be a decision problem in the complexity class $\mathcal{BPP}$ of *bounded-error probabilistic polynomial-time problems*. This means that there is a probabilistic polynomial algorithm A with input $x \in \{0,1\}^*$ and output $A(x) \in \{0,1\}$ which solves the membership decision problem for $\mathcal{L} \subset \{0,1\}^*$, with a bounded error probability. More precisely, there are a positive polynomial P and a constant a, $0 < a < 1$, such that

$$\operatorname{prob}(A(x) = 1) \geq a + \frac{1}{P(|x|)}, \text{ for } x \in \mathcal{L}, \text{ and}$$

$$\operatorname{prob}(A(x) = 1) \leq a - \frac{1}{P(|x|)}, \text{ for } x \notin \mathcal{L}.$$

Let Q be another positive polynomial. Show how to obtain a probabilistic polynomial algorithm $\tilde{A}$, with

$$\operatorname{prob}(\tilde{A}(x) = 1) \geq 1 - \frac{1}{Q(|x|)}, \text{ for } x \in \mathcal{L}, \text{ and}$$

$$\operatorname{prob}(\tilde{A}(x) = 1) \leq \frac{1}{Q(|x|)}, \text{ for } x \notin \mathcal{L}.$$

(Use a similar technique as that used, for example, in the proof of Proposition 5.6.)

3. A decision problem $\mathcal{L} \subset \{0,1\}^*$ belongs to the complexity class $\mathcal{RP}$ of *randomized probabilistic polynomial-time problems* if there exists a probabilistic polynomial algorithm A which on input $x \in \{0,1\}^*$ outputs $A(x) \in \{0,1\}$, and a positive polynomial Q, such that $\operatorname{prob}(A(x) = 1) \geq {}^1\!/\!_{Q(|x|)}$, for $x \in \mathcal{L}$, and $\operatorname{prob}(A(x) = 1) = 0$, for $x \notin \mathcal{L}$.
A decision problem $\mathcal{L}$ belongs to the complexity class $\mathcal{NP}$ if there is a deterministic polynomial algorithm $M(x,y)$ and a polynomial L, such that $M(x,y) = 1$ for some $y \in \{0,1\}^*$, with $|y| \leq L(|x|)$, if and only if $x \in \mathcal{L}$ (y is called a certificate for x). Show that:

a. $\mathcal{RP} \subseteq \mathcal{BPP}$.
b. $\mathcal{RP} \subseteq \mathcal{NP}$.

4. A decision problem $\mathcal{L} \subset \{0,1\}^*$ belongs to the complexity class $\mathcal{ZPP}$ of *zero-sided probabilistic polynomial-time problems* if there exists a Las Vegas algorithm $A(x)$, such that $A(x) = 1$ if $x \in \mathcal{L}$, and $A(x) = 0$ if $x \notin \mathcal{L}$.
Show that $\mathcal{ZPP} \subseteq \mathcal{RP}$.

5. If $S_1, \ldots, S_n$ are independent repetitions of a binary random variable X and $p := \mathrm{prob}(X = 1) = E(X)$, then the *Chernoff bound* holds for $0 < \varepsilon \le p(1-p)$:

$$
\mathrm{prob}\left(\left| \frac{1}{n} \sum_{i=1}^{n} S_i - p \right| < \varepsilon \right) \ge 1 - 2e^{-n\varepsilon^2/2},
$$

(see, e.g., [Rényi70], Section 7.4). Use the Chernoff bound to derive an improved version of Proposition 5.6.

6. One-Way Functions and the Basic Assumptions

In Chapter 3 we introduced the notion of one-way functions. As the examples of RSA encryption and Rabin signatures show, one-way functions play the key role in asymmetric cryptography.

Speaking informally, a *one-way function* is a map $f : X \longrightarrow Y$ which is easy to compute but hard to invert. There is no efficient algorithm that computes pre-images of $y \in Y$. If we want to use a one-way function f for encryption in the straightforward way (applying f to the plaintext, as, for example, in RSA encryption), then f must belong to a special class of one-way functions. Knowing some information ("the trapdoor information": e.g. the factorization of the modulus n in RSA schemes), it must be easy to invert f, and f is one way only if the trapdoor information is kept secret. These functions are called *trapdoor functions*.

Our notion of one-way functions introduced in Chapter 3 was a rather informal one: we did not specify precisely what we mean by "efficiently computable", "infeasible" or "hard to invert". Now, in this chapter, we will clarify these terms and give a precise definition of one-way functions. For example, "efficiently computable" means that the solution can be computed by a probabilistic polynomial algorithm, as defined in Chapter 5.

We discuss three examples in some detail: the discrete exponential function, modular powers and modular squaring. The first is not a trapdoor function. Nevertheless, it has important applications in cryptography (e.g. pseudorandom bit generators, see Chapter 8; ElGamal's encryption and signature scheme and the DSA, see Chapter 3).

Unfortunately, there is no proof that these functions are really one way. However, it is possible to state the basic assumptions precisely, which guarantee the one-way feature. It is widely believed that these assumptions are true.

In order to define the one-way feature (and in a way that naturally matches the examples), we have to consider not only single functions, but, more generally, families of functions defined over appropriate index sets.

In the preliminary (and very short) Section 6.1, we introduce an intuitive notation for probabilities that will be used subsequently.

6.1 A Notation for Probabilities

The notation

$$\text{prob}(B(x) = 1 : x \leftarrow X) := \text{prob}(\{x \in X \mid B(x) = 1\}),$$

introduced for Boolean predicates[1] $B : X \longrightarrow \{0,1\}$ in Appendix B.1 (p. 329), intuitively suggests that we mean the probability of $B(x) = 1$ if x is randomly chosen from X. We will use an analogous notation for probabilistic algorithms.

Let A be a probabilistic algorithm with inputs from X and outputs in Y, and let $B : X \times Y \longrightarrow \{0,1\}$, $(x, y) \longmapsto B(x, y)$ be a Boolean predicate. Let p_X be a probability distribution on X. As in Section 5.1, let A_D be the deterministic extension of A, and let t_x denote the number of coin tosses of A on input x:

$$\text{prob}(B(x, A(x)) = 1 : x \xleftarrow{p_X} X)$$
$$:= \sum_{x \in X} \text{prob}(x) \cdot \text{prob}(B(x, A(x)) = 1)$$
$$= \sum_{x \in X} \text{prob}(x) \cdot \text{prob}(\{r \in \{0,1\}^{t_x} \mid B(x, A_D(x, r)) = 1\})$$
$$= \sum_{x \in X} \text{prob}(x) \cdot \frac{|\{r \in \{0,1\}^{t_x} \mid B(x, A_D(x, r)) = 1\}|}{2^{t_x}}.$$

The notation is typically used in the following situation. A Monte Carlo algorithm A tries to compute a function $f : X \longrightarrow Y$, and $B(x, y) := B_f(x, y) := 1$ if $f(x) = y$, and $B(x, y) := B_f(x, y) := 0$ if $f(x) \neq y$. Then

$$\text{prob}(A(x) = f(x) : x \xleftarrow{p_X} X) := \text{prob}(B_f(x, A(x)) = 1 : x \xleftarrow{p_X} X)$$

is the probability that A succeeds if the input x is randomly chosen from X (according to p_X).

We write $x \xleftarrow{u} X$ if the distribution on X is the uniform one, and often we simply write $x \leftarrow X$ instead of $x \xleftarrow{p_X} X$.

In cryptography, we often consider probabilistic algorithms whose domain X is a joint probability space $X_1 X_2 \ldots X_r$ constructed by iteratively joining fibers $X_{j,x_1 \ldots x_{j-1}}$ to $X_1 \ldots X_{j-1}$ (Appendix B.1, p. 328). In this case, the notation is

$$\text{prob}(B(x_1, \ldots, x_r, A(x_1, \ldots, x_r)) = 1 :$$
$$x_1 \leftarrow X_1, x_2 \leftarrow X_{2,x_1}, x_3 \leftarrow X_{3,x_1 x_2}, \quad \ldots \quad , x_r \leftarrow X_{r,x_1 \ldots x_{r-1}}).$$

Now, the notation suggests that we mean the probability of the event $B(x_1, \ldots, x_r, A(x_1, \ldots, x_r)) = 1$ if first x_1 is randomly chosen, then x_2, then

---

[1] Maps with values in $\{0,1\}$ are called *Boolean predicates*.

x_3, then ....

A typical example is the discrete logarithm assumption in Section 6.2 (Definition 6.1).

The distribution $x_j \leftarrow X_{j,x_1...x_{j-1}}$ is the conditional distribution of $x_j \in X_{j,x_1...x_{j-1}}$, assuming $x_1,\ldots,x_{j-1}$. The probability can be computed as follows (we consider $r = 3$ and the case where A computes a function f and B is the predicate $A(x) = f(x)$):

$$\text{prob}(A(x_1, x_2, x_3) = f(x_1, x_2, x_3) : x_1 \leftarrow X_1, x_2 \leftarrow X_{2,x_1}, x_3 \leftarrow X_{3,x_1 x_2})$$

$$= \sum_{x_1,x_2,x_3} \text{prob}(x_1, x_2, x_3) \cdot \text{prob}(A(x_1, x_2, x_3) = f(x_1, x_2, x_3))$$

$$= \sum_{x_1 \in X_1} \text{prob}(x_1) \cdot \sum_{x_2 \in X_{1,x_1}} \text{prob}(x_2 \mid x_1)$$

$$\cdot \sum_{x_3 \in X_{1,x_1,x_2}} \text{prob}(x_3 \mid x_2, x_1) \cdot \text{prob}(A(x_1, x_2, x_3) = f(x_1, x_2, x_3)).$$

Here $\text{prob}(x_2 \mid x_1)$ (resp. $\text{prob}(x_3 \mid x_2, x_1)$) denotes the conditional probability of x_2 (resp. x_3) assuming x_1 (resp. x_1 and x_2); see Appendix B.1 (p. 328). The last probability, $\text{prob}(A(x_1, x_2, x_3) = f(x_1, x_2, x_3))$, is the probability that the coin tosses of A on input (x_1, x_2, x_3) yield the result $f(x_1, x_2, x_3)$.

In Section 5.1, we introduced the image p_Y of the distribution p_X under a probabilistic algorithm A from X to Y. We have

$$p_Y(y) = \text{prob}(A(x) = y : x \overset{p_X}{\leftarrow} X)$$

for each $y \in Y$.

For each $x \in X$, we have the distribution $p_{A(x)}$ on Y:

$$p_{A(x)}(y) = \text{prob}(A(x) = y).$$

We write $y \leftarrow A(x)$ instead of $y \overset{p_{A(x)}}{\leftarrow} Y$. This notation suggests that y is generated by the random variable $A(x)$. With this notation, we have

$$\text{prob}(A(x) = f(x) : x \leftarrow X) = \text{prob}(f(x) = y : x \leftarrow X, y \leftarrow A(x)).$$

6.2 Discrete Exponential Function

The notion of one-way functions can be precisely defined using probabilistic algorithms. As a first example we consider the discrete exponential function. Let $I := \{(p,g) \mid p \text{ a prime number}, g \in \mathbb{Z}_p^* \text{ a primitive root}\}$. We call the family of discrete exponential functions

$$\text{Exp} := (\text{Exp}_{p,g} : \mathbb{Z}_{p-1} \longrightarrow \mathbb{Z}_p^*, \ x \longmapsto g^x)_{(p,g) \in I}$$

the *Exp family*. Since g is a primitive root, $\mathrm{Exp}_{p,g}$ is an isomorphism between the additive group $\mathbb{Z}_{p-1}$ and the multiplicative group $\mathbb{Z}_p^*$. The family of inverse functions

$$\mathrm{Log} := (\mathrm{Log}_{p,g} : \mathbb{Z}_p^* \longrightarrow \mathbb{Z}_{p-1})_{(p,g) \in I}$$

is called the *Log family*.

The algorithm of modular exponentiation computes $\mathrm{Exp}_{p,g}(x)$ efficiently (see Algorithm A.26). It is unknown whether an efficient algorithm for the computation of the discrete logarithm function exists. All known algorithms have exponential running time, and it is widely believed that, in general, $\mathrm{Log}_{p,g}$ is not efficiently computable. We state this assumption on the one-way property of the Exp family by means of probabilistic algorithms.

Definition 6.1. Let $I_k := \{(p,g) \in I \mid |p| = k\}$, with $k \in \mathbb{N}$,[2] and let $Q(X) \in \mathbb{Z}[X]$ be a positive polynomial. Let $A(p,g,y)$ be a probabilistic polynomial algorithm. Then there exists a $k_0 \in \mathbb{N}$, such that

$$\mathrm{prob}(A(p,g,y) = \mathrm{Log}_{p,g}(y) : (p,g) \xleftarrow{u} I_k, y \xleftarrow{u} \mathbb{Z}_p^*) \leq \frac{1}{Q(k)}$$

for $k \geq k_0$.
This is called the *discrete logarithm assumption*.

Remarks:

1. The probabilistic algorithm A models an attacker who tries to compute the discrete logarithm or, equivalently, to invert the discrete exponential function. The discrete logarithm assumption essentially states that for a sufficiently large size k of the modulus p, the probability of A successfully computing $\mathrm{Log}_{p,g}(y)$ is smaller than $1/Q(k)$. This means that Exp cannot be inverted by A for all but a negligible fraction of the inputs. Therefore, we call Exp a family of *one-way functions*. The term "negligible" is explained more precisely in a subsequent remark.
2. When we use the discrete exponential function in a cryptographic scheme, such as ElGamal's encryption scheme (see Section 3.5.1), selecting a function $\mathrm{Exp}_{p,g}$ from the family means to choose a public key $i = (p,g)$ (actually, i may be only one part of the key).
3. The index set I is partitioned into disjoint subsets: $I = \bigcup_{k \in \mathbb{N}} I_k$. k may be considered as the security parameter of $i = (p,g) \in I_k$. The one-way property requires a sufficiently large security parameter. The security parameter is closely related to the binary length of i. Here, $k = |p|$ is half the length of i.
4. The probability in the discrete logarithm assumption is also taken over the random choice of a key i with a given security parameter k. Hence, the

---

[2] As usual, $|p|$ denotes the binary length of p.

meaning of the probability statement is: choosing both the key $i = (p, g)$ with security parameter k and $y = g^x$ randomly, the probability that A correctly computes the logarithm x from y is small. The statement is not related to a particular key i. In practice, however, a public key is chosen and then fixed for a long time, and it is known to the adversary. Thus, we are interested in the conditional probability of success, assuming a fixed public key i. Even if the security parameter k is very large, there may be keys (p, g) such that A correctly computes $\text{Log}_{p,g}(y)$ with a significant chance. However, as we will see below, the number of such keys (p, g) is negligibly small compared to all keys with security parameter k. Choosing (p, g) at random (and uniformly) from I_k, the probability of obtaining one for which A has a significant chance of success is negligibly small (see Proposition 6.3 for a precise statement). Indeed, if $p - 1$ has only small prime factors, an efficient algorithm developed by Pohlig and Hellman computes the discrete logarithm function (see [PohHel78]).

Remark. In this book, we often consider families $\varepsilon = (\varepsilon_k)_{k \in \mathbb{N}}$ of quantities $\varepsilon_k \in \mathbb{R}$, as the probabilities in the discrete logarithm assumption. We call them *negligible* or *negligibly small* if, for every positive polynomial $Q \in \mathbb{Z}[X]$, there is a $k_0 \in \mathbb{N}$, such that $|\varepsilon_k| \leq 1/Q(k)$ for $k \geq k_0$. "Negligible" means that the absolute value is asymptotically smaller than any polynomial bound.

Remark. In order to simplify, definitions and results are often stated asymptotically (as the discrete logarithm assumption or the notion of negligible quantities). Polynomial running times or negligible probabilities are not specified more precisely, even if it were possible. A typical situation is as follows. A cryptographic scheme is based on a one-way function f (e.g. the Exp family). Let g be a function that describes a property of the cryptographic scheme (e.g. g predicts the next bit of the discrete exponential pseudorandom bit generator; see Chapter 8). It is desirable that this property g cannot be efficiently computed by an adversary. Sometimes, this can be proven. Typically a proof runs by a contradiction. We assume that a probabilistic polynomial algorithm A_1 which successfully computes g with probability ε_1 is given. Then, a probabilistic polynomial algorithm A_2 is constructed which calls A_1 as a subroutine and inverts the underlying one-way function f, with probability ε_2. Such an algorithm is called a *polynomial-time reduction* of f to g. If ε_2 is non-negligible, we get a contradiction to the one-way assumption (e.g. the discrete logarithm assumption).

In our example, a typical statement would be as follows. If discrete logarithms cannot be computed in polynomial time with non-negligible probability (i.e., if the discrete logarithm assumption is true), then a polynomial-time adversary cannot predict, with non-negligible probability, the next bit of the discrete exponential pseudorandom bit generator.

Actually, in many cases the statement could be made more precise, by performing a detailed analysis of the reduction algorithm A_2. The running time of A_2 can be described as an explicit function of $\varepsilon_1, \varepsilon_2$ and the running time of A_1 (see, e.g., the results in Chapter 7).

As in the Exp example, we often meet families of functions indexed on a set of keys which may be partitioned according to a security parameter. Therefore we propose the notion of indexes, whose binary lengths are measured by a security parameter as specified more precisely by the following definition.

Definition 6.2. Let $I = \bigcup_{k \in \mathbb{N}} I_k$ be an infinite index set which is partitioned into finite disjoint subsets I_k. Assume that the indexes are binarily encoded. As always, we denote by $|i|$ the binary length of i.
I is called a *key set with security parameter k* or an *index set with security parameter k*, if:

1. The security parameter k of $i \in I$ can be derived from i by a deterministic polynomial algorithm.
2. There is a constant $m \in \mathbb{N}$, such that
$$k^{1/m} \leq |i| \leq k^m \text{ for } i \in I_k.$$

We usually write $I = (I_k)_{k \in \mathbb{N}}$ instead of $I = \bigcup_{k \in \mathbb{N}} I_k$.

Remarks:

1. The second condition means that the security parameter k is a measure for the binary length $|i|$ of the elements $i \in I_k$. In particular, statements such as:
   (1) "There is a polynomial P with $\ldots \leq P(|i|)$", or
   (2) "For every positive polynomial Q, there is a $k_0 \in \mathbb{N}$, such that $\ldots \leq 1/Q(|i|)$ for $|i| \geq k_0$",
   are equivalent to the corresponding statements in which $|i|$ is replaced by the security parameter k. In almost all of our examples, we have $k \leq |i| \leq 3k$ for $i \in I_k$.
2. The index set I of the Exp family is a key set with security parameter. As with all indexes occurring in this book, the indexes of the Exp family consist of numbers in $\mathbb{N}$ or residues in some residue class ring $\mathbb{Z}_n$. Unless otherwise stated, we consider them as binarily encoded in the natural way (see Appendix A): the binary encoding of $x \in \mathbb{N}$ is its standard encoding as an unsigned number, and the encoding of a residue class $[a] \in \mathbb{Z}_n$ is the encoding of its representative x with $0 \leq x \leq n - 1$.
3. If $I = \bigcup_{k \in \mathbb{N}} I_k$ satisfies only the second condition, then we can easily modify it and turn it into a key set with a security parameter, which also satisfies the first condition. Namely, let $\tilde{I}_k := \{(i, k) \mid i \in I_k\}$ and replace I by $\tilde{I} := \bigcup_{k \in \mathbb{N}} \tilde{I}_k$.

In the discrete logarithm assumption, we do not consider a single fixed key i: the probability is also taken over the random choice of the key. In the following proposition, we relate this average probability to the conditional probabilities, assuming a fixed key.

Proposition 6.3. *Let $I = (I_k)_{k \in \mathbb{N}}$ be a key set with security parameter k. Let $f = (f_i : X_i \longrightarrow Y_i)_{i \in I}$ be a family of functions and A be a probabilistic polynomial algorithm with inputs $i \in I$ and $x \in X_i$ and output in Y_i. Assume that probability distributions are given on I_k and X_i for all k, i (e.g. the uniform distributions). Then the following statements are equivalent:*

1. *For every positive polynomial P, there is a $k_0 \in \mathbb{N}$, such that for all $k \geq k_0$*
$$\mathrm{prob}(A(i, x) = f_i(x) : i \leftarrow I_k, x \leftarrow X_i) \leq \frac{1}{P(k)}.$$

2. *For all positive polynomials Q and R, there is a $k_0 \in \mathbb{N}$, such that for all $k \geq k_0$*
$$\mathrm{prob}\left(\left\{ i \in I_k \,\middle|\, \mathrm{prob}(A(i, x) = f_i(x) : x \leftarrow X_i) > \frac{1}{Q(k)} \right\}\right) \leq \frac{1}{R(k)}.$$

Proof. Let
$$p_i := \mathrm{prob}(A(i, x) = f_i(x) : x \leftarrow X_i)$$

be the conditional probability of success of A assuming a fixed i.

We first prove that statement 2 implies statement 1. Let P be a positive polynomial. By statement 2, there is some $k_0 \in \mathbb{N}$ such that for all $k \geq k_0$

$$\mathrm{prob}\left(\left\{ i \in I_k \,\middle|\, p_i > \frac{1}{2P(k)} \right\}\right) \leq \frac{1}{2P(k)}.$$

Hence

$$
\begin{aligned}
&\mathrm{prob}(A(i, x) = f_i(x) : i \leftarrow I_k, x \leftarrow X_i) \\
&= \sum_{i \in I_k} \mathrm{prob}(i) \cdot p_i \\
&= \sum_{p_i \leq 1/(2P(k))} \mathrm{prob}(i) \cdot p_i + \sum_{p_i > 1/(2P(k))} \mathrm{prob}(i) \cdot p_i \\
&\leq \sum_{p_i \leq 1/(2P(k))} \mathrm{prob}(i) \cdot \frac{1}{2P(k)} + \sum_{p_i > 1/(2P(k))} \mathrm{prob}(i) \cdot 1 \\
&= \mathrm{prob}\left(\left\{ i \in I_k \,\middle|\, p_i \leq \frac{1}{2P(k)} \right\}\right) \cdot \frac{1}{2P(k)} \\
&\qquad + \mathrm{prob}\left(\left\{ i \in I_k \,\middle|\, p_i > \frac{1}{2P(k)} \right\}\right) \\
&\leq \frac{1}{2P(k)} + \frac{1}{2P(k)} = \frac{1}{P(k)},
\end{aligned}
$$

for $k \geq k_0$.

Conversely, assume that statement 1 holds. Let Q and R be positive polynomials. Then there is a $k_0 \in \mathbb{N}$ such that for all $k \geq k_0$

$$\frac{1}{Q(k)R(k)} \geq \text{prob}(A(i,x) = f_i(x) : i \leftarrow I_k, x \leftarrow X_i)$$

$$= \sum_{i \in I_k} \text{prob}(i) \cdot p_i$$

$$\geq \sum_{p_i > 1/Q(k)} \text{prob}(i) \cdot p_i$$

$$> \frac{1}{Q(k)} \cdot \text{prob}\left(\left\{i \in I_k \,\Big|\, p_i > \frac{1}{Q(k)}\right\}\right).$$

This inequality implies statement 2.   □

Remark. A nice feature of the discrete logarithm problem is that it is *random self-reducible*. This means that solving the problem for arbitrary inputs can be reduced to solving the problem for randomly chosen inputs. More precisely, let $(p,g) \in I_k := \{(p,g) \mid p \text{ a prime}, |p| = k, g \in \mathbb{Z}_p^* \text{ a primitive root}\}$. Assume that there is a probabilistic polynomial algorithm A, such that

$$\text{prob}(A(p,g,y) = \text{Log}_{p,g}(y) : y \overset{u}{\leftarrow} \mathbb{Z}_p^*) > \frac{1}{Q(k)} \tag{6.1}$$

for some positive polynomial Q; i.e., $A(p,g,y)$ correctly computes the discrete logarithm with a non-negligible probability if the input y is randomly selected. Since y is chosen uniformly, we may rephrase this statement: $A(p,g,y)$ correctly computes the discrete logarithm for a polynomial fraction of inputs $y \in \mathbb{Z}_p^*$.

Then, however, there is also a probabilistic polynomial algorithm $\tilde{A}$ which correctly computes the discrete logarithm for every input $y \in \mathbb{Z}_p^*$, with an overwhelmingly high probability. Namely, given $y \in \mathbb{Z}_p^*$, we apply a slight modification A_1 of A. On input (p,g,y), A_1 randomly selects $r \overset{u}{\leftarrow} \mathbb{Z}_{p-1}$ and returns

$$A_1(p,g,y) := (A(p,g,yg^r) - r) \bmod (p-1).$$

Then $\text{prob}(A_1(p,g,y) = \text{Log}_{p,g}(y)) > 1/Q(k)$ for every $y \in \mathbb{Z}_p^*$. Now we can apply Proposition 5.7. For every positive polynomial P, we obtain – by repeating the computation of $A_1(p,g,y)$ a polynomial number of times and by checking each time whether the result is correct – a probabilistic polynomial algorithm $\tilde{A}$, with

$$\text{prob}(\tilde{A}(p,g,y) = \text{Log}_{p,g}(y)) > 1 - 2^{-P(k)},$$

for every $y \in \mathbb{Z}_p^*$. The existence of a random self-reduction enhances the credibility of the discrete logarithm assumption. Namely, assume that the discrete logarithm assumption is not true. Then by Proposition 6.3 there is a probabilistic polynomial algorithm A, such that for infinitely many k, the inequality (6.1) holds for a polynomial fraction of keys (p,g); i.e.,

$$\frac{|\{(p,g) \in I_k \mid \text{inequality (6.1) holds}\}|}{|I_k|} > \frac{1}{R(k)}$$

(with R a positive polynomial). For these keys $\tilde{A}$ computes the discrete logarithm for every $y \in \mathbb{Z}_p^*$ with an overwhelmingly high probability, and the probability of obtaining such a key is $> 1/R(k)$ if the keys are selected uniformly at random.

6.3 Uniform Sampling Algorithms

In the discrete logarithm assumption 6.1, the probabilities are taken with respect to the uniform distributions on I_k and $\mathbb{Z}_p^*$. Stating the assumption in this way, we tacitly assumed that it is possible to sample uniformly over I_k (during key generation) and $\mathbb{Z}_p^*$, by using efficient algorithms. In practice it might be difficult to construct a probabilistic polynomial sampling algorithm that selects the elements exactly according to the uniform distribution. However, as in the present case of discrete logarithms (see Proposition 6.6), we are often able to find practical sampling algorithms which sample in a "virtually uniform" way. Then the assumptions stated for the uniform distribution, such as the discrete logarithm assumption, apply. This is shown by the following considerations.

Definition 6.4. Let $J = (J_k)_{k \in \mathbb{N}}$ be an index set with security parameter k (see Definition 6.2). Let $X = (X_j)_{j \in J}$ be a family of finite sets:

1. A probabilistic polynomial algorithm S_X with input $j \in J$ is called a *sampling algorithm* for X if $S_X(j)$ outputs an element in X_j with a probability $\geq 1 - \varepsilon_k$ for $j \in I_k$, where $\varepsilon = (\varepsilon_k)_{k \in \mathbb{N}}$ is negligible; i.e., given a positive polynomial Q, there is a k_0 such that $\varepsilon_k \leq 1/Q(k)$ for $k \geq k_0$.
2. A sampling algorithm S_X for X is called *(virtually) uniform* if the distributions of $S_X(j)$ and the uniform distributions on X_j are polynomially close (see Definition B.22). This means that the statistical distance is negligibly small; i.e., given a positive polynomial Q, there is a k_0 such that the statistical distance (see Definition B.19) between the distribution of $S_X(j)$ and the uniform distribution on X_j is $\leq 1/Q(k)$, for $k \geq k_0$ and $j \in J_k$.

Remark. If S_X is a virtually uniform sampling algorithm for $X = (X_j)_{j \in J}$, we usually do not need to distinguish between the virtually uniform distribution of $S_X(j)$ and the truly uniform distribution when we compute a probability involving $x \leftarrow S_X(j)$. Namely, consider probabilities

$$\text{prob}(B_j(x,y) = 1 : x \leftarrow S_X(j), y \leftarrow Y_{j,x}),$$

where $(Y_{j,x})_{x \in X_j}$ is a family of probability spaces and B_j is a Boolean predicate. Then for every positive polynomial P, there is a $k_0 \in \mathbb{N}$ such that

$$| \operatorname{prob}(B_j(x,y) = 1 : x \leftarrow S_X(j), y \leftarrow Y_{j,x})$$
$$- \operatorname{prob}(B_j(x,y) = 1 : x \xleftarrow{u} X_j, y \leftarrow Y_{j,x}) | < \frac{1}{P(k)},$$

for $k \geq k_0$ and $j \in J_k$ (by Lemmas B.21 and B.24), and we see that the difference between the probabilities is negligibly small.

Therefore, we usually do not distinguish between perfectly and virtually uniform sampling algorithms and simply talk of *uniform sampling algorithms*.

We study an example. Suppose we want to construct a uniform sampling algorithm S for $(\mathbb{Z}_n)_{n \in \mathbb{N}}$. We have $\mathbb{Z}_n \subseteq \{0,1\}^{|n|}$, and could proceed as follows. We toss the coin $|n|$ times and obtain a binary number $x := b_{|n|-1} \ldots b_1 b_0$, with $0 \leq x < 2^{|n|}$. We can easily verify whether $x \in \mathbb{Z}_n$, by checking $x < n$. If the answer is affirmative, we return $S(n) := x$. Otherwise, we repeat the coin tosses. Since S is required to have a polynomial running time, we have to stop after at most $P(|n|)$ iterations (P a polynomial). Thus, $S(n)$ does not always succeed to return an element in $\mathbb{Z}_n$. The probability of a failure is, however, negligibly small.[3]

Our construction, which derives a uniform sampling algorithm for a subset, works, if the membership in this subset can be efficiently tested. It can be applied in many situations. Therefore, we state the following lemma.

Lemma 6.5. *Let $J = (J_k)_{k \in \mathbb{N}}$ be an index set with security parameter k. Let $X = (X_j)_{j \in J}$ and $Y = (Y_j)_{j \in J}$ be families of finite sets with $Y_j \subseteq X_j$ for all $j \in J$. Assume that there is a polynomial Q, such that $|Y_j| \cdot Q(k) \geq |X_j|$ for $j \in J_k$.*

Let S_X be a uniform sampling algorithm for $(X_j)_{j \in J}$ which on input $j \in J$ outputs $x \in X_j$[4] and some additional information $aux(x)$ about x. Let $A(j, x, aux(x))$ be a Monte Carlo algorithm which decides the membership in Y_j; i.e., on input $j \in J, x \in X_j$ and $aux(x)$, it yields 1 if $x \in Y_j$, and 0 if $x \notin Y_j$. Assume that the error probability of A is negligible; i.e., for every positive polynomial P, there is a k_0 such that the error probability is $\leq 1/P(k)$ for $k \geq k_0$.

Then there exists a uniform sampling algorithm S_Y for $(Y_j)_{j \in J}$.

Proof. Let S_Y be the probabilistic polynomial algorithm which on input $j \in J$ repeatedly computes $x := S_X(j)$ until $A(j, x, aux(x)) = 1$. To get a polynomial algorithm, we stop S_Y after at most $\ln(2)kQ(k)$ iterations.

We now show that S_Y has the desired properties. We first assume that $S_X(j) \in X_j$ with certainty and that A has an error probability of 0. By Lemma B.10, we may also assume that S_Y has found an element in Y_j (before

---

[3] We could construct a Las Vegas algorithm in this way, which always succeeds. See Section 5.2.

[4] Here and in what follows we use this formulation, though the sampling algorithm may sometimes yield elements outside of X_j. However, as stated in Definition 6.4, this happens only with a negligible probability.

being stopped), because this event has a probability $\geq 1-(1-1/Q(k))^{kQ(k)} > 1-2^{-k}$, which is exponentially close to 1 (see the proof of Proposition 5.7 for an analogous estimate).

By construction, we have for $V \subset Y_j$ that

$$\mathrm{prob}(S_Y(j) \in V) = \mathrm{prob}(S_X(j) \in V \mid S_X(j) \in Y_j) = \frac{\mathrm{prob}(S_X(j) \in V)}{\mathrm{prob}(S_X(j) \in Y_j)}.$$

Thus, we have for all subsets $V \subset Y_j$ that

$$\mathrm{prob}(S_Y(j) \in V) = \frac{\frac{|V|}{|X_j|} + \varepsilon_j(V)}{\frac{|Y_j|}{|X_j|} + \varepsilon_j(Y_j)},$$

with a negligibly small function ε_j. Then $\mathrm{prob}(S_Y(j) \in V) - \frac{|V|}{|Y_j|}$ is also negligibly small (you can see this immediately by Taylor's formula for the real function $(x,y) \mapsto x/y$). Hence, S_Y is a uniform sampling algorithm for $(Y_j)_{j\in J}$.

The general case, where $S_X(j) \notin X_j$ with a negligible probability and A has a negligible error probability, follows by applying Lemma B.10.   □

Example. Let $Y_n := \mathbb{Z}_n$ or $Y_n := \mathbb{Z}_n^*$. Then Y_n is a subset of $X_n := \{0,1\}^{|n|}, n \in \mathbb{N}$. Obviously, $\{0,1\}^{|n|}$ can be sampled uniformly by $|n|$ coin tosses. The membership of x in $\mathbb{Z}_n$ is checked by $x < n$, and the Euclidean algorithm tells us whether x is a unit. Thus, there are (probabilistic polynomial) uniform sampling algorithms for $(\mathbb{Z}_n)_{n\in\mathbb{N}}$ and $(\mathbb{Z}_n^*)_{n\in\mathbb{N}}$, which on input $n \in \mathbb{N}$ output an element $x \in \mathbb{Z}_n$ (or $x \in \mathbb{Z}_n^*$).
To apply Lemma 6.5 in this example, let $J := \mathbb{N}$ and $J_k := \{n \in \mathbb{N} \mid |n| = k\}$.

Example. Let Primes_k be the set of primes p whose binary length $|p|$ is k; i.e., $\mathrm{Primes}_k := \{p \in \mathrm{Primes} \mid |p| = k\} \subseteq \{0,1\}^k$. The number of primes $< 2^k$ is $\approx 2^k/k\ln(2)$ (Theorem A.68). By iterating a probabilistic primality test (e.g. Miller-Rabin's test, see Appendix A.8), we can, with a probability $> 1 - 2^{-k}$, correctly test the primality of an element x in $\{0,1\}^k$. Thus, there is a (probabilistic polynomial) uniform sampling algorithm S which on input 1^k yields a prime $p \in \mathrm{Primes}_k$.
To apply Lemma 6.5 in this example, let $J_k := \{1^k\}$ and $J := \mathbb{N} = \bigcup_{k\in\mathbb{N}} J_k$, i.e., the index set is the set of natural numbers. However, an index $k \in \mathbb{N}$ is not encoded in the standard way; it is encoded as the constant bit string 1^k (see the subsequent remark on 1^k).

Remark. 1^k denotes the constant bit string $11\ldots1$ of length k. Using it as input for a polynomial algorithm means that the number of steps in the algorithm is bounded by a polynomial in k. If we used k (encoded in the standard way) instead of 1^k as input, the bound would be a polynomial in $\log_2(k)$.

We return to the example of discrete exponentiation.

Proposition 6.6. *Let $I := \{(p,g) \mid p \text{ prime number}, g \in \mathbb{Z}_p^* \text{ primitive root}\}$ and $I_k := \{(p,g) \in I \mid |p| = k\}$. There is a probabilistic polynomial uniform sampling algorithm for $I = (I_k)_{k\in\mathbb{N}}$, which on input 1^k yields an element $(p,g) \in I_k$.*

Proof. We want to apply Lemma 6.5 to the index set $J := \mathbb{N} = \bigcup_{k\in\mathbb{N}} J_k, J_k := \{1^k\}$, and the families of sets $X_k := \text{Primes}_k \times \{0,1\}^k, Y_k := I_k \subseteq X_k (k \in \mathbb{N})$. The number of primitive roots in $\mathbb{Z}_p^*$ is $\varphi(p-1)$, where φ is the Eulerian totient function (see Theorem A.36). For $x \in \mathbb{N}$, we have

$$\varphi(x) = x \prod_{i=1}^{r} \left(1 - \frac{1}{p_i}\right) = x \prod_{i=1}^{r} \frac{p_i - 1}{p_i},$$

where $p_1, \ldots, p_r$ are the primes dividing x (see Corollary A.30). Since $\prod_{i=1}^{r} \frac{p_i-1}{p_i} \geq \prod_{i=2}^{r+1} \frac{i-1}{i} = \frac{1}{r+1}$ and $r + 1 \leq |x|$, we immediately see that $\varphi(x) \cdot |x| \geq x.^5$ In particular, we have $\varphi(p - 1) \cdot k \geq p - 1 \geq 2^{k-1}$ for $p \in \text{Primes}_k$, and hence $2k \cdot |Y_k| \geq |X_k|$.

Given a prime $p \in \text{Primes}_k$ and all prime numbers $q_1, \ldots, q_r$ dividing $p - 1$, we can efficiently verify whether $g \in \{0,1\}^k$ is in $\mathbb{Z}_p^*$ and whether it is a primitive root. Namely, we first test $g < p$ and then apply the criterion for primitive roots (see Algorithm A.39), i.e., we check whether $g^{(p-1)/q} \neq 1$ for all prime divisors q of $p - 1$.

We may apply Lemma 6.5 if there is a probabilistic polynomial uniform sampling algorithm for $(\text{Primes}_k)_{k\in\mathbb{N}}$ which not only outputs a prime p, but also the prime factors of $p - 1$. Bach's algorithm (see [Bach88]) yields such an algorithm: it generates uniformly distributed k-bit integers n, along with their factorization. We may repeatedly generate such numbers n until $n + 1$ is a prime. $\qquad\square$

6.4 Modular Powers

Let $I := \{(n,e) \mid n = pq, p \neq q \text{ primes}, 0 < e < \varphi(n), e \text{ prime to } \varphi(n)\}$. The family

$$\text{RSA} := (\text{RSA}_{n,e} : \mathbb{Z}_n^* \longrightarrow \mathbb{Z}_n^*, \ x \longmapsto x^e)_{(n,e)\in I}$$

is called the *RSA family*.

Consider an $(n,e) \in I$, and let $d \in \mathbb{Z}_{\varphi(n)}^*$ be the inverse of e mod $\varphi(n)$. Then we have $x^{ed} = x^{ed \bmod \varphi(n)} = x$ for $x \in \mathbb{Z}_n^*$, since $x^{\varphi(n)} = 1$ (see Proposition A.25). This shows that $\text{RSA}_{n,e}$ is bijective and that the inverse function is also an RSA function, namely $\text{RSA}_{n,d} : \mathbb{Z}_n^* \longrightarrow \mathbb{Z}_n^*, \ x \longmapsto x^d$.

---

^{5} Actually, $\varphi(x)$ is much closer to x. It can be shown that $\varphi(x) > \frac{x}{6 \log(|x|)}$ (see Appendix A.2).

RSA$_{n,e}$ can be computed by modular exponentiation, an efficient algorithm. d can be easily computed by the extended Euclidean algorithm A.5, if $\varphi(n) = (p-1)(q-1)$ is known. No algorithm to compute RSA$_{n,e}^{-1}$ in polynomial time is known, if p, q and d are kept secret. We call d (or p, q) the *trapdoor information* for the RSA function.

All known attacks to break RSA, if implemented by an efficient algorithm, would deliver an efficient algorithm for factoring n. All known factoring algorithms have exponential running time. Therefore, it is widely believed that *RSA* cannot be efficiently inverted. The following assumption makes this more precise.

Definition 6.7. Let $I_k := \{(n, e) \in I \mid n = pq, |p| = |q| = k\}$, with $k \in \mathbb{N}$, and let $Q(X) \in \mathbb{Z}[X]$ be a positive polynomial. Let $A(n, e, y)$ be a probabilistic polynomial algorithm. Then there exists a $k_0 \in \mathbb{N}$, such that

$$\mathrm{prob}(A(n, e, y) = \mathrm{RSA}_{n,d}(y) : (n, e) \xleftarrow{u} I_k, y \xleftarrow{u} \mathbb{Z}_n^*) \leq \frac{1}{Q(k)}$$

for $k \geq k_0$.
This is called the *RSA assumption*.

Remarks:

1. The assumption states the one-way property of the RSA family. The algorithm A models an adversary, who tries to compute $x = \mathrm{RSA}_{n,d}(y)$ from $y = \mathrm{RSA}_{n,e}(x) = x^e$ (in $\mathbb{Z}_n^*$) without knowing the trapdoor information d. By using Proposition 6.3, we may interpret the RSA assumption in an analogous way to the discrete logarithm assumption (Definition 6.1). The fraction of keys (n, e) in I_k, for which the adversary A has a significant chance to succeed, is negligibly small if the security parameter k is sufficiently large.
2. RSA$_{n,e}$ is bijective, and its range and domain coincide. Therefore, we also speak of a family of *one-way permutations* (or a family of *trapdoor permutations*).
3. Here and in what follows, we restrict the key set I of the RSA family and consider only those functions RSA$_{n,e}$, where $n = pq$ is the product of two primes of the same binary length. Instead, we could also define a stronger version of the assumption, where I_k is the set of pairs (n, e), with $|n| = k$ (the primes may have different length). However, our statement is closer to normal practice. To generate keys with a given security parameter k, usually two primes of length k are chosen and multiplied.
4. The RSA problem – computing x from x^e – is random self-reducible (see the analogous remark on the discrete logarithm problem on p. 154).

Stating the RSA assumption as above, we assume that the set I of keys can be uniformly sampled by an efficient algorithm.

Proposition 6.8. *There is a probabilistic polynomial uniform sampling algorithm for $I = (I_k)_{k \in \mathbb{N}}$, which on input 1^k yields a key $(n, e) \in I_k$ along with the trapdoor information (p, q, d).*

Proof. Above (see the examples after Lemma 6.5), we saw that Primes$_k$ can be uniformly sampled by a probabilistic polynomial algorithm. Thus, there is a probabilistic polynomial uniform sampling algorithm for

$$(X_k := \{n = pq \mid p, q \text{ distinct primes }, |p| = |q| = k\} \times \{0, 1\}^{2k})_{k \in \mathbb{N}}.$$

In the proof of Proposition 6.6, we observed that $|x| \cdot \varphi(x) \geq x$, and we immediately conclude that

$$|\mathbb{Z}^*_{\varphi(n)}| = \varphi(\varphi(n)) \geq \frac{\varphi(n)}{|\varphi(n)|} \geq \frac{n}{|n| \cdot |\varphi(n)|} \geq \frac{n}{4k^2} \geq \frac{2^{2k-2}}{4k^2} = \frac{2^{2k}}{16k^2}.$$

Thus, we can apply Lemma 6.5 to $Y_k := I_k \subseteq X_k$ and obtain the desired sampling algorithm. It yields (p, q, e). The inverse d of e in $\mathbb{Z}^*_{\varphi(n)}$ can be computed using the extended Euclidean algorithm (Algorithm A.5). □

Remarks:

1. The uniform sampling algorithm for $(I_k)_{k \in \mathbb{N}}$ which we derived in the proof of Proposition 6.8 is constructed by the method given in Lemma 6.5. Thus, it chooses triples (p, q, e) uniformly and then tests whether $e < \varphi(n) = (p - 1)(q - 1)$ and whether e is prime to $\varphi(n)$. If this test fails, a new triple (p, q, e) is selected. It would be more natural and more efficient to first choose a pair (p, q) uniformly, and then, with $n = pq$ fixed, to choose an exponent e uniformly from $\mathbb{Z}^*_{\varphi(n)}$. Then, however, the statistical distance between the distribution of the elements (n, e) and the uniform distribution is not negligible. The sampling algorithm is not uniform. Note that even for fixed k, there is a rather large variance of the cardinalities $|\mathbb{Z}^*_{\varphi(n)}|$. Nevertheless, this more natural sampling algorithm is an admissible key generator for the RSA family; i.e., the one-way condition is preserved if the keys are sampled by it (see Definition 6.13, of admissible key generators, and Exercise 1).
   An analogous remark applies to the sampling algorithm, given in the proof of Proposition 6.6.
2. We can generate the primes p and q by uniformly selecting numbers of length k and testing their primality by using a probabilistic prime number test (see the examples after Lemma 6.5). There are also other very efficient algorithms for the generation of uniformly distributed primes (see, e.g., [Maurer95]).

6.5 Modular Squaring

Let $I := \{n \mid n = pq,\ p, q \text{ distinct prime numbers}, |p| = |q|\}$. The family

$$\mathrm{Sq} := (\mathrm{Sq}_n : \mathbb{Z}_n^* \longrightarrow \mathbb{Z}_n^*,\ x \longmapsto x^2)_{n \in I}$$

is called the *Square family*.[6] Sq_n is neither injective nor surjective. If $\mathrm{Sq}_n^{-1}(x) \neq \emptyset$, then $|\mathrm{Sq}_n^{-1}(x)| = 4$ (see Proposition A.62).

Modular squaring can be done efficiently. Square roots modulo p are computable by a probabilistic polynomial algorithm if p is a prime number (see Algorithm A.61). Applying the Chinese Remainder Theorem (Theorem A.29), it is then easy to derive an efficient algorithm that computes square roots in $\mathbb{Z}_n^*$ if $n = pq$ (p and q are distinct prime numbers) and if the factorization of n is known.

Conversely, given an efficient algorithm for computing square roots in $\mathbb{Z}_n^*$, an efficient algorithm for the factorization of n can be derived (see Proposition A.64).

All known factoring algorithms have exponential running time. Therefore, it is widely believed that the factors of n (or, equivalently, square roots modulo n) cannot be computed efficiently. We make this statement more precise by the following assumption.

Definition 6.9. Let $I_k := \{n \in I \mid n = pq, |p| = |q| = k\}$, with $k \in \mathbb{N}$, and let $Q(X) \in \mathbb{Z}[X]$ be a positive polynomial. Let $A(n)$ be a probabilistic polynomial algorithm. Then there exists a $k_0 \in \mathbb{N}$, such that

$$\mathrm{prob}(A(n) = p : n \xleftarrow{u} I_k) \leq \frac{1}{Q(k)}$$

for $k \geq k_0$.
This is called the *factoring assumption*.

Stating the factoring assumption, we again assume that the set I of keys may be uniformly sampled by an efficient algorithm.

Proposition 6.10. *There is a probabilistic polynomial uniform sampling algorithm for* $I = (I_k)_{k \in \mathbb{N}}$, *which on input* 1^k *yields a number* $n \in I_k$, *along with its factors* p *and* q.

Proof. The algorithm chooses at random integers p and q with $|p| = |q| = k$, and applies a probabilistic primality test (see Appendix A.8) to check whether p and q are prime. By repeating the probabilistic primality test sufficiently

---

[6] As above in the RSA family, we only consider moduli n which are the product of two primes of equal binary length; see the remarks after the RSA assumption (Definition 6.7).

often, we can, with a probability $> 1 - 2^{-k}$, correctly test the primality of an element x in $\{0,1\}^k$. This sampling algorithm is uniform (Lemma 6.5).   □

Restricting the range and the domain to the set QR_n of squares modulo n (called the quadratic residues modulo n, see Definition A.48), the modular squaring function can be made bijective in many cases. Of course, each $x \in \mathrm{QR}_n$ has a square root. If p and q are distinct primes with $p, q \equiv 3 \bmod 4$ and $n := pq$, then exactly one of the four square roots of $x \in \mathrm{QR}_n$ is an element in QR_n (see Proposition A.66). Taking as key set

$$I := \{n \mid n = pq,\ p, q \text{ distinct prime numbers}, |p| = |q|, p, q \equiv 3 \bmod 4\},$$

we get a family

$$\mathrm{Square} := (\mathrm{Square}_n : \mathrm{QR}_n \longrightarrow \mathrm{QR}_n,\ x \longmapsto x^2)_{n \in I}$$

of bijective functions, also called the *Square family*. Since the range and domain are of the same set, we speak of a family of permutations. The family of inverse maps is denoted by

$$\mathrm{Sqrt} := (\mathrm{Sqrt}_n : \mathrm{QR}_n \longrightarrow \mathrm{QR}_n)_{n \in I}.$$

Sqrt_n maps x to the square root of x which is an element of QR_n.

The same considerations as those on $\mathrm{Sq}_n : \mathbb{Z}_n^* \longrightarrow \mathbb{Z}_n^*$ above show that Square_n is efficiently computable, and that computing Sqrt_n is equivalent to factoring n. Square is a family of trapdoor permutations with trapdoor information p and q.

6.6 Quadratic Residuosity Property

If p is a prime and $x \in \mathbb{Z}_p^*$, the Legendre symbol $\left(\frac{x}{p}\right)$ tells us whether x is a quadratic residue modulo p: $\left(\frac{x}{p}\right) = 1$ if $x \in \mathrm{QR}_p$, and $\left(\frac{x}{p}\right) = -1$ if $x \notin \mathrm{QR}_p$ (Definition A.51). The Legendre symbol can be easily computed using Euler's criterion (Proposition A.52): $\left(\frac{x}{p}\right) = x^{(p-1)/2} \bmod p$.

Now, let p and q be distinct prime numbers and $n := pq$. The Jacobi symbol $\left(\frac{x}{n}\right)$ is defined as $\left(\frac{x}{n}\right) := \left(\frac{x}{p}\right) \cdot \left(\frac{x}{q}\right)$ (Definition A.55). It is efficiently computable for every element $x \in \mathbb{Z}_n^*$ – without knowing the prime factors p and q of n (see Algorithm A.59). The Jacobi symbol cannot be used to decide whether $x \in \mathrm{QR}_n$. If $\left(\frac{x}{n}\right) = -1$, then x is not in QR_n. However, if $\left(\frac{x}{n}\right) = 1$, both cases $x \in \mathrm{QR}_n$ and $x \notin \mathrm{QR}_n$ are possible. $x \in \mathbb{Z}_n^*$ is a quadratic residue if and only if both $x \bmod p \in \mathbb{Z}_p^*$ and $x \bmod q \in \mathbb{Z}_q^*$ are quadratic residues, which is equivalent to $\left(\frac{x}{p}\right) = \left(\frac{x}{q}\right) = 1$.

Let $I := \{n \mid n = pq,\ p, q \text{ distinct prime numbers}, |p| = |q|\}$ and let

$$J_n^{+1} := \left\{ x \in \mathbb{Z}_n^* \ \middle| \ \left(\frac{x}{n}\right) = +1 \right\}$$

be the elements with Jacobi symbol $+1$. QR_n is a proper subset of J_n^{+1}. Consider the functions

$$PQR_n : J_n^{+1} \longrightarrow \{0,1\}, PQR_n(x) := \begin{cases} 1 \text{ if } x \in QR_n, \\ 0 \text{ otherwise.} \end{cases}$$

The family $PQR := (PQR_n)_{n \in I}$ is called the *quadratic residuosity family*.

It is believed that there is no efficient algorithm which, without knowing the factors of n, is able to decide whether $x \in J_n^{+1}$ is a quadratic residue. We make this precise in the following assumption.

Definition 6.11. Let $I_k := \{ n \in I \mid n = pq, |p| = |q| = k \}$, with $k \in \mathbb{N}$, and let $Q(X) \in \mathbb{Z}[X]$ be a positive polynomial. Let $A(n,x)$ be a probabilistic polynomial algorithm. Then there exists a $k_0 \in \mathbb{N}$, such that

$$\text{prob}(A(n,x) = PQR_n(x) : n \xleftarrow{u} I_k, x \xleftarrow{u} J_n^{+1}) \leq \frac{1}{2} + \frac{1}{Q(k)}$$

for $k \geq k_0$.
This is called the *quadratic residuosity assumption*.

Remark. The assumption states that there is not a significant chance of computing the predicate PQR_n if the factors of n are secret. It differs a little from the previous assumptions: the adversary algorithm A now has to compute a predicate. Since exactly half of the elements in J_n^{+1} are quadratic residues (see Proposition A.65), A can always predict the correct value with probability $1/2$, simply by tossing a coin. However, her probability of success is at most negligibly more than $1/2$.

Remark. The factoring assumption follows from the RSA assumption and also from the quadratic residuosity assumption. Hence, each of these two assumptions is stronger than the factoring assumption.

6.7 Formal Definition of One-Way Functions

As our examples show, one-way functions actually are families of functions. We give a formal definition of such families.

Definition 6.12. Let $I = (I_k)_{k \in \mathbb{N}}$ be a key set with security parameter k. Let K be a probabilistic polynomial sampling algorithm for I, which on input 1^k outputs $i \in I_k$.

A family

$$f = (f_i : D_i \longrightarrow R_i)_{i \in I}$$

of functions between finite sets D_i and R_i is a *family of one-way functions* (or, for short, a *one-way function*) with *key generator* K if and only if:

1. f can be computed by a Monte Carlo algorithm $F(i, x)$.
2. There is a uniform sampling algorithm S for $D := (D_i)_{i \in I}$, which on input $i \in I$ outputs $x \in D_i$.
3. f is not invertible by any efficient algorithm if the keys are generated by K. More precisely, for every positive polynomial $Q \in \mathbb{Z}[X]$ and every probabilistic polynomial algorithm $A(i, y)$ ($i \in I, y \in R_i$), there is a $k_0 \in \mathbb{N}$, such that for all $k \geq k_0$

$$\text{prob}(f_i(A(i, f_i(x))) = f_i(x) : i \leftarrow K(1^k), x \xleftarrow{u} D_i) \leq \frac{1}{Q(k)}.$$

If K is a uniform sampling algorithm for I, then we call f a family of one-way functions (or a one-way function), without explicitly referring to a key generator.

If f_i is bijective for all $i \in I$, then f is called a *bijective one-way function*, and if, in addition, the domain D_i coincides with the range R_i for all i, we call f a *family of one-way permutations* (or simply a *one-way permutation*).

Examples: The examples studied earlier in this chapter are families of one-way functions, provided Assumptions 6.1, 6.7 and 6.9 are true:

1. Discrete exponential function.
2. Modular powers.
3. Modular squaring.

Our considerations on the "random generation of the key" above (Propositions 6.6 and 6.8, 6.10) show that there are uniform key generators for these families. There are uniform sampling algorithms S for the domains $\mathbb{Z}_{p-1}$ and $\mathbb{Z}_n^*$, as we have seen in the examples after Lemma 6.5. Squaring a uniformly selected $x \in \mathbb{Z}_n^*$, $x \mapsto x^2$, we get a uniform sampling algorithm for the domains QR_n of the Square family.

Modular powers is a one-way permutation, as well as the modular squaring function Square. The discrete exponential function is a bijective one-way function.

Remarks:

1. Selecting an index i, for example, to use f_i as an encryption function, is equivalent to choosing a public key. Recall from Definition 6.2 that for $i \in I_k$, the security parameter k is a measure of the key length in bits.
2. Condition 3 means that pre-images of $y := f_i(x)$ cannot be computed in polynomial time if x is randomly and uniformly chosen from the domain (all inputs have the same probability), or equivalently, if y is random with respect to the image distribution induced by f. f is only called a one-way function if the random and uniform choice of elements in the domain can be accomplished by a probabilistic algorithm in polynomial time (condition 2).

Definition 6.12 can be immediately generalized to the case of one-way functions, where the inputs $x \in D_i$ are generated by any probabilistic polynomial – not necessarily uniform – sampling algorithm S (see, e.g., [Goldreich01]). The distribution $x \stackrel{u}{\leftarrow} D_i$ is replaced by $x \leftarrow S(i)$.

In this book, we consider only families of one-way functions with uniformly distributed inputs. The keys generated by the key generator K may be distributed in a non-uniform way.

3. The definition can be easily extended to formally define *families of trapdoor functions* (or, for short, *trapdoor functions*). We only sketch this definition. A bijective one-way function $f = (f_i)_{i \in I}$ is a trapdoor function if the inverse family $f^{-1} := (f_i^{-1})_{i \in I}$ can be computed by a Monte Carlo algorithm $F^{-1}(i, t_i, y)$, which takes as inputs the (public) key i, the (secret) trapdoor information t_i for f_i and a function value $y := f_i(x)$. It is required that the key generator K generates the trapdoor information t_i along with i.

   The RSA and the Square families are examples of trapdoor functions (see above).

4. The probability of success of the adversary A in the "one-way condition" (condition 3) is taken over the random choice of a key of a given security parameter k. It says that over all possibly generated keys, A has on average only a small probability of success. An adversary A usually knows the public key i when performing her attack. Thus, in a concrete attack the probability of success is given by the conditional probability assuming a fixed i, and this conditional probability might be high even if the average probability is negligibly small, as stated in condition 3. However, according to Proposition 6.3, the probability of such insecure keys is negligibly small. Thus, when randomly generating a key i by K, the probability of obtaining one for which A has a significant chance of succeeding is negligibly small (see Proposition 6.3 for a precise statement).

5. Condition 2 implies, in particular, that the binary length of the elements in D_i is bounded by the running time of $S(i)$, and hence is $\leq P(|i|)$ if P is a polynomial bound for the running time of S.

6. In all our examples, the one-way function can be computed using a deterministic polynomial algorithm. Computable by a Monte Carlo algorithm (see Definition 5.3) means that there is a probabilistic polynomial algorithm $F(i, x)$ with

$$\text{prob}(F(i, x) = f_i(x)) \geq 1 - 2^{-k} \quad (i \in I_k)$$

   (see Proposition 5.6 and Exercise 5 in Chapter 5).

7. Families of one-way functions, as defined here, are also called collections of strong one-way functions. They may be considered as a single one-way function $\{0, 1\}^* \longrightarrow \{0, 1\}^*$, defined on the infinite domain $\{0, 1\}^*$ (see [GolBel01]; [Goldreich01]). For the notion of weak one-way functions, see Exercise 3.

The key generator of a one-way function f is not uniquely determined: there are more suitable key generation algorithms (see Proposition 6.14 below). We call them "admissible generators".

Definition 6.13. Let $f = (f_i : D_i \longrightarrow R_i)_{i \in I}, I = (I_k)_{k \in \mathbb{N}}$, be a family of one-way functions with key generator K. A probabilistic polynomial algorithm $\tilde{K}$ that on input 1^k outputs a key $i \in I_k$ is called an *admissible key generator* for f if the one-way condition 3 of Definition 6.12 is satisfied for $\tilde{K}$.

Proposition 6.14. *Let* $f = (f_i : D_i \longrightarrow R_i)_{i \in I}, I = (I_k)_{k \in \mathbb{N}}$, *be a family of one-way functions with key generator K. Let $\tilde{K}$ be a probabilistic polynomial sampling algorithm for I, which on input 1^k yields $i \in I_k$. Assume that the family of distributions $i \leftarrow \tilde{K}(1^k)$ is polynomially bounded by the family of distributions $i \leftarrow K(1^k)$ (see Definition B.25).*
Then $\tilde{K}$ is also an admissible key generator for f.

Proof. This is a consequence of Proposition B.26. Namely, apply Proposition B.26 to $J := \mathbb{N} := \bigcup_{k \in \mathbb{N}} J_k, J_k := \{1^k\}, X_k := \{(i, x) \mid i \in I_k, x \in D_i\}$ and the probability distributions $(i \leftarrow \tilde{K}(1^k), x \xleftarrow{u} D_i)$ and $(i \leftarrow K(1^k), x \xleftarrow{u} D_i)$, $k \in \mathbb{N}$. The first family of distributions is polynomially bounded by the second. Assume as event $\mathcal{E}_k$ that $f_i(A(i, f_i(x))) = f_i(x)$. $\square$

Example. Let $f = (f_i : D_i \longrightarrow R_i)_{i \in I}$ be a family of one-way functions (with uniform key generation), and let $J \subseteq I$ with $|J_k| \cdot Q(k) \geq |I_k|$, for some polynomial Q. Let $\tilde{K}$ be a uniform sampling algorithm for J. Then $i \leftarrow \tilde{K}(1^k)$ is polynomially bounded by the uniform distribution $i \xleftarrow{u} I_k$. Thus, $\tilde{K}$ is an admissible key generator for f. This fact may be restated as:
$f = (f_i : D_i \longrightarrow R_i)_{i \in J}$ is also a one-way function.

Example. As a special case of the previous example, consider the RSA one-way function (Section 6.4). Take as keys only pairs $(n, e) \in I_k$ (notation as above), with e a prime number in $\mathbb{Z}^*_{\varphi(n)}$. Since the number of primes in $\mathbb{Z}_{\varphi(n)}$ is of the same order as $\varphi(n)/k$ (by the Prime Number Theorem, Theorem A.68), we get an admissible key generator in this way. In other words, the classical RSA assumption (Assumption 6.7) implies an RSA assumption, with $(n, e) \xleftarrow{u} I_k$ replaced by $(n, e) \xleftarrow{u} J_k$, where $J_k := \{(n, e) \in I_k \mid e \text{ prime}\}$.

Example. As already mentioned in Section 6.4, the key generator that first uniformly chooses an $n = pq$ and then, in a second step, uniformly chooses an exponent $e \in \mathbb{Z}^*_{\varphi(n)}$ is an admissible key generator for the RSA one-way function. The distribution given by this generator is polynomially bounded by the uniform distribution.

Similarly, we get an admissible key generator for the discrete exponential function (Section 6.2) if we first uniformly generate a prime p (together with a factorization of $p - 1$) and then, for this fixed p, repeatedly select $g \xleftarrow{u} \mathbb{Z}^*_p$ until g happens to be a primitive root (see Exercise 1).

6.8 Hard-Core Predicates

Given a one-way function f, it is impossible to compute a pre-image x from $y = f(x)$ using an efficient algorithm. Nevertheless, it is often easy to derive single bits of the pre-image x from $f(x)$. For example, if f is the discrete exponential function, the least-significant bit of x is derived from $f(x)$ in a straightforward way (see Chapter 7). On the other hand, since f is one way there should be other bits of the pre-image x, or more generally, properties of x stated as Boolean predicates, which are very hard to derive from $f(x)$. Examples of such hard-core predicates are studied thoroughly in Chapter 7.

Definition 6.15. Let $I = (I_k)_{k \in \mathbb{N}}$ be a key set with security parameter k. Let $f = (f_i : D_i \longrightarrow R_i)_{i \in I}$ be a family of one-way functions with key generator K, and let $B = (B_i : D_i \longrightarrow \{0,1\})_{i \in I}$ be a family of Boolean predicates. B is called a *family of hard-core predicates* (or, for short, a *hard-core predicate*) of f if and only if:

1. B can be computed by a Monte Carlo algorithm $A_1(i, x)$.
2. $B(x)$ is not computable from $f(x)$ by an efficient algorithm; i.e., for every positive polynomial $Q \in \mathbb{Z}[X]$ and every probabilistic polynomial algorithm $A_2(i, y)$, there is a $k_0 \in \mathbb{N}$ such that for all $k \geq k_0$

$$\text{prob}(A_2(i, f_i(x)) = B_i(x) : i \leftarrow K(1^k), x \xleftarrow{u} D_i) \leq \frac{1}{2} + \frac{1}{Q(k)}.$$

Remarks:

1. As above in the one-way conditions, we do not consider a single fixed key in the probability statement of the definition. The probability is also taken over the random generation of a key i, with a given security parameter k. Hence, the meaning of statement 2 is: choosing both the key i with security parameter k and $x \in D_i$ randomly, the adversary A_2 does not have a significantly better chance of finding out the bit $B_i(x)$ from $f_i(x)$ than by simply tossing a coin, provided the security parameter is sufficiently large. In practice, the public key is known to the adversary, and we are interested in the conditional probability of success assuming a fixed public key i. Even if the security parameter k is very large, there may be keys i, such that A_2 has a significantly better chance than $1/2$ of determining $B_i(x)$. The probability of such insecure keys is, however, negligibly small. This is shown in Proposition 6.17.
2. Considering a family of one-way functions with hard-core predicate B, we call a key generator K admissible only if it guarantees the "hard-core" condition 2 of Definition 6.15, in addition to the one-way condition 3 of Definition 6.12. Proposition 6.14 remains valid for one-way functions with hard-core predicates. This immediately follows from Proposition 6.17.

The *inner product bit* yields a generic hard-core predicate for all one-way functions.

Theorem 6.16. *Let* $f = (f_i : D_i \longrightarrow R_i)_{i \in I}$ *be a family of one-way functions,* $D_i \subset \{0,1\}^*$ *for all* $i \in I$. *Extend the functions* f_i *to functions* $\tilde{f}_i$ *which on input* $x \in D_i$ *and* $y \in \{0,1\}^{|x|}$ *return the concatenation* $f_i(x)\|y$. *Let*

$$B_i(x,y) := \left(\sum_{j=1}^{l} x_j \cdot y_j \right) \bmod 2,$$

where $l = |x|$, $x = x_1 \ldots x_l$, $y = y_1 \ldots y_l$, $x_j, y_j \in \{0,1\}$,

be the inner product modulo 2. Then $B := (B_i)_{i \in I}$ *is a hard-core predicate of* $\tilde{f} := (\tilde{f}_i)_{i \in I}$.

For a proof, see [GolLev89], [Goldreich99] or [Luby96].

Proposition 6.17. *Let* $I = (I_k)_{k \in \mathbb{N}}$ *be a key set with security parameter* k. *Let* $f = (f_i : X_i \longrightarrow Y_i)_{i \in I}$ *be a family of functions between finite sets, and let* $B = (B_i : X_i \longrightarrow \{0,1\})_{i \in I}$ *be a family of Boolean predicates. Assume that both* f *and* B *can be computed by a Monte Carlo algorithm, with inputs* $i \in I$ *and* $x \in X_i$. *Let probability distributions be given on* I_k *and* X_i *for all* k *and* i. *Assume there is a probabilistic polynomial sampling algorithm* S *for* $X := (X_i)_{i \in I}$ *which, on input* $i \in I$, *randomly chooses an* $x \in X_i$ *with respect to the given distribution on* X_i, *i.e.,* $\mathrm{prob}(S(i) = x) = \mathrm{prob}(x)$. *Then the following statements are equivalent:*

1. *For every probabilistic polynomial algorithm* A *with inputs* $i \in I$ *and* $y \in Y_i$ *and output in* $\{0,1\}$, *and every positive polynomial* P, *there is a* $k_0 \in \mathbb{N}$, *such that for all* $k \geq k_0$

$$\mathrm{prob}(A(i, f_i(x)) = B_i(x) : i \leftarrow I_k, x \leftarrow X_i) \leq \frac{1}{2} + \frac{1}{P(k)}.$$

2. *For every probabilistic polynomial algorithm* A *with inputs* $i \in I$ *and* $x \in X_i$ *and output in* $\{0,1\}$, *and all positive polynomials* Q *and* R, *there is a* $k_0 \in \mathbb{N}$, *such that for all* $k \geq k_0$

$$\mathrm{prob}\left(\left\{ i \in I_k \,\middle|\, \mathrm{prob}(A(i, f_i(x)) = B_i(x) : x \leftarrow X_i) > \frac{1}{2} + \frac{1}{Q(k)} \right\} \right)$$

$$\leq \frac{1}{R(k)}.$$

Statement 2 implies statement 1, even if we omit "for every algorithm" in both statements and instead consider a fixed probabilistic polynomial algorithm A.

Proof. An analogous computation as in the proof of Proposition 6.3 shows that statement 2 implies statement 1.

Now, assume that statement 1 holds, and let $A(i, y)$ be a probabilistic polynomial algorithm with output in $\{0, 1\}$. Let Q and R be positive polynomials. We abbreviate the probability of success of A, conditional on a fixed i, by p_i:

$$p_i := \text{prob}(A(i, f_i(x)) = B_i(x) : x \leftarrow X_i).$$

Assume that for k in an infinite subset $\mathcal{K} \subseteq \mathbb{N}$,

$$\text{prob}\left(\left\{i \in I_k \,\middle|\, p_i > \frac{1}{2} + \frac{1}{Q(k)}\right\}\right) > \frac{1}{R(k)}. \qquad (6.2)$$

If all $p_i, i \in I_k$, were $\geq 1/2$, we could easily conclude that their average also significantly exceeds $1/2$ and obtain a contradiction to statement 1. Unfortunately, it might happen that many of the probabilities p_i are significantly larger than $1/2$ and many are smaller than $1/2$, whereas their average is close to $1/2$. The basic idea now is to modify A and replace A by an algorithm $\tilde{A}$. We want to replace the output $A(i, y)$ by its complementary value $1 - A(i, y)$ if $p_i < 1/2$. In this way we could force all the probabilities to be $\geq 1/2$. At this point we face another problem. We want to define $\tilde{A}$ as

$$(i, y) \mapsto \begin{cases} A(i, y) & \text{if } p_i \geq 1/2, \\ 1 - A(i, y) & \text{if } p_i < 1/2, \end{cases}$$

and see that we have to determine by a polynomial algorithm whether $p_i \geq 1/2$. At least we can compute the correct answer to this question with a high probability in polynomial time. By Proposition 6.18, there is a probabilistic polynomial algorithm $C(i)$, such that

$$\text{prob}\left(|C(i) - p_i| < \frac{1}{4Q(k)R(k)}\right) \geq 1 - \frac{1}{4Q(k)R(k)}.$$

We define that $Sign(i) := +1$ if $C(i) \geq 1/2$, and $Sign(i) := -1$ if $C(i) < 1/2$. Then $Sign$ computes the sign $\sigma_i \in \{+1, -1\}$ of $(p_i - 1/2)$ with a high probability if the distance of p_i from $1/2$ is not too small:

$$\text{prob}(Sign(i) = \sigma_i) \geq 1 - \frac{1}{4Q(k)R(k)} \text{ for all } i \text{ with } \left|p_i - \frac{1}{2}\right| \geq \frac{1}{4Q(k)R(k)}.$$

Now the modified algorithm $\tilde{A} = \tilde{A}(i, y)$ is defined as follows. Let

$$\tilde{A}(i, y) := \begin{cases} A(i, y) & \text{if } Sign(i) = +1, \\ 1 - A(i, y) & \text{if } Sign(i) = -1, \end{cases}$$

for $i \in I$ and $y \in Y_i$. Similarly as before, we write

$$\tilde{p}_i := \text{prob}(\tilde{A}(i, f_i(x)) = B_i(x) : x \leftarrow X_i)$$

for short. By the definition of $\tilde{A}$, $\tilde{p}_i = p_i$ if $Sign(i) = +1$, and $\tilde{p}_i = 1 - p_i$ if $Sign(i) = -1$. Hence $|p_i - 1/2| = |\tilde{p}_i - 1/2|$. Moreover, we have $\tilde{p}_i \geq 1/2$, if $Sign(i) = \sigma_i$, and $\mathrm{prob}(Sign(i) \neq \sigma_i) \leq 1/(4Q(k)R(k))$ if $|p_i - 1/2| \geq 1/(4Q(k)R(k))$.

Let

$$I_{1,k} := \left\{ i \in I_k \,\middle|\, p_i - \frac{1}{2} > \frac{1}{Q(k)} \right\},$$

$$I_{2,k} := \left\{ i \in I_k \,\middle|\, \left|p_i - \frac{1}{2}\right| > \frac{1}{4Q(k)R(k)} \right\}.$$

We compute

$$\mathrm{prob}(\tilde{A}(i, f_i(x)) = B_i(x) : i \leftarrow I_k, x \leftarrow X_i) - \frac{1}{2}$$

$$= \sum_{i \in I_k} \mathrm{prob}(i) \cdot \left(\tilde{p}_i - \frac{1}{2} \right)$$

$$= \sum_{i \in I_{2,k}} \mathrm{prob}(i) \cdot \left(\tilde{p}_i - \frac{1}{2} \right) + \sum_{i \notin I_{2,k}} \mathrm{prob}(i) \cdot \left(\tilde{p}_i - \frac{1}{2} \right)$$

$$=: (1).$$

For $i \notin I_{2,k}$, we have $|\tilde{p}_i - 1/2| = |p_i - 1/2| \leq 1/(4Q(k)R(k))$ and hence $\tilde{p}_i - 1/2 \geq -1/(4Q(k)R(k))$. Thus

$$(1) \geq \sum_{i \in I_{2,k}} \mathrm{prob}(i) \cdot \left(\tilde{p}_i - \frac{1}{2} \right) - \frac{1}{4Q(k)R(k)}$$

$$= \sum_{i \in I_{2,k}} \mathrm{prob}(Sign(i) = \sigma_i) \cdot \mathrm{prob}(i) \cdot \left(\tilde{p}_i - \frac{1}{2} \right)$$

$$+ \sum_{i \in I_{2,k}} \mathrm{prob}(Sign(i) \neq \sigma_i) \cdot \mathrm{prob}(i) \cdot \left(\tilde{p}_i - \frac{1}{2} \right) - \frac{1}{4Q(k)R(k)}$$

$$=: (2).$$

We observed before that $\tilde{p}_i \geq 1/2$ if $Sign(i) = \sigma_i$, and $\mathrm{prob}(Sign(i) \neq \sigma_i) \leq 1/(4Q(k)R(k))$ for $i \in I_{2,k}$. Moreover, we obviously have $I_{1,k} \subseteq I_{2,k}$, and our assumption (6.2) means that

$$\mathrm{prob}(I_{1,k}) = \sum_{i \in I_{1,k}} \mathrm{prob}(i) > \frac{1}{R(k)},$$

for the infinitely many $k \in \mathcal{K}$. Therefore, we may continue our computation for $k \in \mathcal{K}$ in the following way:

$$(2) \geq \left(1 - \frac{1}{4Q(k)R(k)}\right) \sum_{i \in I_{2,k}} \text{prob}(i) \cdot \left|p_i - \frac{1}{2}\right| - \frac{1}{4Q(k)R(k)} - \frac{1}{4Q(k)R(k)}$$

$$\geq \left(1 - \frac{1}{4Q(k)R(k)}\right) \sum_{i \in I_{1,k}} \text{prob}(i) \cdot \left(p_i - \frac{1}{2}\right) - \frac{1}{4Q(k)R(k)} - \frac{1}{4Q(k)R(k)}$$

$$\geq \left(1 - \frac{1}{4Q(k)R(k)}\right) \cdot \frac{1}{R(k)} \cdot \frac{1}{Q(k)} - \frac{1}{4Q(k)R(k)} - \frac{1}{4Q(k)R(k)}$$

$$= \frac{1}{2Q(k)R(k)} - \frac{1}{4Q^2(k)R^2(k)}$$

$$\geq \frac{1}{4Q(k)R(k)}.$$

Since $\mathcal{K}$ is infinite, we obtain a contradiction to statement 1 applied to $\tilde{A}$ and $P = 4QR$, and the proof that statement 1 implies statement 2 is finished. $\square$

As in the preceding proof, we sometimes want to compute probabilities with a probabilistic polynomial algorithm, at least approximately.

Proposition 6.18. *Let $B(i, x)$ be a probabilistic polynomial algorithm that on input $i \in I$ and $x \in X_i$ outputs a bit $b \in \{0, 1\}$. Assume that a probability distribution is given on X_i, for every $i \in I$. Further assume that there is a probabilistic polynomial sampling algorithm S which on input $i \in I$ randomly chooses an element $x \in X_i$ with respect to the distribution given on X_i, i.e., $\text{prob}(S(i) = x) = \text{prob}(x)$. Let P and Q be positive polynomials.*

Let $p_i := \text{prob}(B(i, x) = 1 : x \leftarrow X_i)$ be the probability of $B(i, x) = 1$, assuming that i is fixed. Then there is a probabilistic polynomial algorithm A that approximates the probabilities $p_i, i \in I$, with high probability; i.e.,

$$\text{prob}\left(|A(i) - p_i| < \frac{1}{P(|i|)}\right) \geq 1 - \frac{1}{Q(|i|)}.$$

Proof. We first observe that $\text{prob}(B(i, S(i)) = 1) = p_i$ for every $i \in I$, since $S(i)$ samples (by use of its coin tosses) according to the distribution given on X_i. We define the algorithm A on input i as follows:

1. Let t be the smallest $n \in \mathbb{N}$ with $n \geq 1/4 \cdot P(|i|)^2 \cdot Q(|i|)$.
2. Compute $B(i, S(i))$ t times, and obtain the results
   $b_1, \ldots, b_t \in \{0, 1\}$.
3. Let

$$A(i) := \frac{1}{t} \sum_{i=1}^{t} b_i.$$

Applying Corollary B.17 to the t independent computations of the random variable $B(i, S(i))$, we get

$$\text{prob}\left(|A(i) - p_i| < \frac{1}{P(|i|)}\right) \geq 1 - \frac{P(|i|)^2}{4t} \geq 1 - \frac{1}{Q(|i|)},$$

as desired. $\square$

Exercises

1. Let S be the key generator for
   a. the RSA family, which on input 1^k first generates random prime numbers p and q of binary length k, and then repeatedly generates random exponents $e \in \mathbb{Z}_{\varphi(n)}, n := pq$, until it finds an e prime to $\varphi(n)$.
   b. the Exp family, which on input 1^k first generates a random prime p of binary length k, together with the prime factors of $p-1$, and then randomly chooses elements $g \in \mathbb{Z}_p^*$, until it finds a primitive root g (see the proof of Proposition 6.6).

   Show that S is an admissible key generator for the RSA family and the Exp family (see the remark after Proposition 6.8).

2. Compare the running times of the uniform key generator for RSA keys constructed in Proposition 6.8 and the (more efficient) admissible key generator S in Exercise 1.

3. Consider the Definition 6.12 of "strong" one-way functions. If we replace the probability statement in condition 3 by "There is a positive polynomial Q, such that for all probabilistic polynomial algorithms A

$$\mathrm{prob}(f_i(A(i, f_i(x))) = f_i(x) : i \leftarrow K(1^k), x \xleftarrow{u} D_i) \leq 1 - \frac{1}{Q(k)},$$

   for sufficiently large k", then $f = (f_i)_{i \in I}$ is called a family of *weak one-way functions*.[7]
   Let $X_j := \{2, 3, \ldots, 2^j - 1\}$ be the set of numbers > 1 of binary length $\leq j$. Let $D_n := \bigcup_{j=2}^{n-2} X_j \times X_{n-j}$ (disjoint union), and let $R_n := X_n$.
   Show that
   $$f := (f_n : D_n \longrightarrow R_n, \ (x, y) \longmapsto x \cdot y)_{n \in \mathbb{N}}$$
   is a weak – not a strong – one-way function, if the factoring assumption (Definition 6.9) is true.

4. We consider the representation problem (Section 4.5.3).
   Let $A(p, q, g_1, \ldots, g_r)$ be a probabilistic polynomial algorithm ($r \geq 2$). The inputs are primes p and q, such that q divides $p - 1$, and elements $g_1, \ldots, g_r \in \mathbb{Z}_p^*$ of order q. A outputs two tuples $(a_1, \ldots, a_r)$ and $(a'_1, \ldots, a'_r)$ of integer exponents. Assume that r is polynomially bounded by the binary length of p, i.e., $r \leq T(|p|)$ for some polynomial T. We denote by G_q the subgroup of order q of $\mathbb{Z}_p^*$. Recall that G_q is cyclic and each $g \in G_q, g \neq 1$, is a generator (Lemma A.40). Let P be a positive polynomial, and let $\mathcal{K}$ be the set of pairs (p, q), such that

---

[7] A weak one-way function f yields a strong one-way function by the following construction: $(x_1, \ldots, x_{2kQ(k)}) \mapsto (f_i(x_1), \ldots, f_i(x_{2kQ(k)}))$ (where $i \in I_k$). See [Luby96] for a proof.

$$\text{prob}\big(A(p,q,g_1,\ldots,g_r) = ((a_1,\ldots,a_r),(a'_1,\ldots,a'_r)),$$

$$(a_1,\ldots,a_r) \neq (a'_1,\ldots,a'_r),\quad \prod_{j=1}^{r} g_j^{a_j} = \prod_{j=1}^{r} g_j^{a'_j}:$$

$$g_j \xleftarrow{u} G_q \setminus \{1\}, 1 \leq j \leq r\big)$$

$$\geq {}^1\!/P(|p|).$$

Show that for every positive polynomial Q there is a probabilistic polynomial algorithm $\tilde{A} = \tilde{A}(p,q,g,y)$, such that

$$\text{prob}(\tilde{A}(p,q,g,y) = \text{Log}_{p,g}(y)) \geq 1 - 2^{-Q(|p|)},$$

for all $(p,q) \in \mathcal{K}$, $g \in G_q \setminus \{1\}$ and $y \in G_q$.

5. Let $J_k := \{n \in \mathbb{N} \mid n = pq, p, q \text{ distinct primes}, |p| = |q| = k\}$ be the set of RSA moduli with security parameter k. For a prime $\tilde{p}$, we denote by $J_{k,\tilde{p}} := \{n \in J_k \mid \tilde{p} \text{ does not divide } \varphi(n)\}$ the set of those moduli for which $\tilde{p}$ may serve as an RSA exponent. Let $\text{Primes}_{\leq 2k}$ be the primes of binary length $\leq 2k$.

   Show that the RSA assumption remains valid if we first choose a prime exponent and then a suitable modulus. More precisely, show that the classical RSA assumption (Definition 6.7) implies that:

   For every probabilistic polynomial algorithm A and every positive polynomial P there is a $k_0 \in \mathbb{N}$, such that for all $k \geq k_0$

   $$\text{prob}(A(n,\tilde{p},x^{\tilde{p}}) = x \mid \tilde{p} \xleftarrow{u} \text{Primes}_{\leq 2k}, n \xleftarrow{u} J_{k,\tilde{p}}, x \xleftarrow{u} \mathbb{Z}_n^*) \leq \frac{1}{P(k)}.$$

6. Let $f = (f_i : D_i \longrightarrow R_i)_{i \in I}$, $I = (I_k)_{k \in \mathbb{N}}$, be a family of one-way functions with key generator K, and let $B = (B_i : D_i \longrightarrow \{0,1\})_{i \in I}$ be a hard-core predicate for f.

   Show that for every positive polynomial P, there is a $k_0 \in \mathbb{N}$ such that for all $k \geq k_0$

   $$\left| \text{prob}(B_i(x) = 0 : i \leftarrow K(1^k), x \xleftarrow{u} D_i) - \frac{1}{2} \right| \leq \frac{1}{P(k)}.$$

7. Let $f = (f_i : D_i \longrightarrow R_i)_{i \in I}$, $I = (I_k)_{k \in \mathbb{N}}$, be a family of one-way functions with key generator K, and let $B = (B_i : D_i \longrightarrow \{0,1\})_{i \in I}$ be a family of predicates which is computable by a Monte Carlo algorithm. Show that B is a hard-core predicate of f if and only if for every probabilistic polynomial algorithm $A(i,x,y)$ and every positive polynomial P there is a k_0, such that for all $k \geq k_0$

   $$|\, \text{prob}(A(i,f_i(x),B_i(x)) = 1 : i \leftarrow K(1^k), x \xleftarrow{u} D_i)$$

   $$- \text{prob}(A(i,f_i(x),z) = 1 : i \leftarrow K(1^k), x \xleftarrow{u} D_i, z \xleftarrow{u} \{0,1\}) \,| \leq \frac{1}{P(k)}.$$

If the functions f_i are bijective, then the latter statement means that the family $(\{(f_i(x), B_i(x)) : x \xleftarrow{u} D_i\})_{i \in I}$ of distributions cannot be distinguished from the uniform distributions on $(R_i \times \{0,1\})_{i \in I}$ by a probabilistic polynomial algorithm.

8. Let $I = (I_k)_{k \in \mathbb{N}}$ be a key set with security parameter k. Consider probabilistic polynomial algorithms A which on input $i \in I$ and $x \in X_i$ compute a Boolean value $A(i, x) \in \{0, 1\}$. Assume that one family of probability distributions is given on $(I_k)_{k \in \mathbb{N}}$, and two families of probability distributions $p := (p_i)_{i \in I}$ and $q := (q_i)_{i \in I}$ are given on $X := (X_i)_{i \in I}$. Further assume that there are probabilistic polynomial sampling algorithms $S_1(i)$ and $S_2(i)$ which randomly choose an $x \in X_i$ with $\mathrm{prob}(S_1(i) = x) = p_i(x)$ and $\mathrm{prob}(S_2(i) = x) = q_i(x)$, for all $i \in I$ and $x \in X_i$.

Prove a result that is analogous to the statements of Propositions 6.3 and 6.17 for

$$| \mathrm{prob}(A(i, x) = 1 : i \leftarrow I_k, x \xleftarrow{p_i} X_i)$$

$$- \mathrm{prob}(A(i, x) = 1 : i \leftarrow I_k, x \xleftarrow{q_i} X_i)| \leq \frac{1}{P(k)}.$$

9. Let $n := pq$, with distinct primes p and q. The quadratic residuosity assumption (Definition 6.11) states that it is infeasible to decide whether a given $x \in J_n^{+1}$ is a quadratic residue or not. If p and q are kept secret, no efficient algorithm for selecting a quadratic non-residue modulo n is known. Thus, it might be easier to decide quadratic residuosity if, additionally, a random quadratic non-residue with Jacobi symbol 1 is revealed. In [GolMic84] (Section 6.2) it is shown that this is not the case. More precisely, let $I_k := \{n \mid n = pq, p, q \text{ distinct primes }, |p| = |q| = k\}$ and $I := (I_k)_{k \in \mathbb{N}}$. Let $\mathrm{QNR}_n := \mathbb{Z}_n^* \setminus \mathrm{QR}_n$ be the set of quadratic non-residues (see Definition A.48) and $\mathrm{QNR}_n^{+1} := \mathrm{QNR}_n \cap J_n^{+1}$. Then the following statements are equivalent:

a. For all probabilistic polynomial algorithms A, with inputs $n \in I$ and $x \in J_n^{+1}$ and output in $\{0, 1\}$, and every positive polynomial P, there exists a $k_0 \in \mathbb{N}$ such that for all $k \geq k_0$

$$\mathrm{prob}(A(n, x) = \mathrm{PQR}_n(x) : n \xleftarrow{u} I_k, x \xleftarrow{u} J_n^{+1}) \leq \frac{1}{2} + \frac{1}{P(k)}.$$

b. For all probabilistic polynomial algorithms A, with inputs $n \in I$ and $z, x \in J_n^{+1}$ and output in $\{0, 1\}$, and every positive polynomial P, there exists a $k_0 \in \mathbb{N}$ such that for all $k \geq k_0$

$$\mathrm{prob}(A(n, z, x) = \mathrm{PQR}_n(x) : n \xleftarrow{u} I_k, z \xleftarrow{u} \mathrm{QNR}_n^{+1}, x \xleftarrow{u} J_n^{+1})$$

$$\leq \frac{1}{2} + \frac{1}{P(k)}.$$

Study [GolMic84] and give a proof.

7. Bit Security of One-Way Functions

Let $f : X \longrightarrow Y$ be a bijective one-way function and let $x \in X$. Sometimes it is possible to compute some bits of x from $f(x)$ without inverting f. A function f does not necessarily hide everything about x, even if f is one way. Let b be a bit of x. We call b a *secure bit* of f if it is as difficult to compute b from $f(x)$ as it is to compute x from $f(x)$. We prove that the most-significant bit of x is a secure bit of Exp, and that the least-significant bit is a secure bit of RSA and Square.

We show how to compute x from $\text{Exp}(x)$, assuming that we can compute the most-significant bit of x from $\text{Exp}(x)$. Then we show the same for the least-significant bit and RSA or Square. First we assume that deterministic algorithms are used to compute the most- or least-significant bit. In this much easier case, we demonstrate the basic ideas. Then we study the probabilistic case. We assume that we can compute the most- or least-significant bit with a probability $p \geq 1/2 + 1/P(|x|)$, for some positive polynomial P, and derive that then x can be computed with an overwhelmingly high probability.

As a consequence, the discrete logarithm assumption implies that the most-significant bit is a hard-core predicate for the Exp family. Given the RSA or the factoring assumption, the least-significant bit yields a hard-core predicate for the RSA or the Square family.

Bit security is not only of theoretical interest. Bleichenbacher's 1-Million-Chosen-Ciphertext Attack against PKCS#1-based schemes shows that a leaking secure bit can lead to dangerous practical attacks (see Section 3.3.3).

Let $n, r \in \mathbb{N}$ such that $2^{r-1} \leq n < 2^r$. As usual, the *binary encoding* of $x \in \mathbb{Z}_n$ is the binary encoding in $\{0,1\}^r$ of the representative of x between 0 and $n - 1$ as an unsigned number (see Appendix A.2). This defines an embedding $\mathbb{Z}_n \subset \{0,1\}^r$. Bits of x and properties of x that depend on a representative of x are defined relative to this embedding. The property "x is even", for example, is such a property.

7.1 Bit Security of the Exp Family

Let p be an odd prime number and let g be a primitive root in $\mathbb{Z}_p^*$. We consider the discrete exponential function

$$\text{Exp}_{p,g} : \mathbb{Z}_{p-1} \longrightarrow \mathbb{Z}_p^*, \ x \longmapsto g^x,$$

and its inverse

$$\text{Log}_{p,g} : \mathbb{Z}_p^* \longrightarrow \mathbb{Z}_{p-1},$$

which is the discrete logarithm function.

$\text{Log}_{p,g}(x)$ is even if and only if x is a square (Lemma A.49). The square property modulo a prime can be computed efficiently by Euler's criterion (Proposition A.52). Hence, we have an efficient algorithm that computes the least-significant bit of x from $\text{Exp}_{p,g}(x)$. The most-significant bit, however, is as difficult to compute as the discrete logarithm.

Definition 7.1. Let p be an odd prime, and let g be a primitive root in $\mathbb{Z}_p^*$:

1. The predicate Msb_p is defined by

$$\text{Msb}_p : \mathbb{Z}_{p-1} \longrightarrow \{0,1\}, x \longmapsto \begin{cases} 0 \text{ if } 0 \le x < \frac{p-1}{2}, \\ 1 \text{ if } \frac{p-1}{2} \le x < p-1. \end{cases}$$

2. The predicate $\text{B}_{p,g}$ is defined by

$$\text{B}_{p,g} : \mathbb{Z}_p^* \longrightarrow \{0,1\}, \text{B}_{p,g}(x) = \text{Msb}_p(\text{Log}_{p,g}(x)).$$

Remark. If $p-1$ is a power of 2, then the predicate Msb_p is the most-significant bit of the binary encoding of x.

Let $x \in \text{QR}_p$, $x \neq 1$. There is a probabilistic polynomial algorithm which computes the square roots of x (Algorithm A.61). Let $y := y_n \dots y_0, y_i \in \{0,1\}$, be the binary representation of $y := \text{Log}_{p,g}(x)$. As we observed before, $y_0 = 0$. Therefore, $w_1 := g^{\tilde{y}}$ with $\tilde{y} := y_n \dots y_1$ is a root of x with $\text{B}_{p,g}(w_1) = 0$. The other root w_2 is $g^{\tilde{y}+(p-1)/2}$, and obviously $\text{B}_{p,g}(w_2) = 1$. Thus, exactly one of the two square roots w satisfies the condition $\text{B}_{p,g}(w) = 0$.

Definition 7.2. Let $x \in \text{QR}_p$. The square root w of x which satisfies the condition $\text{B}_{p,g}(w) = 0$ is called the *principal square root* of x. The map $\text{PSqrt}_{p,g}$ is defined by

$$\text{PSqrt}_{p,g} : \text{QR}_p \longrightarrow \mathbb{Z}_p^*, \ x \longmapsto \text{principal square root of } x.$$

Remark. The ability to compute the principal square root in polynomial time is equivalent to the ability to compute the predicate $\text{B}_{p,g}$ in polynomial time. Namely, let A be a deterministic polynomial algorithm for the computation of $\text{PSqrt}_{p,g}$. Then the algorithm

Algorithm 7.3.
```
int B(int p, g, x)
1   if A(p, g, x²) = x
2       then return 0
3       else return 1
```

computes $B_{p,g}$ and is deterministic polynomial.

Conversely, let B be a deterministic polynomial algorithm which computes $B_{p,g}$. If Sqrt is a polynomial algorithm for the computation of square roots modulo p, then the polynomial algorithm

Algorithm 7.4.
   int $A(\text{int } p, g, x)$
  1   $\{u, v\} \leftarrow \text{Sqrt}(x, p)$
  2   if $B(p, g, u) = 0$
  3   then return u
  4   else return v

computes $\text{PSqrt}_{p,g}$. It is deterministic if Sqrt is deterministic. We have a polynomial algorithm Sqrt (Algorithm A.61). It is deterministic for $p \equiv 3 \bmod 4$. In the other case, the only non-deterministic step is to find a quadratic non-residue modulo p, which is an easy task. If we select t numbers in $\{1, p-1\}$ at random, then the probability of finding a quadratic non-residue is $1 - 1/2^t$. Thus, the probability of success of A can be made almost 1, independent of the size of the input x.

In the following proposition and theorem, we show how to reduce the computation of the discrete logarithm function to the computation of $\text{PSqrt}_{p,g}$. Given an algorithm A_1 for computing $\text{PSqrt}_{p,g}$, we develop an algorithm A_2 for the discrete logarithm which calls A_1 as a subroutine.[1] The resulting algorithm A_2 has the same complexity as A_1. Therefore, $\text{PSqrt}_{p,g}$ is also believed to be not efficiently computable.

Proposition 7.5. *Let A_1 be a deterministic polynomial algorithm, such that*

$$A_1(p, g, x) = \text{PSqrt}_{p,g}(x) \text{ for all } x \in \text{QR}_p,$$

with p an odd prime and g a primitive root in $\mathbb{Z}_p^*$. Then there is a deterministic polynomial algorithm A_2, such that

$$A_2(p, g, x) = \text{Log}_{p,g}(x) \text{ for all } x \in \text{QR}_p.$$

Proof. The basic idea of the reduction is the following:
1. Let $x = g^y \in \mathbb{Z}_p^*$ and $y = y_k \ldots y_0$, $y_i \in \{0, 1\}$, $i = 0, \ldots, k$, be the binary encoding of y. We compute the bits of y from right to left. Bit y_0 is 0 if and only if $x \in \text{QR}_p$ (Lemma A.49). This condition can be tested by use of Euler's criterion for quadratic residuosity (Proposition A.52).
2. To get the next bit y_1, we replace x by $xg^{-1} = g^{y_k \cdots y_1 0}$ if $y_0 = 1$. Then bit y_1 can be obtained from $\text{PSqrt}_{p,g}(x) = g^{y_k \cdots y_1}$ as in step 1.

The following algorithm A_2, which calls A_1 as a subroutine, computes $\text{Log}_{p,g}$.

---

[1] In our construction we use A_1 as an "oracle" for $\text{PSqrt}_{p,g}$. Therefore, algorithms such as A_1 are sometimes called *"oracle algorithms"*.

Algorithm 7.6.

```
int A₂(int p, g, x)
1   y ← empty word, k ← |p|
2   for c ← 0 to k − 1 do
3       if x ∈ QRₚ
4           then y ← y‖0
5           else y ← y‖1
6               x ← xg⁻¹
7       x ← A₁(p, g, x)
8   return y
```

This completes the proof.   □

Theorem 7.7. *Let $P, Q \in \mathbb{Z}[X]$ be positive polynomials and A_1 be a probabilistic polynomial algorithm, such that*

$$\mathrm{prob}(A_1(p, g, x) = \mathrm{PSqrt}_{p,g}(x) : x \overset{u}{\leftarrow} \mathrm{QR}_n) \geq \frac{1}{2} + \frac{1}{P(k)},$$

where p is an odd prime number, g is a primitive root in $\mathbb{Z}_p^*$ and $k = |p|$ is the binary length of p. Then there is a probabilistic polynomial algorithm A_2, such that for every $x \in \mathbb{Z}_p^*$,

$$\mathrm{prob}(A_2(p, g, x) = \mathrm{Log}_{p,g}(x)) \geq 1 - 2^{-Q(k)}.$$

Proof. Let $\varepsilon := {}^1\!/P(k)$. In order to reduce $\mathrm{Log}_{p,g}$ to $\mathrm{PSqrt}_{p,g}$, we intend to proceed as in the deterministic case (Proposition 7.5). There, A_1 is applied k times by A_2. A_2 correctly yields the desired logarithm if PSqrt is correctly computed by A_1 in each step. Now the algorithm A_1 computes the function $\mathrm{PSqrt}_{p,g}$ with a probability of success of only $\geq {}^1\!/2 + \varepsilon$. Thus, the probability of success of A_2 is $\geq (1/2 + \varepsilon)^k$. This value is exponentially close to 0, and hence is too small.

The basic idea now is to replace A_1 by an algorithm B which computes $\mathrm{PSqrt}_{p,g}(x)$ with a high probability of success, for a polynomial fraction of inputs.

Lemma 7.8. *Under the assumptions of the theorem, let $t := k\varepsilon^{-2}$. Then there is a probabilistic polynomial algorithm B, such that*

$$\mathrm{prob}(B(x) = \mathrm{PSqrt}_{p,g}(x)) \geq 1 - \frac{1}{k}, \text{ for } x = g^{2s}, 0 \leq s \leq \frac{p-1}{2t}.$$

Proof (of the Lemma). Let $x = g^{2s}, 0 \leq s \leq (p-1)/2t$. In our algorithm B we want to increase the probability of success on input x by computing $A_1(x)$ repeatedly. By assumption, we have

$$\mathrm{prob}(A_1(p, g, x) = \mathrm{PSqrt}_{p,g}(x) : x \overset{u}{\leftarrow} \mathrm{QR}_n) \geq \frac{1}{2} + \varepsilon.$$

Here the probability is also taken over the random choice of $x \in \mathrm{QR}_n$. Therefore, we must modify x randomly each time we apply A_1. For this purpose, we iteratively select $r \in \left\{0, \ldots, \frac{(p-1)}{2} - 1\right\}$ at random and compute $A_1(p, g, xg^{2r})$. If $A_1(p, g, xg^{2r})$ successfully computes $\mathrm{PSqrt}_{p,g}(xg^{2r})$, then $\mathrm{PSqrt}_{p,g}(x) = A_1(p, g, xg^{2r}) \cdot g^{-r}$, at least if $s + r < (p-1)/2$. The latter happens with a high probability, since s is assumed to be small. Since the points xg^{2r} are sampled randomly and independently, we can then compute the principal square root $\mathrm{PSqrt}_{p,g}(x)$ with a high probability, using a majority decision on the values $A_1(p, g, xg^{2r}) \cdot g^{-r}$. The probability of success increases linearly with the number of sampled points, and it can be computed by Corollary B.18, which is a consequence of the weak law of large numbers.

Algorithm 7.9.
```
int B(int x)
 1   C₀ ← 0, C₁ ← 0
 2   {u, v} ← Sqrt(x)
 3   for i ← 1 to t do
 4       select r ∈ {0, ..., (p-1)/2 - 1} at random
 5       if A₁(p, g, xg²ʳ) = ugʳ
 6           then C₀ ← C₀ + 1
 7           else  C₁ ← C₁ + 1
 8   if C₀ > C₁
 9       then return u
10       else  return v
```

We now show that

$$\mathrm{prob}(B(x) = \mathrm{PSqrt}_{p,g}(x)) \geq 1 - \frac{1}{k}, \text{ for } x = g^{2s}, 0 \leq s \leq \frac{p-1}{2t}.$$

Let r_i be the randomly selected elements r and $x_i = xg^{2r_i}$, $i = 1, \ldots, t$. Every element $z \in \mathrm{QR}_p$ has a unique representation $z = xg^{2r}$, $0 \leq r < (p-1)/2$. Therefore, the uniform and random choice of r_i implies the uniform and random choice of x_i. Let $x = g^{2s}, 0 \leq s \leq (p-1)/2t$, and $r < (t-1)(p-1)/2t$. Then

$$2s + 2r < \frac{p-1}{t} + \frac{(t-1)(p-1)}{t} = p - 1,$$

and hence

$$\mathrm{PSqrt}_{p,g}(x) \cdot g^r = \mathrm{PSqrt}_{p,g}(xg^{2r}).$$

Let E_1 be the event $r < (t-1)(p-1)/2t$ and E_2 be the event $A_1(p, g, xg^{2r}) = \mathrm{PSqrt}_{p,g}(xg^{2r})$. We have $\mathrm{prob}(E_1) \geq 1 - 1/t$ and $\mathrm{prob}(E_2) \geq 1/2 + \varepsilon$, and we correctly compute $\mathrm{PSqrt}_{p,g}(x) \cdot g^r$, if both events E_1 and E_2 occur. Thus, we have (denoting by $\overline{E_2}$ the complement of event E_2)

$$\mathrm{prob}(A_1(p, g, xg^{2r}) = \mathrm{PSqrt}_{p,g}(x) \cdot g^r)$$

$$\geq \text{prob}(E_1 \text{ and } E_2) \geq \text{prob}(E_1) - \text{prob}(\overline{E_2})$$

$$\geq 1 - \frac{1}{t} - \left(\frac{1}{2} - \varepsilon\right) = \frac{1}{2} + \varepsilon - \frac{1}{k}\varepsilon^2 \geq \frac{1}{2} + \frac{1}{2}\varepsilon.$$

In each of the t iterations of B, $\text{PSqrt}_{p,g}(x) \cdot g^r$ is computed with probability $> 1/2$. Taking the most frequent result, we get $\text{PSqrt}_{p,g}(x)$ with a very high probability. More precisely, we can apply Corollary B.18 to the independent random variables S_i, $i = 1, \ldots, t$, defined by

$$S_i = \begin{cases} 1 \text{ if } A_1(p, g, xg^{2r_i}) = \text{PSqrt}_{p,g}(x) \cdot g^{r_i}, \\ 0 \text{ otherwise.} \end{cases}$$

We have $E(S_i) = \text{prob}(S_i = 1) \geq 1/2 + \varepsilon$.

If $u = \text{PSqrt}_{p,g}(x)$, then we conclude by Corollary B.18 that

$$\text{prob}(B(x) = \text{PSqrt}_{p,g}(x)) = \text{prob}\left(C_0 > \frac{t}{2}\right) \geq 1 - \frac{4}{4t\varepsilon^2} = 1 - \frac{1}{k}.$$

The case $v = \text{PSqrt}_{p,g}(x)$ follows analogously. This completes the proof of the lemma.   □

We continue with the proof of the theorem. The following algorithm computes the discrete logarithm.

Algorithm 7.10.
   int A(int p, g, x)
    1   $y \leftarrow$ empty word, $k \leftarrow |p|, t \leftarrow k\varepsilon^{-2}$
    2   guess $j \in \{0, \ldots, t-1\}$ satisfying $j\frac{p-1}{t} \leq \text{Log}_{p,g}(x) < (j+1)\frac{p-1}{t}$
    3   $x = xg^{-\lceil j(p-1)/t \rceil}$
    4   for $c \leftarrow 1$ to k do
    5   if $x \in \text{QR}_p$
    6   then $y \leftarrow y\|0$
    7   else $y \leftarrow y\|1$
    8            $x \leftarrow xg^{-1}$
    9      $x \leftarrow B(x)$
  10 return $y + \left\lceil \frac{j(p-1)}{t} \right\rceil$

In algorithm A we use "to guess" as a basic operation. To guess the right alternative means to find out the right choice by computation.

Here, to *guess* the correct j means to carry out lines 3–9 and then test whether $y + \left\lceil \frac{j(p-1)}{t} \right\rceil$ is equal to $\text{Log}_{p,g}(x)$, for $j = 0, 1, \ldots$. The test is done by modular exponentiation. We stop if the test succeeds, i.e., if $\text{Log}_{p,g}(x)$ is computed. We have to consider at most $t = k\varepsilon^{-2} = kP^2(k)$ many intervals. Hence, in this way we get a polynomial algorithm. This notion of "to guess" will also be used in the subsequent sections.

We have $A(p, g, x) = \text{Log}_{p,g}(x)$ if B correctly computes $\text{PSqrt}_{p,g}(x)$ in each iteration of the for loop. Thus, we get

$$\text{prob}(A(p, g, x) = \text{Log}_{p,g}(x)) \geq \left(1 - \frac{1}{k}\right)^k.$$

As $(1 - 1/k)^k$ increases monotonously (it converges to e^{-1}), this implies

$$\text{prob}(A(p, g, x) = \text{Log}_{p,g}(x)) \geq \left(1 - \frac{1}{2}\right)^2 = \frac{1}{4}.$$

By Proposition 5.7, we get, by repeating the computation $A(p, g, x)$ independently, a probabilistic polynomial algorithm $A_2(p, g, x)$ with

$$\text{prob}(A_2(p, g, x) = \text{Log}_{p,g}(x)) > 1 - 2^{-Q(k)}.$$

This concludes the proof of the theorem.   □

Remarks:

1. The expectation is that A will compute $\text{Log}_{p,g}(x)$ after four repetitions (see Lemma B.12). Thus, we expect that we have to call A_1 at most $4k^3\varepsilon^{-4}$ times to compute $\text{Log}_{p,g}(x)$.
2. In [BluMic84], Blum and Micali introduced the idea to reduce the discrete logarithm problem to the problem of computing principal square roots. They developed the techniques we used to prove Theorem 7.7. In that paper they also constructed cryptographically strong pseudo-random bit generators using hard-core bits. They proved that the most-significant bit of the discrete logarithm is unpredictable and achieved as an application the discrete exponential generator (see Section 8.1).

Corollary 7.11. *Let $I = \{(p, g) \mid p \text{ prime}, g \in \mathbb{Z}_p^* \text{ a primitive root}\}$. Provided the discrete logarithm assumption is true,*

$$\text{Msb} := \left(\text{Msb}_p : \mathbb{Z}_{p-1} \longrightarrow \{0, 1\}, x \longmapsto \begin{cases} 0 \text{ if } 0 \leq x < \frac{p-1}{2}, \\ 1 \text{ if } \frac{p-1}{2} \leq x < p - 1 \end{cases}\right)_{(p,g)\in I}$$

is a family of hard-core predicates for the Exp family

$$\text{Exp} = (\text{Exp}_{p,g} : \mathbb{Z}_{p-1} \longrightarrow \mathbb{Z}_p^*, x \longmapsto g^x \bmod p)_{(p,g)\in I}.$$

Proof. Assume Msb is not a family of hard-core predicates for Exp. Then, there is a positive polynomial $P \in \mathbb{Z}[X]$ and an algorithm A_1, such that for infinitely many k

$$\text{prob}(A_1(p, g, g^x) = \text{Msb}_p(x) : (p, g) \xleftarrow{u} I_k, x \xleftarrow{u} \mathbb{Z}_{p-1}) > \frac{1}{2} + \frac{1}{P(k)}.$$

By Proposition 6.17, there are positive polynomials Q, R, such that

$$\text{prob}\left(\left\{(p,g) \in I_k \,\middle|\, \text{prob}(A_1(p,g,g^x) = \text{Msb}_p(x) : x \xleftarrow{u} \mathbb{Z}_{p-1})\right.\right.$$
$$\left.\left. > \frac{1}{2} + \frac{1}{Q(k)}\right\}\right) > \frac{1}{R(k)},$$

for infinitely many k. From the theorem and the remark after Definition 7.2 above, we conclude that there is an algorithm A_2 and a positive polynomial S, such that

$$\text{prob}\left(\left\{(p,g) \in I_k \,\middle|\, \text{prob}(A_2(p,g,g^x) = x : x \xleftarrow{u} \mathbb{Z}_{p-1}) \geq 1 - \frac{1}{S(k)}\right\}\right)$$
$$> \frac{1}{R(k)}$$

for infinitely many k. By Proposition 6.3, there is a positive polynomial T such that

$$\text{prob}(A_2(p,g,g^x) = x : (p,g) \leftarrow I_k, x \xleftarrow{u} \mathbb{Z}_{p-1}) > \frac{1}{T(k)}$$

for infinitely many k, a contradiction to the discrete logarithm assumption (Definition 6.1). $\square$

Remark. Suppose that $p - 1 = 2^t a$, where a is odd. The t least significant bits of x can be easily computed from $g^x = \text{Exp}_{p,g}(x)$ (see Exercise 3 of this chapter). But all the other bits of x are secure bits, i.e., each of them yields a hard-core predicate, as shown by Håstad and Näslund ([HåsNäs98]; [HåsNäs99]).

7.2 Bit Security of the RSA Family

Let $\text{Lsb}(x) := x \bmod 2$ be the least-significant bit of $x \in \mathbb{Z}, x \geq 0$, with respect to the binary encoding of x as an unsigned number. In order to compute $\text{Lsb}(x)$ for an element x in $\mathbb{Z}_n$, we apply Lsb to the representative of x between 0 and $n - 1$.

In this section, we study the RSA function

$$\text{RSA}_{n,e} : \mathbb{Z}_n^* \longrightarrow \mathbb{Z}_n^*, \; x \longmapsto x^e$$

and its inverse $\text{RSA}_{n,e}^{-1}$, for $n = pq$, with p and q odd, distinct primes, and e prime to $\varphi(n)$.

To compute $\text{Lsb}(x)$ from $y = x^e$ is as difficult as to compute x from y. The following Proposition 7.12 and Theorem 7.14 make this statement more precise. The proofs show how to reduce the computation of x from y to the computation of $\text{Lsb}(x)$ from y.

First, we study the deterministic case where $\text{Lsb}(x)$ can be computed from y by a deterministic algorithm in polynomial time.

Proposition 7.12. *Let A_1 be a deterministic polynomial algorithm, such that*

$$A_1(n, e, x^e) = \mathrm{Lsb}(x) \text{ for all } x \in \mathbb{Z}_n^*,$$

where $n := pq$, p and q are odd distinct prime numbers, and e is relatively prime to $\varphi(n)$. Then there is a deterministic polynomial algorithm A_2, such that

$$A_2(n, e, x^e) = x \text{ for all } x \in \mathbb{Z}_n^*.$$

Proof. Let $x \in \mathbb{Z}_n^*$ and $y = x^e$. The basic idea of the inversion algorithm is to compute $a \in \mathbb{Z}_n^*$ and a rational number $u \in \mathbb{Q}, 0 \le u < 1$, with

$$|ax \bmod n - un| < \frac{1}{2}.$$

Then we have $ax \bmod n = \lfloor un + 1/2 \rfloor$, and hence $x = a^{-1} \lfloor un + 1/2 \rfloor \bmod n$. This method to invert the RSA function is called *rational approximation*. We approximate $ax \bmod n$ by the rational number un.

For $z \in \mathbb{Z}$, let $\overline{z} := z \bmod n$. Let 2^{-1} denote the inverse element of 2 mod n in $\mathbb{Z}_n^*$. We start with $u_0 = 0$ and $a_0 = 1$ to get an approximation for $\overline{a_0 x}$ with

$$|\overline{a_0 x} - u_0 n| < n.$$

We define

$$a_t := 2^{-1} a_{t-1} \text{ and}$$

$$u_t := \frac{1}{2}(u_{t-1} + \mathrm{Lsb}(\overline{a_{t-1}x}))$$

(the last computation is done in $\mathbb{Q}$). In each step, we replace a_{t-1} by a_t and u_{t-1} by u_t, and we observe that

$$\overline{a_t x} = \overline{2^{-1} a_{t-1} x} = \begin{cases} \frac{1}{2}\overline{a_{t-1}x} & \text{if } \overline{a_{t-1}x} \text{ is even,} \\ \frac{1}{2}(\overline{a_{t-1}x} + n) & \text{if } \overline{a_{t-1}x} \text{ is odd,} \end{cases}$$

and hence

$$|\overline{a_t x} - u_t n| = \frac{1}{2}|\overline{a_{t-1}x} - u_{t-1}n|.$$

After $r = |n| + 1$ steps, we reach

$$|\overline{a_r x} - u_r n| < \frac{n}{2^r} < \frac{1}{2}.$$

Since $\mathrm{Lsb}(\overline{a_t x}) = A_1(n, e, a_t^e y \bmod n)$, we can decide whether $\overline{a_t x}$ is even without knowing x. Thus, we can compute a_t and u_t in each step, and finally get x. The following algorithm inverts the RSA function.

Algorithm 7.13.

   int $A_2(\text{int } n, e, y)$
    1   $a \leftarrow 1, u \leftarrow 0, k \leftarrow |n|$
    2   for $t \leftarrow 0$ to k do
    3       $u \leftarrow \frac{1}{2}(u + A_1(n, e, a^e y \bmod n))$
    4       $a \leftarrow 2^{-1}a \bmod n$
    5   return $a^{-1} \left\lfloor un + \frac{1}{2} \right\rfloor \bmod n$

This completes the proof of the Proposition. $\qquad\qquad\qquad\qquad$ □

Next we study the probabilistic case. Now the algorithm A_1 does not compute the predicate $\mathrm{Lsb}(x)$ deterministically, but only with a probability slightly better than guessing it at random. Nevertheless, RSA can be inverted with a high probability.

Theorem 7.14. *Let $P, Q \in \mathbb{Z}[X]$ be positive polynomials and A_1 be a probabilistic polynomial algorithm, such that*

$$\mathrm{prob}(A_1(n, e, x^e) = \mathrm{Lsb}(x) : x \xleftarrow{u} \mathbb{Z}_n^*) \geq \frac{1}{2} + \frac{1}{P(k)},$$

where $n := pq$, $k := |n|$, p and q are odd distinct primes and e is relatively prime to $\varphi(n)$. Then there is a probabilistic polynomial algorithm A_2, such that

$$\mathrm{prob}(A_2(n, e, x^e) = x) \geq 1 - 2^{-Q(k)} \text{ for all } x \in \mathbb{Z}_n^*.$$

Proof. Let $y := x^e$ and let $\varepsilon := 1/P(k)$. As in the deterministic case, we use rational approximation to invert the RSA function. We try to approximate $ax \bmod n$ by a rational number un. To invert RSA correctly, we have to compute $\mathrm{Lsb}(ax)$ correctly in each step. However, now we only know that the probability for k correct computations of Lsb is $\geq (1/2 + \varepsilon)^k$, which is exponentially close to 0 and thus too small. In order to increase the probability of success, we develop the algorithm L. The probability of success of L is sufficiently high. This is the statement of the following lemma.

Lemma 7.15. *Under the assumptions of the theorem, there is a probabilistic polynomial algorithm L with the following properties: given $y := x^e$, randomly chosen $a, b \in \mathbb{Z}_n^*$,[2] $\alpha := \mathrm{Lsb}(ax \bmod n)$, $\beta := \mathrm{Lsb}(bx \bmod n)$, $u \in \mathbb{Q}$ with $|ax \bmod n - un| \leq \varepsilon^3 n/8$ and $v \in \mathbb{Q}$ with $|bx \bmod n - vn| \leq \varepsilon n/8$, then L successively computes values $l_t, t = 0, 1, 2, \ldots, k$, such that*

---

[2] Actually we randomly choose $a, b \in \mathbb{Z}_n$. In the rare case that $a, b \notin \mathbb{Z}_n^*$, we can factor n using Euclid's algorithm and compute x from x^e.

$$\text{prob}(l_t = \text{Lsb}(a_t x \bmod n)| \wedge_{j=0}^{t-1} l_j = \text{Lsb}(a_j x \bmod n) : a, b \xleftarrow{u} \mathbb{Z}_n^*) \geq 1 - \frac{1}{2k},$$

where $a_0 := a, a_t := 2^{-1} a_{t-1}$.

Proof (of the Lemma). Let $m := \min(2^t \varepsilon^{-2}, 2k\varepsilon^{-2})$. We may assume that both primes p and q are $> m$. Namely, if one of the primes is $\leq m$, then we may factorize n in polynomial time (and then easily compute the inverse of the RSA function), simply by checking whether one of the polynomially many numbers $\leq m$ divide n.

To compute $\text{Lsb}(a_t x \bmod n)$, L will apply A_1 m times. In each step, a least-significant bit is computed and the return value of L is the more frequently occurring bit. In the theorem, we assume that

$$\text{prob}(A_1(n, e, x^e) = \text{Lsb}(x) : x \xleftarrow{u} \mathbb{Z}_n^*) \geq \frac{1}{2} + \varepsilon.$$

The probability is also taken over the random choice of $x \in \mathbb{Z}_n^*$. Thus, we cannot simply repeat the execution of A_1 with the same input $(a_t x)^e$, but have to modify the input randomly. We use the modifiers $a_t + i a_{t-1} + b$, $i \in \mathbb{Z}, -m/2 \leq i \leq m/2 - 1$, and compute $\text{Lsb}((a_t + i a_{t-1} + b)x \bmod n)$.[3] As we will see below, the assumptions of the lemma guarantee that we can infer $\text{Lsb}(a_t x \bmod n)$ from $\text{Lsb}((a_t + i a_{t-1} + b)x \bmod n)$ with high probability (here sufficiently good rational approximations u_t of $a_t x \bmod n$ and v of $bx \bmod n$ are needed). a and b are chosen independently and at random, because then the modifiers $a_t + i a_{t-1} + b$ are pairwise independent. Then Corollary B.18, which is a consequence of the weak law of large numbers, applies. This implies that the probability of success of L is as large as desired.

We now describe how L works on input of $y = x^e, a, b, \alpha, \beta, u$ and v to compute $l_t, t = 0, 1, \ldots, k$. In its computation, L uses the variable a_t to store the current a_t, and the variable a_{t-1} to store the a_t from the preceding iteration. Analogously, we use variables u_t and u_{t-1}. $u_t n$ is a rational approximation of $a_t x$. We have $u_0 = u$ and $u_t = 1/2 (u_{t-1} + l_{t-1})$.

It is the goal of L to return $l_t = \text{Lsb}(a_t x \bmod n)$ for $t = 0, 1, \ldots, k$. The first iteration $t = 0$ is easy: l_0 is the given α. From now on, the variable α is used to store the last computed l_t.

Before L starts to compute $l_1, l_2, \ldots, l_k$, its variables are initialized by

$$a_{t-1} := a_0 = a, u_{t-1} := u.$$

To compute $l_t, t \geq 1$, L repeats the following subroutine.

---

[3] If $a_t + i a_{t-1} + b \notin \mathbb{Z}_n^*$, we factor n using Euclid's algorithm and compute x from x^e.

Algorithm 7.16.

$\underline{L'(\ )}$

```
1   C_0 ← 0; C_1 ← 0
2   a_t ← 2^{-1} a_{t-1}; u_t ← ½(u_{t-1} + α)
3   for i ← -m/2 to m/2 - 1 do
4       A ← a_t + i a_{t-1} + b
5       W ← ⌊u_t + i u_{t-1} + v⌋
6       B ← (iα + β + W) mod 2
7       if A_1(n, e, A^e y mod n) ⊕ B = 0
8           then C_0 ← C_0 + 1
9           else  C_1 ← C_1 + 1
10      u_{t-1} ← u_t, a_{t-1} ← a_t
11  if C_0 > C_1
12      then α ← 0
13      else  α ← 1
14  return α
```

For $z \in \mathbb{Z}$, we denote by $\bar{z}$ the remainder of z modulo n.

For $i \in \{-m/2, \ldots, m/2 - 1\}$, let

$A_{t,i} := a_t + i a_{t-1} + b,$

$W'_{t,i} := u_t + i u_{t-1} + v, W_{t,i} = \lfloor W'_{t,i} \rfloor,$

$B_{t,i} := (i \cdot \mathrm{Lsb}(\overline{a_{t-1}x}) + \mathrm{Lsb}(\overline{bx}) + \mathrm{Lsb}(W_{t,i})) \bmod 2.$

We want to compute $\mathrm{Lsb}(\overline{a_t x})$ from $\mathrm{Lsb}(\overline{A_{t,i}x})$, $\mathrm{Lsb}(\overline{a_{t-1}x})$ and $\mathrm{Lsb}(\overline{bx})$. For this purpose, let $\lambda_{t,i} := \overline{a_t x} + i \cdot \overline{a_{t-1}x} + \overline{bx} = qn + \overline{A_{t,i}x}$ with $q = \lfloor \lambda_{t,i}/n \rfloor$. Then $\mathrm{Lsb}(\lambda_{t,i}) = (\mathrm{Lsb}(\overline{a_t x}) + i \cdot \mathrm{Lsb}(\overline{a_{t-1}x}) + \mathrm{Lsb}(\overline{bx})) \bmod 2$ and

$$\mathrm{Lsb}(\overline{A_{t,i}x}) = (\mathrm{Lsb}(\lambda_{t,i}) + \mathrm{Lsb}(q)) \bmod 2$$
$$= (\mathrm{Lsb}(\overline{a_t x}) + i \cdot \mathrm{Lsb}(\overline{a_{t-1}x}) + \mathrm{Lsb}(\overline{bx}) + \mathrm{Lsb}(q)) \bmod 2,$$

and we obtain

$$\mathrm{Lsb}(\overline{a_t x}) = (\mathrm{Lsb}(\overline{A_{t,i}x}) + i \cdot \mathrm{Lsb}(\overline{a_{t-1}x}) + \mathrm{Lsb}(\overline{bx}) + \mathrm{Lsb}(q)) \bmod 2.$$

The problem is to get q and $\mathrm{Lsb}(q)$. We will show that $W_{t,i}$ is equal to q with a high probability, and $W_{t,i}$ is easily computed from the rational approximations u_t of $\overline{a_t x}$, u_{t-1} of $\overline{a_{t-1}x}$ and v of $\overline{bx}$. If $W_{t,i} = q$, we have

$$\mathrm{Lsb}(\overline{a_t x}) = \mathrm{Lsb}(\overline{A_{t,i}x}) \oplus B_{t,i}.$$

We assume from now on that L computed the least-significant bit correctly in the preceding steps:

$$\mathrm{Lsb}(\overline{a_j x}) = l_j, 0 \le j \le t - 1.$$

Next, we give a lower bound for the probability that $W_{t,i} = q$. Let $Z = |\lambda_{t,i} - W'_{t,i}n|$.

$$Z = |\overline{a_t x} - u_t n + i(\overline{a_{t-1} x} - u_{t-1} n) + \overline{bx} - vn|$$

$$\leq \left| \frac{1}{2}(\overline{a_{t-1} x} - u_{t-1} n)(1 + 2i) \right| + |\overline{bx} - vn|$$

$$\leq \frac{n}{2^t} \frac{\varepsilon^3}{8} |1 + 2i| + \frac{\varepsilon}{8} n \leq \frac{\varepsilon}{8} n \left(\frac{\varepsilon^2 m}{2^t} + 1 \right) \leq \frac{\varepsilon}{4} n.$$

Note that $|\overline{a_j x} - u_j n| = 1/2 |(\overline{a_{j-1} x} - u_{j-1} n)|$ for $1 \leq j \leq t$ under our assumption $l_j = \mathrm{Lsb}(\overline{a_j x})$, $0 \leq j \leq t - 1$ (see the proof of Proposition 7.12). Moreover, $|1 + 2i| \leq m$, because $-m/2 \leq i \leq m/2 - 1$, and $m = \min(2^t \varepsilon^{-2}, 2k\varepsilon^{-2})$.

Now $W_{t,i} \neq q$ if and only if there is a multiple of n between $\lambda_{t,i}$ and $W'_{t,i} n$. There is no multiple of n between $\lambda_{t,i}$ and $W'_{t,i} n$ if

$$\frac{\varepsilon}{4} n < \overline{\lambda_{t,i}} = \overline{A_{t,i} x} < n - \frac{\varepsilon}{4} n,$$

because $Z \leq \varepsilon/4 \, n$.

Since $a, b \in \mathbb{Z}_n^*$ are selected uniformly and at random, the remainders $\overline{\lambda_{t,i}} = (a_t + i a_{t-1} + b) x \bmod n = ((2^{-1} + i) a_{t-1} + bx) \bmod n$ are also selected uniformly and at random.

This implies

$$\mathrm{prob}(W_{t,i} = q) \geq \mathrm{prob}\left(\frac{\varepsilon}{4} n < \overline{A_{t,i} x} < n - \frac{\varepsilon}{4} n \right) \geq 1 - \frac{\varepsilon}{2}.$$

We are now ready to show

$$\mathrm{prob}(l_t = \mathrm{Lsb}(\overline{a_t x})| \wedge_{j=0}^{t-1} l_j = \mathrm{Lsb}(\overline{a_j x})\}) \geq 1 - \frac{1}{2k}.$$

Let $E_{1,i}$ be the event $A_1(n, e, \overline{A_{t,i}^e y}) = \mathrm{Lsb}(\overline{A_{t,i} x})$ (here recall that $y = x^e$), and let $E_{2,i}$ be the event that $\overline{A_{t,i} x}$ satisfies the condition

$$\frac{\varepsilon}{4} n < \overline{A_{t,i} x} < n - \frac{\varepsilon}{4} n.$$

We have $\mathrm{prob}(E_{1,i}) \geq 1/2 + \varepsilon$ and $\mathrm{prob}(E_{2,i}) = 1 - \varepsilon/2$. We define random variables

$$S_i := \begin{cases} 1 \text{ if } E_{1,i} \text{ and } E_{2,i} \text{ occur,} \\ 0 \text{ otherwise.} \end{cases}$$

We do not err in computing $\mathrm{Lsb}(a_t x)$ in the i-th step of our algorithm L if both events $E_{1,i}$ and $E_{2,i}$ occur, i.e., if $S_i = 1$.

We have (denoting by $\overline{E_{1,i}}$ the complement of event $E_{1,i}$)

$$\mathrm{prob}(S_i = 1) = \mathrm{prob}(E_{1,i} \text{ and } E_{2,i}) \geq \mathrm{prob}(E_{2,i}) - \mathrm{prob}(\overline{E_{1,i}})$$

$$> \left(1 - \frac{\varepsilon}{2} \right) - \left(\frac{1}{2} - \varepsilon \right) = \frac{1}{2} + \frac{\varepsilon}{2}.$$

Let $i \neq j$. The probabilities $\mathrm{prob}(S_i = d)$ and $\mathrm{prob}(S_j = d)$ $(d \in \{0,1\})$ are taken over the random choice of $a, b \in \mathbb{Z}_n^*$ and the coin tosses of $A_1(n, e, \overline{A_{t,i}^e y})$ and $A_1(n, e, \overline{A_{t,j}^e y})$. The elements $a_0 = a$ and b are chosen independently (and uniformly), and we have $(\overline{A_{t,i}}, \overline{A_{t,j}}) = (a_{t-1}, b)\Delta = (2^{-t+1}a, b)\Delta$ with the invertible matrix

$$\Delta = \begin{pmatrix} 2^{-1}+i & 2^{-1}+j \\ 1 & 1 \end{pmatrix}$$

over $\mathbb{Z}_n^*$. Thus, $\overline{A_{t,i}}$ and $\overline{A_{t,j}}$ are also independent. This implies that the events $E_{2,i}$ and $E_{2,j}$ are independent and that the inputs $\overline{A_{t,i}^e y}$ and $\overline{A_{t,j}^e y}$ are independent random elements. Since the coin tosses during an execution of A_1 are independent of all other random events (see Chapter 5), the events $E_{1,i}$ and $E_{1,j}$ are also independent. We see that S_i and S_j are indeed independent. Note that the determinant $i - j$ of Δ is in $\mathbb{Z}_n^*$, since $|i - j| < m < \min\{p, q\}$.

The number of i, $-m/2 \leq i \leq m/2 - 1$, and hence the number of random variables S_i, is m. By Corollary B.18, we conclude

$$\mathrm{prob}\left(\sum_i S_i > \frac{m}{2}\right) \geq 1 - \frac{1}{m\varepsilon^2} \geq 1 - \frac{1}{2k}.$$

Recall that we do not err in computing $\mathrm{Lsb}(a_t x)$ in the i-th step of our algorithm L, if both events $E_{1,i}$ and $E_{2,i}$ occur, i.e., if $S_i = 1$. Thus, we have $C_0 \geq \sum_i S_i$ and hence $\mathrm{prob}(C_0 > C_1) \geq 1 - 1/2k$ if $\mathrm{Lsb}(\overline{a_t x}) = 0$, and $C_1 \geq \sum_i S_i$ and hence $\mathrm{prob}(C_1 > C_0) \geq 1 - 1/2k$ if $\mathrm{Lsb}(\overline{a_t x}) = 1$. Therefore, we have shown that

$$\mathrm{prob}(l_t = \mathrm{Lsb}(\overline{a_t x}) | \wedge_{j=0}^{t-1} l_j = \mathrm{Lsb}(\overline{a_j x})) \geq 1 - \frac{1}{2k}.$$

The proof of the lemma is complete.   □

We continue in the proof of the theorem. The following algorithm A inverts the RSA function by the method of rational approximation. The basic structure of A is the same as that of Algorithm 7.13. Now we call L to compute $\mathrm{Lsb}(ax)$. Therefore, we must meet the assumptions of Lemma 7.15. This is done in lines 1–4.

Algorithm 7.17.

int $A(\text{int } n, e, y)$

1   select $a, b \in \mathbb{Z}_n^*$ at random
2   guess $u, v \in \mathbb{Q} \cap [0, 1[$ satisfying
3   $|ax \bmod n - un| \leq \frac{\varepsilon^3 n}{8}, |bx \bmod n - vn| \leq \frac{\varepsilon n}{8}$
4   guess $\alpha \leftarrow \text{Lsb}(ax \bmod n), \text{guess } \beta \leftarrow \text{Lsb}(bx \bmod n)$
5   Compute $l_0, l_1, \ldots, l_k$ by L
6   for $t \leftarrow 0$ to k do
7       $u \leftarrow \frac{1}{2}(u + l_t)$
8       $a \leftarrow 2^{-1}a \bmod n$
9   return $a^{-1} \lfloor un + \frac{1}{2} \rfloor \bmod n$

Algorithm L computes $l_0, \ldots, l_k$ in advance. Lines 7 and 8 of A also appear in L. In a real and efficient implementation, it is possible to avoid this redundancy.

As above in Algorithm 7.10, we can "guess" the right alternative. This means we can find out the right alternative in polynomial time. There are only a polynomial number of alternatives, and both the computation for each alternative as well as checking the result can be done in polynomial time. In order to guess u or v, we have to consider $8/\varepsilon^3 = 8P(k)^3$ and $8/\varepsilon = 8P(k)$ many intervals. There are only two alternatives for $\text{Lsb}(\overline{ax})$ and $\text{Lsb}(\overline{bx})$.

$A(n, e, y) = \text{RSA}_{n,e}^{-1}(y) = x$ for $y = x^e$ if L correctly computes $l_t = \text{Lsb}(\overline{a_t x})$ for $t = 1, \ldots, k$. Thus, we have

$$\text{prob}(A(n, e, y) = x) \geq \left(1 - \frac{1}{2k}\right)^k.$$

Since $(1 - 1/2k)^k$ increases monotonously (converging to $e^{-1/2}$), we conclude

$$\text{prob}(A(n, e, y) = x) \geq \frac{1}{2}.$$

Repeating the computation $A(n, e, y)$ independently, we get, by Proposition 5.7, a probabilistic polynomial algorithm $A_2(n, e, y)$ with

$$\text{prob}(A_2(n, e, y) = \text{RSA}_{n,e}^{-1}(y)) \geq 1 - 2^{-Q(k)},$$

and Theorem 7.14 is proven.   □

Remarks:

1. The expectation is that A will compute $\text{RSA}_{n,e}^{-1}(y)$ after two repetitions (see Lemma B.12). The input of A_1 does not depend on the guessed elements u, v, α and β (it only depends on a and b). Thus, we can also use the return values of A_1, computed for the first guess of u, v, α and β, for all subsequent guesses. Then we expect that we have to call A_1 at most $4k^2\varepsilon^{-2}$ times to compute $\text{RSA}_{n,e}^{-1}(y)$.

2. The bit security of the RSA family was first studied in [GolMicTon82], in which a method for inverting RSA by guessing the least-significant bit was introduced (see Exercise 11).

The problem of inverting RSA, if the least-significant bit is predicted only with probability $\geq 1/2 + 1/P(k)$, is studied in [SchnAle84], [VazVaz84], [AleChoGolSch88] and [FisSch2000]. The technique we used to prove Theorem 7.14 is from [FisSch2000].

Corollary 7.18. *Let*
$$I := \{(n, e) \mid n = pq, \ p \text{ and } q \text{ odd distinct primes}, \ |p| = |q|, \ e \text{ prime to } \varphi(n)\}.$$
Provided the RSA assumption is true, then
$$\text{Lsb} = (\text{Lsb}_{n,e} : \mathbb{Z}_n^* \longrightarrow \{0, 1\}, \ x \longmapsto \text{Lsb}(x))_{(n,e) \in I}$$
is a family of hard-core predicates for the RSA family
$$\text{RSA} = (\text{RSA}_{n,e} : \mathbb{Z}_n^* \longrightarrow \mathbb{Z}_n^*, \ x \longmapsto x^e)_{(n,e) \in I}.$$

Proof. The proof is analogous to the proof of Corollary 7.11.   □

Remark. Håstad and Näslund have shown that all the plaintext bits are secure bits of the RSA function, i.e., each of them yields a hard-core predicate ([HåsNäs98]; [HåsNäs99]).

7.3 Bit Security of the Square Family

Let $n := pq$, with p and q distinct primes, and $p, q \equiv 3 \bmod 4$. We consider the (bijective) modular squaring function
$$\text{Square}_n : \text{QR}_n \longrightarrow \text{QR}_n, \ x \longmapsto x^2$$
and its inverse, the modular square root function
$$\text{Sqrt}_n : \text{QR}_n \longrightarrow \text{QR}_n, \ y \longmapsto \text{Sqrt}_n(y)$$
(see Section 6.5). The computation of $\text{Sqrt}_n(y)$ can be reduced to the computation of the least-significant bit $\text{Lsb}(\text{Sqrt}_n(y))$ of the square root. This is shown in Proposition 7.19 and Theorem 7.22. In the proposition, the algorithm that computes the least-significant bit is assumed to be deterministic polynomial. Then the algorithm which we obtain by the reduction is also deterministic polynomial. It computes $\text{Sqrt}_n(y)$ for all $y \in \text{QR}_n$ (under the additional assumption $n \equiv 1 \bmod 8$).

Proposition 7.19. *Let A_1 be a deterministic polynomial algorithm, such that*
$$A_1(n, x^2) = \text{Lsb}(x) \text{ for all } x \in \text{QR}_n,$$
where $n = pq$, p and q are distinct primes, $p, q \equiv 3 \bmod 4$ and $n \equiv 1 \bmod 8$. Then there exists a deterministic polynomial algorithm A_2, such that
$$A_2(n, y) = \text{Sqrt}_n(y) \text{ for all } y \in \text{QR}_n.$$

Proof. As in the RSA case, we use rational approximation to invert the Square function. Let $y = x^2, x \in QR_n$. Since n is assumed to be $\equiv 1 \bmod 8$, either $2 \in QR_n$ or $-2 \in QR_n$ (see the remark following this proof).

First, let $2 \in QR_n$. Then $2^{-1} \in QR_n$. We define $a_0 := 1$, $a_t := 2^{-1}a_{t-1} \bmod n$ for $t \geq 1$, and $u_0 := 1$, $u_t := 1/2\,(u_{t-1} + \mathrm{Lsb}(a_{t-1}x \bmod n))$ for $t \geq 1$. Since $2^{-1} \in QR_n$, we have $a_t \in QR_n$ and hence $a_t x \in QR_n$ for all $t \geq 1$. Thus, $\mathrm{Sqrt}_n(a_t^2 x^2) = a_t x$, and hence we can compute $\mathrm{Lsb}(a_t x \bmod n) = A_1(n, a_t^2 x^2) = A_1(n, a_t^2 y)$ by A_1 for all $t \geq 1$. The rational approximation works as in the RSA case.

Now, let $-2 \in QR_n$. Then $-2^{-1} \in QR_n$. We modify the method of rational approximation and define $a_0 = 1$, $a_t = -2^{-1}a_{t-1} \bmod n$ for $t \geq 1$, and $u_0 = 1$, $u_t = 1/2\,(2 - \mathrm{Lsb}(a_{t-1}x \bmod n) - u_{t-1})$ for $t \geq 1$. Then, we get

$$|a_t x \bmod n - u_t n| = \frac{1}{2}\,|(a_{t-1}x \bmod n - u_{t-1}n)|,$$

because

$$a_t x \bmod n = -2^{-1}a_{t-1}x \bmod n = 2^{-1}(n - a_{t-1}x \bmod n)$$

$$= \begin{cases} \frac{1}{2}(n - a_{t-1}x \bmod n) & \text{if } a_{t-1}x \bmod n \text{ is odd,} \\ \frac{1}{2}(n - a_{t-1}x \bmod n + n) & \text{otherwise.} \end{cases}$$

After $r = |n| + 1$ steps we reach

$$|a_r x - u_r n| \leq \frac{n}{2^r} < \frac{1}{2}.$$

Since $-2^{-1} \in QR_n$, we have $a_t \in QR_n$ and hence $a_t x \in QR_n$ for $t \geq 1$. Thus, $\mathrm{Sqrt}_n(a_t^2 x^2) = a_t x$, and hence we can compute $\mathrm{Lsb}(a_t x \bmod n) = A_1(n, a_t^2 x^2) = A_1(n, a_t^2 y)$ by A_1 for all $t \geq 1$. $\square$

Remarks:

1. As $p, q \equiv 3 \bmod 4$ is assumed, p and q are $\equiv 3 \bmod 8$ or $\equiv -1 \bmod 8$, and hence either $n \equiv 1 \bmod 8$ or $n \equiv 5 \bmod 8$. We do not consider the case $n \equiv 5 \bmod 8$ in the proposition.

2. The proof actually works if we have $2 \in QR_n$ or $-2 \in QR_n$. This is equivalent to $n \equiv 1 \bmod 8$, as follows from Theorem A.53. We have $2 \in QR_n$ if and only if $2 \in QR_p$ and $2 \in QR_q$, and this in turn is equivalent to $p \equiv q \equiv -1 \bmod 8$ (by Theorem A.53). On the other hand, $-2 \in QR_n$ if and only if $-2 \in QR_p$ and $-2 \in QR_q$, and this in turn is equivalent to $p \equiv q \equiv 3 \bmod 8$ (by Theorem A.53).

Studying the Square function, we face, in the probabilistic case, an additional difficulty compared to the reduction in the RSA case. Membership in the domain of the Square function – the set QR_n of quadratic residues – is not efficiently decidable without knowing the factorization of n. To overcome this difficulty, we develop a probabilistic polynomial reduction of the quadratic residuosity property to the predicate defined by the least-significant bit of Sqrt. Then the same reduction as in the RSA case also works for the Sqrt function.

Let $J_n^{+1} = \{x \in \mathbb{Z}_n^* \mid \left(\frac{x}{n}\right) = +1\}$. The predicate

$$\mathrm{PQR}_n : J_n^{+1} \longrightarrow \{0,1\}, \mathrm{PQR}_n(x) = \begin{cases} 1 \text{ if } x \in \mathrm{QR}_n, \\ 0 \text{ otherwise,} \end{cases}$$

is believed to be a trapdoor predicate (see Definition 6.11). In Proposition 7.20, we show how to reduce the computation of $\mathrm{PQR}_n(x)$ to the computation of the least-significant bit of a $\mathrm{Sqrt}_n(x)$.

Proposition 7.20. *Let $P, Q \in \mathbb{Z}[X]$ be positive polynomials, and let A_1 be a probabilistic polynomial algorithm, such that*

$$\mathrm{prob}(A_1(n, x^2) = \mathrm{Lsb}(x) : x \xleftarrow{u} \mathrm{QR}_n) \geq \frac{1}{2} + \frac{1}{P(k)},$$

where $n = pq$, p and q are distinct primes, $p, q \equiv 3 \bmod 4$, and $k = |n|$. Then there exists a probabilistic polynomial algorithm A_2, such that

$$\mathrm{prob}(A_2(n, x) = \mathrm{PQR}_n(x)) \geq 1 - \frac{1}{Q(k)} \text{ for all } x \in J_n^{+1}.$$

Proof. Let $x \in J_n^{+1}$. If $x \in \mathrm{QR}_n$, then $x = \mathrm{Sqrt}_n(x^2)$ and therefore $\mathrm{Lsb}(x) = \mathrm{Lsb}(\mathrm{Sqrt}_n(x^2))$. If $x \notin \mathrm{QR}_n$, then $-x \bmod n = n - x = \mathrm{Sqrt}_n(x^2)$ and $\mathrm{Lsb}(x) \neq \mathrm{Lsb}(\mathrm{Sqrt}_n(x^2))$ (note that $-1 \in J_n^{+1} \setminus \mathrm{QR}_n$ for $p, q \equiv 3 \bmod 4$, by Theorem A.53). Consequently, we get

$$\mathrm{PQR}_n(x) = \mathrm{Lsb}(x) \oplus \mathrm{Lsb}(\mathrm{Sqrt}_n(x^2)) \oplus 1.$$

Since for each $y \in \mathrm{QR}_n$ there are exactly two elements $x \in J_n^{+1}$ with $x^2 = y$ and because $|J_n^{+1}| = 2|\mathrm{QR}_n|$, we conclude

$$\mathrm{prob}(A_1(n, x^2) = \mathrm{Lsb}(\mathrm{Sqrt}_n(x^2)) : x \xleftarrow{u} J_n^{+1})$$
$$= \mathrm{prob}(A_1(n, y) = \mathrm{Lsb}(\mathrm{Sqrt}_n(y)) : y \xleftarrow{u} \mathrm{QR}_n).$$

Hence, we get

$$\mathrm{prob}(\mathrm{PQR}_n(x) = A_1(n, x^2) \oplus \mathrm{Lsb}(x) \oplus 1 : x \xleftarrow{u} J_n^{+1}) \geq \frac{1}{2} + \frac{1}{P(k)}.$$

We construct an algorithm A_2, such that for every $x \in J_n^{+1}$

$$\mathrm{prob}(A_2(n, x) = \mathrm{PQR}_n(x)) \geq 1 - \frac{1}{Q(k)}.$$

Algorithm 7.21.
int $A_2(\text{int } n, x)$

1    $c \leftarrow 0, l \leftarrow \left\lceil \frac{Q(k)P^2(k)}{4} \right\rceil$

2   for $i \leftarrow 1$ to l do

3   select $r \in \text{QR}_n$ at random

4            $c \leftarrow c + (A_1(n, (rx)^2 \bmod n) \oplus \text{Lsb}(rx \bmod n) \oplus 1)$

5   if $c > \frac{l}{2}$

6   then return 1

7   else return 0

Let $x \in J_n^{+1}$. $\text{PQR}_n(x) = \text{PQR}_n(rx)$ is computed l times by applying A_1 to the l independent random inputs rx. We compute $\text{PQR}_n(rx)$ in each step with a probability $> 1/2$. The weak law of large numbers guarantees that, for sufficiently large l, we can compute $\text{PQR}_n(x)$ by the majority of the results, with a high probability. More precisely, let the random variable S_i, $i = 1, \ldots, l$, be defined by

$$S_i = \begin{cases} 1 \text{ if } \text{PQR}_n(rx) = A_1(n, (rx)^2) \oplus \text{Lsb}(rx) \oplus 1, \\ 0 \text{ otherwise,} \end{cases}$$

with $r \in \text{QR}_n$ randomly chosen, as in the algorithm. Then we have $\text{prob}(S_i = 1) \geq 1/2 + 1/P(k)$. The random variables S_i, $i = 1, \ldots, l$, are independent. If $\text{PQR}_n(x) = 1$, we get by Corollary B.18

$$\text{prob}(c > \frac{l}{2}) \geq 1 - \frac{P^2(k)}{4l} \geq 1 - \frac{1}{Q(k)} .$$

The case $\text{PQR}_n(x) = 0$ follows analogously. Thus, we have shown

$$\text{prob}(A_2(n, x) = \text{PQR}_n(x)) \geq 1 - \frac{1}{Q(k)} ,$$

as desired.   □

Theorem 7.22. *Let $P, Q \in \mathbb{Z}[X]$ be positive polynomials and A_1 be a probabilistic polynomial algorithm, such that*

$$\text{prob}(A_1(n, x^2) = \text{Lsb}(x) : x \overset{u}{\leftarrow} \text{QR}_n) \geq \frac{1}{2} + \frac{1}{P(k)} ,$$

where $n := pq$, p and q are distinct primes, $p, q \equiv 3 \bmod 4$, and $k := |n|$. Then there is a probabilistic polynomial algorithm A_2, such that

$$\text{prob}(A_2(n, x) = \text{Sqrt}_n(x)) \geq 1 - 2^{-Q(k)} \text{ for all } x \in \text{QR}_n .$$

Proof. The proof runs in the same way as the proof of Theorem 7.14. We only describe the differences to this proof. Here, the algorithm A_1 is only applicable to quadratic residues. However, it is easy to compute $\left(\frac{x}{n}\right)$ for $x \in \mathbb{Z}_n^*$, and we can use algorithm A_2 from Proposition 7.20 to check whether a given element $x \in J_n^{+1}$ is a quadratic residue. Assume that $\mathrm{prob}(A_2(n,x) = \mathrm{PQR}_n(x)) \geq 1 - 1/P^2(k)$.

If $p, q \equiv 3 \bmod 4$, we have $-1 \notin \mathrm{QR}_n$ (see Theorem A.53). Therefore, either a or $-a \in \mathrm{QR}_n$ for $\left(\frac{a}{n}\right) = 1$. We are looking for m multipliers $\tilde{a}$ of the form $a_t + i a_{t-1} + b$ with $\left(\frac{\tilde{a}}{n}\right) = 1$, where $m := \min(2^t \varepsilon^{-2}, 2k\varepsilon^{-2})$. If $\tilde{a} \in \mathrm{QR}_n$, $\mathrm{Lsb}(\tilde{a}x)$ can be computed with algorithm A_1, and if $-\tilde{a} = n - \tilde{a} \in \mathrm{QR}_n$, $\mathrm{Lsb}(\tilde{a}x) = 1 - \mathrm{Lsb}(-\tilde{a}x)$ and $\mathrm{Lsb}(-\tilde{a}x)$ can be computed with algorithm A_1. $\mathrm{Lsb}(\tilde{a}x)$ is correctly computed if A_1 correctly computes the predicate Lsb and A_2 from Proposition 7.20 correctly computes the predicate PQR_n. Both events are independent. Thus $\mathrm{Lsb}(\tilde{a}x)$ is computed correctly with a probability $> \left(1/2 + 1/P(k)\right)\left(1 - 1/P^2(k)\right) > \left(1/2 + 1/2P(k)\right)$. Thus we set $\varepsilon = 1/2P(k)$.

With i varying in an interval, the fraction of the multipliers $a_t + i a_{t-1} + b$ which are in J_n^{+1} differs from $1/2$ only negligibly, because J_n^{+1} is nearly uniformly distributed in $\mathbb{Z}_n^*$ (see [Peralta92]).

We double the range for i, take $i \in [-m, m-1]$, and halve the distances of the initial rational approximations:

$$|ax \bmod n - un| \leq \frac{\varepsilon^3 n}{16} \text{ and } |bx \bmod n - vn| \leq \frac{\varepsilon n}{16}.$$

Now we obtain the same estimates as in the proof of Theorem 7.14.   □

Corollary 7.23. *Let $I := \{n \mid n = pq,\ p \text{ and } q \text{ distinct primes},\ |p| = |q|, p, q \equiv 3 \bmod 4\}$. Provided that the factorization assumption is true,*

$$\mathrm{QRLsb} = (\mathrm{QRLsb}_n : \mathrm{QR}_n \longrightarrow \{0,1\},\ x \longmapsto \mathrm{Lsb}(x))_{n \in I}$$

is a family of hard-core predicates for

$$\mathrm{Square} = \left(\mathrm{Square}_n : \mathrm{QR}_n \longrightarrow \mathrm{QR}_n,\ x \longmapsto x^2\right)_{n \in I}.$$

Proof. The proof is analogous to the proof of Corollary 7.11. Observe that the ability to compute square roots modulo n is equivalent to the ability to compute the prime factors of n (Proposition A.64).   □

Exercises

1. Compute $\mathrm{Log}_{p,g}(17)$ using Algorithm 7.6, for $p := 19$ and $g := 2$ (note that 2, 4 and 13 are principal square roots).

2. Let p be an odd prime number, and g be a primitive root in $\mathbb{Z}_p^*$. For $y := \mathrm{Exp}_{p,g}(x)$, we have

$$B_{p,g}(y) = 0 \text{ if and only if } 0 \le x < \frac{p-1}{2} ,$$

$$B_{p,g}(y^2) = 0 \text{ if and only if } 0 \le x < \frac{p-1}{4} \text{ and } \frac{p-1}{2} \le x < \frac{3(p-1)}{4} ,$$

and so on. Let A_1 be a a deterministic polynomial algorithm such that $A_1(p,g,y) = B_{p,g}(y)$ for all $y \in \mathbb{Z}_p^*$.
By using a binary search technique, prove that there is a deterministic polynomial algorithm A_2 such that

$$A_2(p,g,y) = \mathrm{Log}_{p,g}(y)$$

for all $y \in \mathbb{Z}_p^*$.

3. Let p be an odd prime number. Suppose that $p-1 = 2^t a$, where a is odd. Let g be a primitive root in $\mathbb{Z}_p^*$:
   a. Show how the t least-significant bits (bits at the positions 0 to $t-1$) of $x \in \mathbb{Z}_{p-1}$ can be easily computed from $g^x = \mathrm{Exp}_{p,g}(x)$.
   b. Denote by $\mathrm{Lsb}_t(x)$ the t-th least-significant bit of x (bit at position t counted from the right, beginning with 0). Let A_1 be a deterministic polynomial algorithm such that

$$A_1(p,g,g^x) = \mathrm{Lsb}_t(x)$$

   for all $x \in \mathbb{Z}_{p-1}$. By using A_1, construct a deterministic polynomial algorithm A_2 such that

$$A_2(p,g,y) = \mathrm{Log}_{p,g}(y)$$

   for all $y \in \mathbb{Z}_p^*$. (Here assume that a deterministic algorithm for computing square roots exists.)
   c. Show that Lsb_t yields a hard-core predicate for the Exp family.

4. As in the preceding exercise, Lsb_j denotes the j-th least-significant bit.
   a. Let A_1 be a deterministic polynomial algorithm such that for all $x \in \mathbb{Z}_{p-1}$

$$A_1(p,g,g^x,\mathrm{Lsb}_t(x),\ldots,\mathrm{Lsb}_{t+j-1}(x)) = \mathrm{Lsb}_{t+j}(x),$$

   where p is an odd prime, $g \in \mathbb{Z}_p^*$ is a primitive root, $p-1 = 2^t a$, a is odd, $k = |p|$ and $j \in \{0,\ldots,\lfloor \log_2(k) \rfloor\}$.
   Construct a deterministic polynomial algorithm A_2 such that

$$A_2(p, g, y) = \mathrm{Log}_{p,g}(y)$$

for all $y \in \mathbb{Z}_p^*$.

b. Let $P, Q \in \mathbb{Z}[X]$ be positive polynomials and A_1 be a probabilistic polynomial algorithm such that

$$\mathrm{prob}(A_1(p, g, g^x, \mathrm{Lsb}_t(x), \ldots, \mathrm{Lsb}_{t+j-1}(x))$$
$$= \mathrm{Lsb}_{t+j}(x) : x \overset{u}{\leftarrow} \mathbb{Z}_{p-1}) \geq \frac{1}{2} + \frac{1}{P(k)},$$

where p is a an odd prime, $g \in \mathbb{Z}_p^*$ is a primitive root, $p - 1 = 2^t a$, a is odd, $k = |p|$ and $j \in \{0, \ldots, \lfloor \log_2(k) \rfloor\}$.

Construct a probabilistic polynomial algorithm A_2 such that

$$\mathrm{prob}(A_2(p, g, y) = \mathrm{Log}_{p,g}(y)) \geq 1 - 2^{-Q(k)}$$

for all $y \in \mathbb{Z}_p^*$.

5. Let $I := \{(p, g) \mid p \text{ an odd prime}, g \in \mathbb{Z}_p^* \text{ a primitive root}\}$ and $I_k := \{(p, g) \in I \mid |p| = k\}$. Assume that the discrete logarithm assumption (see Definition 6.1) is true.

Show that for every probabilistic polynomial algorithm A, with inputs $p, g, y, b_0, \ldots, b_{j-1}$, $1 \leq j \leq \lfloor \log_2(k) \rfloor$, and for every positive polynomial P, there is a $k_0 \in \mathbb{N}$ such that

$$\mathrm{prob}(A(p, g, g^x, \mathrm{Lsb}_t(x), \ldots, \mathrm{Lsb}_{t+j-1}(x))$$
$$= \mathrm{Lsb}_{t+j}(x) : (p, g) \overset{u}{\leftarrow} I_k, x \overset{u}{\leftarrow} \mathbb{Z}_{p-1}) \leq \frac{1}{2} + \frac{1}{P(k)}$$

for $k \geq k_0$ and for all $j \in \{0, \ldots, \lfloor \log_2(k) \rfloor\}$. t is defined by $p - 1 = 2^t a$, with a odd.

In particular, the predicates $\mathrm{Lsb}_{t+j}, 0 \leq j \leq \lfloor \log_2(k) \rfloor$, are hard-core predicates for Exp.

6. Compute the rational approximation (a, u) for $13 \in \mathbb{Z}_{29}$.

7. Let $p := 17, q := 23, n := pq$ and $e := 3$. List the least-significant bits that A_1 will return if you compute $\mathrm{RSA}_{n,e}^{-1}(49)$ using Algorithm 7.12 (note that $49 = 196^3 \bmod n$).

8. Let $n := pq$, with p and q distinct primes and e relatively prime to $\varphi(n)$, $x \in \mathbb{Z}_n^*$ and $y := \mathrm{RSA}_{n,e}(x) = x^e \bmod n$. Msb is defined analogously to Definition 7.1. Show that you can compute $\mathrm{Msb}(x)$ from y if and only if you can compute $\mathrm{Lsb}(x)$ from y.

9. Show that the most-significant bit of x is a hard-core predicate for the RSA family, provided that the RSA assumption is true.

10. Use the most significant bit and prove Proposition 7.12 using a binary search technique (analogous to Exercise 2).

11. RSA inversion by binary division ([GolMicTon82]). Let $y := \mathrm{RSA}_{n,e}(x) = x^e \bmod n$. Let A be an algorithm that on input y outputs $\mathrm{Lsb}(x)$. Let $k := |n|$, and let $2^{-e} \in \mathbb{Z}_n^*$ be the inverse of $2^e \in \mathbb{Z}_n^*$. Compute x from y by using the bit vector $(b_{k-1}, \ldots, b_0)$, defined by

$$y_0 = y,$$
$$b_0 = A(y_0),$$
$$y_i = \begin{cases} y_{i-1}2^{-e} \bmod n & \text{if } b_{i-1} = 0, \\ (n - y_{i-1})2^{-e} \bmod n & \text{otherwise,} \end{cases}$$
$$b_i = A(y_i), \text{ for } 1 \le i \le k.$$

Describe the algorithm inverting RSA and show that it really does invert RSA.

12. a. Let A_1 be a deterministic polynomial algorithm such that

$$A_1(n, e, x^e, \mathrm{Lsb}_0(x), \ldots, \mathrm{Lsb}_{j-1}(x)) = \mathrm{Lsb}_j(x)$$

for all $x \in \mathbb{Z}_n^*$, where $n := pq$, p and q are odd distinct primes, e is relatively prime to $\varphi(n)$, $k := |n|$ and $j \in \{0, \ldots, \lfloor \log_2(k) \rfloor\}$. Construct a deterministic polynomial algorithm A_2 such that

$$A_2(n, e, x^e) = x$$

for all $x \in \mathbb{Z}_n^*$.

   b. Let $P, Q \in \mathbb{Z}[X]$ be positive polynomials and A_1 be a probabilistic polynomial algorithm, such that

$$\mathrm{prob}(A_1(n, e, x^e, \mathrm{Lsb}_0(x), \ldots, \mathrm{Lsb}_{j-1}(x))$$
$$= \mathrm{Lsb}_j(x) : x \overset{u}{\leftarrow} \mathbb{Z}_n^*) \ge \frac{1}{2} + \frac{1}{P(k)},$$

where $n := pq$, p and q are odd distinct primes, e is relatively prime to $\varphi(n)$, $k := |n|$ and $j \in \{0, \ldots, \lfloor \log_2(k) \rfloor\}$. Construct a probabilistic polynomial algorithm A_2 such that

$$\mathrm{prob}(A_2(n, e, x^e) = x \ge 1 - 2^{-Q(k)}$$

for all $x \in \mathbb{Z}_n^*$.

13. Let $I := \{(n, e) \mid n = pq, p \ne q \text{ prime numbers}, |p| = |q|, e < \varphi(n), e \text{ prime to } \varphi(n)\}$ and $I_k := \{(n, e) \in I \mid n = pq, |p| = |q| = k\}$. Assume that the RSA assumption (see Definition 6.7) is true.
Show that for every probabilistic polynomial algorithm A, with inputs $n, e, y, b_0, \ldots, b_{j-1}, 0 \le j \le \lfloor \log_2(|n|) \rfloor$, and every positive polynomial P, there is a $k_0 \in \mathbb{N}$ such that

$$\text{prob}(A(n, e, x^e, \text{Lsb}_0(x), \dots, \text{Lsb}_{j-1}(x))$$

$$= \text{Lsb}_j(x) : (n, e) \xleftarrow{u} I_k, x \xleftarrow{u} \mathbb{Z}_n^*) \leq \frac{1}{2} + \frac{1}{P(k)}$$

for $k \geq k_0$ and for all $j \in \{0, \dots, \lfloor \log_2(|n|) \rfloor\}$. In particular, the predicates $\text{Lsb}_j, 0 \leq j \leq \lfloor \log_2(|n|) \rfloor$, are hard-core predicates for RSA.

8. One-Way Functions and Pseudorandomness

There is a close relationship between encryption and randomness. The security of encryption algorithms usually depends on the random choice of keys and bit sequences. A famous example is Shannon's result. Ciphers with perfect secrecy require randomly chosen key strings that are of the same length as the encrypted message. In Chapter 9, we will study the classical Shannon approach to provable security, together with more recent notions of security. One main problem is that truly random bit sequences of sufficient length are not available in most practical situations. Therefore, one works with pseudorandom bit sequences. They appear to be random, but actually they are generated by an algorithm. Such algorithms are called *pseudorandom bit generators*. They output, given a short random input value (called the *seed*), a long pseudorandom bit sequence. Classical techniques for the generation of pseudorandom bits or numbers (see [Knuth98]) yield well-distributed sequences. Therefore, they are well-suited for Monte Carlo simulations. However, they are often cryptographically insecure. For example, in linear congruential pseudorandom number generators or linear feedback shift registers (see, e.g., [MenOorVan96]), the secret parameters and hence the complete pseudorandom sequence can be efficiently computed from a small number of outputs.

It turns out that computationally perfect (hence cryptographically secure) pseudorandom bit generators can be derived from one-way permutations with hard-core predicates. We will discuss this close relation in this chapter. The pseudorandom bit generators G studied are families of functions whose indexes vary over a set of keys. Before we can use G, we have to select such a key with a sufficiently large security parameter.

Of course, even applying a perfect pseudorandom generator requires starting with a truly random seed. Thus, in any case you need some "natural" source of random bits, such as independent fair coin tosses (see Chapter 5).

8.1 Computationally Perfect Pseudorandom Bit Generators

In the definition of pseudorandom generators, we use the notion of polynomial functions.

Definition 8.1. We call a function $l : \mathbb{N} \longrightarrow \mathbb{N}$ a *polynomial function* if it is computable by a polynomial algorithm and if there is a polynomial $Q \in \mathbb{Z}[X]$, such that $l(k) \leq Q(k)$ for all $k \in \mathbb{N}$.

Now we are ready to define pseudorandom bit generators.

Definition 8.2. Let $I = (I_k)_{k \in \mathbb{N}}$ be a key set with security parameter k, and let K be a probabilistic polynomial sampling algorithm for I, which on input 1^k outputs an $i \in I_k$. Let l be a polynomial function.

A *pseudorandom bit generator* with *key generator* K and *stretch function* l is a family $G = (G_i)_{i \in I}$ of functions

$$G_i : X_i \longrightarrow \{0,1\}^{l(k)} \qquad (i \in I_k),$$

such that

1. G is computable by a deterministic polynomial algorithm G:
   $G(i, x) = G_i(x)$ for all $i \in I$ and $x \in X_i$.
2. There is a uniform sampling algorithm S for $X := (X_i)_{i \in I}$, which on input $i \in I$ outputs $x \in X_i$.

The generator G is *computationally perfect* (or *cryptographically secure*), if the pseudorandom sequences generated by G cannot be distinguished from true random sequences by an efficient algorithm; i.e., for every positive polynomial $P \in \mathbb{Z}[X]$ and every probabilistic polynomial algorithm A with inputs $i \in I_k, z \in \{0,1\}^{l(k)}$ and output in $\{0,1\}$, there is a $k_0 \in \mathbb{N}$ such that for all $k \geq k_0$

$$|\operatorname{prob}(A(i, z) = 1 : i \leftarrow K(1^k), z \xleftarrow{u} \{0,1\}^{l(k)})$$
$$- \operatorname{prob}(A(i, G_i(x)) = 1 : i \leftarrow K(1^k), x \xleftarrow{u} X_i)| \leq \frac{1}{P(k)}.$$

Remarks:

1. The probabilistic polynomial algorithm A in the definition may be considered as a statistical test trying to compute some property which distinguishes truly random sequences in $\{0,1\}^{l(k)}$ from the pseudorandom sequences generated by G. Classical statistical tests for randomness, such as the Chi-square test ([Knuth98], Chapter 3), can be considered as such tests and can be implemented as polynomial algorithms. Thus, "computationally perfect" means that no statistical test – which can be implemented as a probabilistic algorithm with polynomial running time – can significantly distinguish between true random sequences and sequences generated by G, provided a sufficiently large key i is chosen.
2. By condition 2, we can randomly generate uniformly distributed seeds $x \in X_i$ for the generator G. We could (but do not) generalize the definition and allow non-uniform seed generators S (see the analogous remark after

the formal definition of one-way functions – Definition 6.12, remark 2). The constructions and proofs given below also work in this more general case.

3. We study only computationally perfect pseudorandom generators. Therefore, we do not specify any other level of pseudorandomness for the sequences generated by G. In the literature, the term "pseudorandom generator" is sometimes only used for generators that are computationally perfect (see, e.g., [Goldreich99]; [Goldreich01]).

4. Our definition of computationally perfect pseudorandom generators is a definition in the "public-key model". The key i is an input to the statistical tests A (which are the adversaries). Thus, the key i is assumed to be public and available to everyone. Definition 8.2 can be adapted to the "private-key model", where the selected key i is kept secret, and hence is not known to the adversaries. The input i of the statistical tests A has to be omitted. We only discuss the public-key model.

5. *Admissible key generators* can be analogously defined as in Definition 6.13.

6. The probability in the definition is also taken over the random generation of a key i, with a given security parameter k. Even for very large k, there may be keys i such that A can successfully distinguish pseudorandom from truly random sequences. However, when generating a key i by K, the probability of obtaining one for which A has a significant chance of success is negligibly small (see Exercise 8 in Chapter 6).

7. As is common, we require that the pseudorandom generator can be implemented by a deterministic algorithm. However, if the sequences can be computed probabilistically in polynomial time, then we can also compute them almost deterministically: for every positive polynomial Q, there is a probabilistic polynomial algorithm $G(i, x)$ with

$$\text{prob}(G(i, x) = G_i(x)) \geq 1 - 2^{-Q(k)} \quad (i \in I_k)$$

(see Proposition 5.6 and Exercise 5 in Chapter 5). Thus, a modified definition, which relaxes condition 1 to "Monte-Carlo computable", would also work. In all our examples, the pseudorandom sequences can be efficiently computed by deterministic algorithms.

We will now derive pseudorandom bit generators from one-way permutations with hard-core predicates (see Definition 6.15). These generators turn out to be computationally perfect. The construction was introduced by Blum and Micali ([BluMic84]).

Definition 8.3. Let $I = (I_k)_{k \in \mathbb{N}}$ be a key set with security parameter k, and let $Q \in \mathbb{Z}[X]$ be a positive polynomial.
Let $f = (f_i : D_i \longrightarrow D_i)_{i \in I}$ be a family of one-way permutations with hard-core predicate $B = (B_i : D_i \longrightarrow \{0, 1\})_{i \in I}$ and key generator K. Then we have the following pseudorandom bit generator with stretch function Q and key generator K:

$$G := G(f, B, Q) := (G_i : D_i \longrightarrow \{0,1\}^{Q(k)})_{k \in \mathbb{N}, i \in I_k},$$

$$x \in D_i \longmapsto (B_i(x), B_i(f_i(x)), B_i(f_i^2(x)), \dots, B_i(f_i^{Q(k)-1}(x))).$$

We call this generator the *pseudorandom bit generator induced by* f, B *and* Q.

Remark. We obtain the pseudorandom bit sequence by a very simple construction: choose some random seed $x \xleftarrow{u} D_i$. Compute the first pseudorandom bit as $B_i(x)$, apply f_i and get $y := f_i(x)$. Compute the next pseudorandom bit as $B_i(y)$, apply f_i to y and get a new $y := f_i(y)$. Compute the next pseudorandom bit as $B_i(y)$, and so on.

Examples:

1. Provided that the discrete logarithm assumption is true, the discrete exponential function

$$\mathrm{Exp} = (\mathrm{Exp}_{p,g} : \mathbb{Z}_{p-1} \longrightarrow \mathbb{Z}_p^*, \ x \longmapsto g^x)_{(p,g) \in I}$$

with $I := \{(p,g) \mid p \text{ prime}, g \in \mathbb{Z}_p^* \text{ a primitive root}\}$ is a bijective one-way function, and the most-significant bit $\mathrm{Msb}_p(x)$ defined by

$$\mathrm{Msb}_p(x) = \begin{cases} 0 \text{ for } 0 \leq x < \frac{p-1}{2}, \\ 1 \text{ for } \frac{p-1}{2} \leq x < p-1, \end{cases}$$

is a hard-core predicate for Exp (see Section 7.1). Identifying $\mathbb{Z}_{p-1}$ with $\mathbb{Z}_p^*$ in the straightforward way,[1] we may consider Exp as a one-way permutation. The induced pseudorandom bit generator is called the *discrete exponential generator* (or *Blum-Micali generator*).

2. Provided that the RSA assumption is true, the RSA family

$$\mathrm{RSA} = (\mathrm{RSA}_{n,e} : \mathbb{Z}_n^* \longrightarrow \mathbb{Z}_n^*, \ x \longmapsto x^e)_{(n,e) \in I}$$

with $I := \{(n,e) \mid n = pq, p, q \text{ distinct primes}, |p| = |q|, e \text{ prime to } \varphi(n)\}$ is a one-way permutation, and the least-significant bit $\mathrm{Lsb}_n(x)$ is a hard-core predicate for RSA (see Section 7.2). The induced pseudorandom bit generator is called the *RSA generator*.

3. Provided that the factorization assumption is true, the modular squaring function

$$\mathrm{Square} = (\mathrm{Square}_n : \mathrm{QR}_n \longrightarrow \mathrm{QR}_n, \ x \longmapsto x^2)_{n \in I}$$

with $I := \{n \mid n = pq, p, q \text{ distinct primes}, |p| = |q|, p, q \equiv 3 \bmod 4\}$ is a one-way permutation, and the least-significant bit $\mathrm{Lsb}_n(x)$ is a hard-core predicate for Square (see Section 7.3). The induced pseudorandom bit

---

[1] $\mathbb{Z}_{p-1} = \{0, \dots, p-2\} \longrightarrow \mathbb{Z}_p^* = \{1, \dots, p-1\}, 0 \mapsto p-1, x \mapsto x$ for $1 \leq x \leq p-2$.

generator is called the *(x² mod n) generator* or *Blum-Blum-Shub gener-*
ator.
Here the computation of the pseudorandom bits is particularly simple:
choose a random seed $x \in \mathbb{Z}_n^*$, repeatedly square x and reduce it by
modulo n, take the least-significant bit after each step.

We will now show that the pseudorandom bit generators, induced by one-
way permutations, are computationally perfect. For later applications, it is
useful to prove a slightly more general statement that covers the pseudorandom bits and the seed, which is encrypted using $f_i^{Q(k)}$.

Theorem 8.4. *Let $I = (I_k)_{k \in \mathbb{N}}$ be a key set with security parameter k, and
let $Q \in \mathbb{Z}[X]$ be a positive polynomial. Let $f = (f_i : D_i \longrightarrow D_i)_{i \in I}$ be a family
of one-way permutations with hard-core predicate $B = (B_i : D_i \longrightarrow \{0,1\})_{i \in I}$
and key generator K. Let $G := G(f, B, Q)$ be the induced pseudorandom bit
generator.*
*Then, for every probabilistic polynomial algorithm A with inputs $i \in I_k, z \in
\{0,1\}^{Q(k)}, y \in D_i$ and output in $\{0,1\}$, and every positive polynomial $P \in
\mathbb{Z}[X]$, there is a $k_0 \in \mathbb{N}$ such that for all $k \geq k_0$*

$$| \operatorname{prob}(A(i, G_i(x), f_i^{Q(k)}(x)) = 1 : i \leftarrow K(1^k), x \xleftarrow{u} D_i)$$
$$- \operatorname{prob}(A(i, z, y) = 1 : i \leftarrow K(1^k), z \xleftarrow{u} \{0,1\}^{Q(k)}, y \xleftarrow{u} D_i)| \leq \frac{1}{P(k)}.$$

Remark. The theorem states that for sufficiently large keys, the probability
of distinguishing successfully between truly random sequences and pseudorandom sequences – using a given efficient algorithm – is negligibly small,
even if the encryption $f_i^{Q(k)}(x)$ of the seed x is known.

Proof. Assume that there is a probabilistic polynomial algorithm A, such
that the inequality is false for infinitely many k. Replacing A by $1 - A$ if
necessary, we may drop the absolute value and assume that

$$\operatorname{prob}(A(i, G_i(x), f_i^{Q(k)}(x)) = 1 : i \leftarrow K(1^k), x \xleftarrow{u} D_i)$$
$$- \operatorname{prob}(A(i, z, y) = 1 : i \leftarrow K(1^k), z \xleftarrow{u} \{0,1\}^{Q(k)}, y \xleftarrow{u} D_i) > \frac{1}{P(k)},$$

for k in an infinite subset $\mathcal{K}$ of $\mathbb{N}$.
For $k \in \mathcal{K}$ and $i \in I_k$, we consider the following sequence of distributions
$p_{i,0}, p_{i,1}, \ldots, p_{i,Q(k)}$ on $Z_i := \{0,1\}^{Q(k)} \times D_i$:[2]

---

[2] We use the notation for image distributions introduced in Appendix B.1 (p.
330): $p_{i,r}$ is the direct product of the uniform distribution on $\{0,1\}^{Q(k)-r}$
with the image of the uniform distribution on D_i under the mapping $x \mapsto
(B_i(x), B_i(f_i(x)), \ldots, B_i(f_i^{r-1}(x)), f_i^r(x))$.

$$p_{i,0} := \{(b_1, \ldots, b_{Q(k)}, y) : (b_1, \ldots, b_{Q(k)}) \overset{u}{\leftarrow} \{0,1\}^{Q(k)}, y \overset{u}{\leftarrow} D_i\}$$

$$p_{i,1} := \{(b_1, \ldots, b_{Q(k)-1}, B_i(x), f_i(x)) :$$
$$(b_1, \ldots, b_{Q(k)-1}) \overset{u}{\leftarrow} \{0,1\}^{Q(k)-1}, x \overset{u}{\leftarrow} D_i\}$$

$$p_{i,2} := \{(b_1, \ldots, b_{Q(k)-2}, B_i(x), B_i(f_i(x)), f_i^2(x)) :$$
$$(b_1, \ldots, b_{Q(k)-2}) \overset{u}{\leftarrow} \{0,1\}^{Q(k)-2}, x \overset{u}{\leftarrow} D_i\}$$

$$\vdots$$

$$p_{i,r} := \{(b_1, \ldots, b_{Q(k)-r}, B_i(x), B_i(f_i(x)), \ldots, B_i(f_i^{r-1}(x)), f_i^r(x)) :$$
$$(b_1, \ldots, b_{Q(k)-r}) \overset{u}{\leftarrow} \{0,1\}^{Q(k)-r}, x \overset{u}{\leftarrow} D_i\}$$

$$\vdots$$

$$p_{i,Q(k)} := \{(B_i(x), B_i(f_i(x)), \ldots, B_i(f_i^{Q(k)-1}(x)), f_i^{Q(k)}(x)) : x \overset{u}{\leftarrow} D_i\}.$$

We start with truly random bit sequences. In each step, we replace one more truly random bit from the right with a pseudorandom bit. The seed x encrypted by f_i^r is always appended on the right. Note that the image $\{f_i(x) : x \overset{u}{\leftarrow} D_i\}$ of the uniform distribution under f_i is again the uniform distribution, since f_i is bijective. Finally, in $p_{i,Q(k)}$ we have the distribution of the pseudorandom sequences supplemented by the encrypted seed. We observe that

$$\mathrm{prob}(A(i,z,y) = 1 : i \leftarrow K(1^k), z \overset{u}{\leftarrow} \{0,1\}^{Q(k)}, y \overset{u}{\leftarrow} D_i)$$
$$= \mathrm{prob}(A(i,z,y) = 1 : i \leftarrow K(1^k), (z,y) \overset{p_{i,0}}{\leftarrow} Z_i)$$

and

$$\mathrm{prob}(A(i, G_i(x), f_i^{Q(k)}(x)) = 1 : i \leftarrow K(1^k), x \overset{u}{\leftarrow} D_i)$$
$$= \mathrm{prob}(A(i,z,y) = 1 : i \leftarrow K(1^k), (z,y) \overset{p_{i,Q(k)}}{\leftarrow} Z_i).$$

Thus, our assumption says that for $k \in \mathcal{K}$, the algorithm A is able to distinguish between the distribution $p_{i,Q(k)}$ (of pseudorandom sequences) and the (uniform) distribution $p_{i,0}$. Hence, A must be able to distinguish between two subsequent distributions $p_{i,r}$ and $p_{i,r+1}$, for some r.

Since f_i is bijective, we have the following equation (8.1):

$$p_{i,r} = \{(b_1, \ldots, b_{Q(k)-r}, B_i(x), B_i(f_i(x)), \ldots, B_i(f_i^{r-1}(x)), f_i^r(x)) :$$
$$(b_1, \ldots, b_{Q(k)-r}) \overset{u}{\leftarrow} \{0,1\}^{Q(k)-r}, x \overset{u}{\leftarrow} D_i\}$$
$$= \{(b_1, \ldots, b_{Q(k)-r}, B_i(f_i(x)), B_i(f_i^2(x)), \ldots, B_i(f_i^r(x)), f_i^{r+1}(x)) :$$
$$(b_1, \ldots, b_{Q(k)-r}) \overset{u}{\leftarrow} \{0,1\}^{Q(k)-r}, x \overset{u}{\leftarrow} D_i\}.$$

We see that $p_{i,r}$ differs from $p_{i,r+1}$ only at one position, namely at position $Q(k) - r$. There, the hard-core bit $B_i(x)$ is replaced by a truly random bit.

Therefore, algorithm A, which distinguishes between $p_{i,r}$ and $p_{i,r+1}$, can also be used to compute $B_i(x)$ from $f_i(x)$.

More precisely, we will derive a probabilistic polynomial algorithm $\tilde{A}(i, y)$ from A that on inputs $i \in I_k$ and $y := f_i(x)$ computes $B_i(x)$ with probability $> 1/2 + 1/P(k)Q(k)$, for the infinitely many $k \in \mathcal{K}$. This contradiction to the hard-core property of B will finish the proof of the theorem.

For $k \in \mathcal{K}$, we have

$$\frac{1}{P(k)} < \text{prob}(A(i, z, y) = 1 : i \leftarrow K(1^k), (z, y) \overset{p_{i,Q(k)}}{\leftarrow} Z_i)$$

$$- \text{prob}(A(i, z, y) = 1 : i \leftarrow K(1^k), (z, y) \overset{p_{i,0}}{\leftarrow} Z_i)$$

$$= \sum_{r=0}^{Q(k)-1} (\text{prob}(A(i, z, y) = 1 : i \leftarrow K(1^k), (z, y) \overset{p_{i,r+1}}{\leftarrow} Z_i)$$

$$- \text{prob}(A(i, z, y) = 1 : i \leftarrow K(1^k), (z, y) \overset{p_{i,r}}{\leftarrow} Z_i)).$$

Randomly choosing r, we expect that the r-th term in the sum is $> 1/P(k)Q(k)$.

On inputs $i \in I_k, y \in D_i$, the algorithm $\tilde{A}$ works as follows:

1. Choose r, with $0 \leq r < Q(k)$, uniformly at random.
2. Independently choose random bits $b_1, b_2, \ldots, b_{Q(k)-r-1}$ and another random bit b.
3. For $y = f_i(x) \in D_i$, let

$$\tilde{A}(i, y) = \tilde{A}(i, f_i(x))$$

$$:= \begin{cases} b & \text{if } A(i, b_1, \ldots, b_{Q(k)-r-1}, b, \\ & \qquad B_i(f_i(x)), \ldots, B_i(f_i^r(x)), f_i^{r+1}(x)) = 1, \\ 1 - b & \text{otherwise.} \end{cases}$$

If A distinguishes between $p_{i,r}$ and $p_{i,r+1}$, it yields 1 with higher probability if the $(Q(k) - r)$-th bit of its input is $B_i(x)$ and not a random bit. Therefore, we guess in our algorithm that the randomly chosen b is the desired hard-core bit if A outputs 1.

We now check that $\tilde{A}$ indeed computes the hard-core bit with a non-negligible probability. Let R be the random variable describing the choice of r in the first step of the algorithm. Since r is selected with respect to the uniform distribution, we have $\text{prob}(R = r) = 1/Q(k)$ for all r. Applying Lemma B.13, we get

$$\mathrm{prob}(\tilde{A}(i, f_i(x)) = B_i(x) : i \leftarrow K(1^k), x \xleftarrow{u} D_i)$$

$$= \frac{1}{2} + \mathrm{prob}(\tilde{A}(i, f_i(x)) = b \,|\, B_i(x) = b) - \mathrm{prob}(\tilde{A}(i, f_i(x)) = b)$$

$$= \frac{1}{2} + \sum_{r=0}^{Q(k)-1} \mathrm{prob}(R = r) \cdot (\mathrm{prob}(\tilde{A}(i, f_i(x)) = b \,|\, B_i(x) = b, R = r)$$

$$- \mathrm{prob}(\tilde{A}(i, f_i(x)) = b \,|\, R = r))$$

$$= \frac{1}{2} + \frac{1}{Q(k)} \sum_{r=0}^{Q(k)-1} (\mathrm{prob}(\tilde{A}(i, f_i(x)) = b \,|\, B_i(x) = b : i \leftarrow K(1^k), x \xleftarrow{u} D_i)$$

$$- \mathrm{prob}(\tilde{A}(i, f_i(x)) = b : i \leftarrow K(1^k), x \xleftarrow{u} D_i))$$

$$= \frac{1}{2} + \frac{1}{Q(k)} \sum_{r=0}^{Q(k)-1} (\mathrm{prob}(A(i, z, y) = 1 : i \leftarrow K(1^k), (z, y) \xleftarrow{p_{i,r+1}} Z_i)$$

$$- \mathrm{prob}(A(i, z, y) = 1 : i \leftarrow K(1^k), (z, y) \xleftarrow{p_{i,r}} Z_i))$$

$$> \frac{1}{2} + \frac{1}{Q(k)P(k)},$$

for the infinitely many $k \in \mathcal{K}$. The probabilities in lines 2 and 3 are computed with respect to $i \leftarrow K(1^k)$ and $x \xleftarrow{u} D_i$ (and the random choice of the elements b_i, b, r). Since r is chosen independently, we can omit the conditions $R = r$. Taking the probability $\mathrm{prob}(\tilde{A}(i, f_i(x)) = b \,|\, B_i(x) = b)$ conditional on $B_i(x) = b$ just means that the inputs to A in step 3 of the algorithm $\tilde{A}$ are distributed according to $p_{i,r+1}$. Finally, recall equation (8.1) for $p_{i,r}$ from above.

Since B is a hard-core predicate, our computation yields the desired contradiction, and the proof of Theorem 8.4 is complete. $\square$

Corollary 8.5 *(Theorem of Blum and Micali). Pseudorandom bit generators induced by one-way permutations with hard-core predicates are computationally perfect.*

Proof. Let $A(i, z)$ be a probabilistic polynomial algorithm with inputs $i \in I_k, z \in \{0,1\}^{Q(k)}$ and output in $\{0,1\}$. We define $\tilde{A}(i, z, y) := A(i, z)$, and observe that

$$\mathrm{prob}(\tilde{A}(i, z, y) = 1 : i \leftarrow I_k, z \leftarrow \{0,1\}^{Q(k)}, y \leftarrow D_i)$$
$$= \mathrm{prob}(A(i, z) = 1 : i \leftarrow I_k, z \leftarrow \{0,1\}^{Q(k)}),$$

and the corollary follows from Theorem 8.4 applied to $\tilde{A}$. $\square$

8.2 Yao's Theorem

Computationally perfect pseudorandom bit generators such as the ones induced by one-way permutations are characterized by another unique feature: it is not possible to predict the next bit in the pseudorandom sequence from the preceding bits.

Definition 8.6. Let $I = (I_k)_{k \in \mathbb{N}}$ be a key set with security parameter k, and let $G = (G_i : X_i \longrightarrow \{0,1\}^{l(k)})_{i \in I}$ be a pseudorandom bit generator with polynomial stretch function l and key generator K:

1. A *next-bit predictor* for G is a probabilistic polynomial algorithm $A(i, z_1 \ldots z_r)$ which, given $i \in I_k$, outputs a bit ("the next bit") from r input bits z_j $(0 \leq r < l(k))$.
2. G *passes all next-bit tests* if and only if for every next-bit predictor A and every positive polynomial $P \in \mathbb{Z}[X]$, there is a $k_0 \in \mathbb{N}$ such that for all $k \geq k_0$ and all $0 \leq r < l(k)$

$$\text{prob}(A(i, G_{i,1}(x) \ldots G_{i,r}(x)) = G_{i,r+1}(x) : i \leftarrow K(1^k), x \xleftarrow{u} X_i)$$

$$\leq \frac{1}{2} + \frac{1}{P(k)}.$$

Here and in what follows we denote by $G_{i,j}, 1 \leq j \leq l(k)$, the j-th bit generated by G_i:

$$G_i(x) = (G_{i,1}(x), G_{i,2}(x), \ldots, G_{i,l(k)}(x)).$$

Remarks:

1. A next-bit predictor has two inputs: the key i and a bit string $z_1 \ldots z_r$ of variable length.
2. As usual, the probability in the definition is also taken over the random choice of a key i with security parameter k. This means that when randomly generating a key i, the probability of obtaining one for which A has a significant chance of predicting a next bit is negligibly small (see Proposition 6.17).

Theorem 8.7 *(Yao's Theorem). Let $I = (I_k)_{k \in \mathbb{N}}$ be a key set with security parameter k, and let $G = (G_i : X_i \longrightarrow \{0,1\}^{l(k)})_{i \in I}$ be a pseudorandom bit generator with polynomial stretch function l and key generator K.*
Then G is computationally perfect if and only if G passes all next-bit tests.

Proof. Assume that G is computationally perfect and does not pass all next-bit tests. Then there is a next-bit predictor A and a positive polynomial P, such that for k in an infinite subset $\mathcal{K}$ of $\mathbb{N}$, we have a position r_k, $0 \leq r_k < l(k)$, with $q_{k,r_k} > 1/2 + 1/P(k)$, where

$q_{k,r} := \mathrm{prob}(A(i, G_{i,1}(x) \ldots G_{i,r}(x)) = G_{i,r+1}(x) : i \leftarrow K(1^k), x \xleftarrow{u} X_i)$.

By Proposition 6.18, we can compute the probabilities $q_{k,r}$, $r = 0, 1, 2, \ldots$, approximately with high probability, and we conclude that there is a probabilistic polynomial algorithm R which on input 1^k finds a position where the next-bit predictor is successful:

$$\mathrm{prob}\left(q_{k,R(1^k)} > \frac{1}{2} + \frac{1}{2P(k)} \right) \geq 1 - \frac{1}{4P(k)}.$$

We define a probabilistic polynomial algorithm $\tilde{A}$ (a statistical test) for inputs $i \in I$ and $z = (z_1, \ldots, z_{l(k)}) \in \{0,1\}^{l(k)}$ as follows: Let $r := R(1^k)$ and set

$$\tilde{A}(i, z) := \begin{cases} 1 & \text{if } z_{r+1} = A(i, z_1 \ldots z_r), \\ 0 & \text{otherwise.} \end{cases}$$

For truly random sequences, it is not possible to predict a next bit with probability $> 1/2$. Thus, we have for the uniform distribution on $\{0,1\}^{l(k)}$ that

$$\mathrm{prob}(\tilde{A}(i, z) = 1 : i \leftarrow K(1^k), z \xleftarrow{u} \{0,1\}^{l(k)}) \leq \frac{1}{2}.$$

We obtain, for the infinitely many $k \in \mathcal{K}$, that

$$|\mathrm{prob}(\tilde{A}(i, G_i(x)) = 1 : i \leftarrow K(1^k), x \xleftarrow{u} X_i) \ -$$
$$\mathrm{prob}(\tilde{A}(i, z) = 1 : i \leftarrow K(1^k), z \xleftarrow{u} \{0,1\}^{l(k)})|$$
$$> \left(1 - \frac{1}{4P(k)}\right) \cdot \left(\frac{1}{2} + \frac{1}{2P(k)} - \frac{1}{2}\right) - \frac{1}{4P(k)}$$
$$= \frac{1}{4P(k)} - \frac{1}{8P^2(k)} \geq \frac{1}{8P(k)},$$

which is a contradiction to the assumption that G is computationally perfect.

Conversely, assume that the sequences generated by G pass all next-bit tests, but can be distinguished from truly random sequences by a statistical test A. This means that

$$|\mathrm{prob}(A(i, G_i(x)) = 1 : i \leftarrow K(1^k), x \xleftarrow{u} X_i)$$
$$- \mathrm{prob}(A(i, z) = 1 : i \leftarrow K(1^k), z \xleftarrow{u} \{0,1\}^{l(k)})| > \frac{1}{P(k)},$$

for some positive polynomial P and k in an infinite subset $\mathcal{K}$ of $\mathbb{N}$. Replacing A by $1 - A$, if necessary, we may drop the absolute value.

The proof now runs in a similar way to the proof of Theorem 8.4. For $k \in \mathcal{K}$ and $i \in I_k$, we consider a sequence $p_{i,0}, p_{i,1}, \ldots, p_{i,l(k)}$ of distributions on $\{0,1\}^{l(k)}$:

$$p_{i,0} := \{(b_1, \ldots, b_{l(k)}) : (b_1, \ldots, b_{l(k)}) \xleftarrow{u} \{0,1\}^{l(k)}\}$$

$$p_{i,1} := \{(G_{i,1}(x), b_2, \ldots, b_{l(k)}) : (b_2, \ldots, b_{l(k)}) \xleftarrow{u} \{0,1\}^{l(k)-1}, x \xleftarrow{u} X_i\}$$

$$p_{i,2} := \{(G_{i,1}(x), G_{i,2}(x), b_3, \ldots, b_{l(k)}) :$$
$$(b_3, \ldots, b_{l(k)}) \xleftarrow{u} \{0,1\}^{l(k)-2}, x \xleftarrow{u} X_i\}$$

$$\vdots$$

$$p_{i,r} := \{(G_{i,1}(x), G_{i,2}(x), \ldots, G_{i,r}(x), b_{r+1}, \ldots, b_{l(k)}) :$$
$$(b_{r+1}, \ldots, b_{l(k)}) \xleftarrow{u} \{0,1\}^{l(k)-r}, x \xleftarrow{u} X_i\}$$

$$\vdots$$

$$p_{i,l(k)} := \{(G_{i,1}(x), G_{i,2}(x), \ldots, G_{i,l(k)}(x)) : x \xleftarrow{u} X_i\}.$$

We start with truly random bit sequences, and in each step we replace one more truly random bit from the left by a pseudorandom bit. Finally, in $p_{i,l(k)}$ we have the distribution of the pseudorandom sequences.

Our assumption says that for $k \in \mathcal{K}$, algorithm A is able to distinguish between the distribution $p_{i,l(k)}$ (of pseudorandom sequences) and the (uniform) distribution $p_{i,0}$. Again, the basic idea of the proof now is that A must be able to distinguish between two subsequent distributions $p_{i,r}$ and $p_{i,r+1}$, for some r. However, $p_{i,r+1}$ differs from $p_{i,r}$ in one position only, and there a truly random bit is replaced by the next bit $G_{i,r+1}(x)$ of the pseudorandom sequence. Therefore, algorithm A can also be used to predict $G_{i,r+1}(x)$.

More precisely, we will derive a probabilistic polynomial algorithm $\tilde{A}(i, z_1, \ldots, z_r)$ that successfully predicts the next bit $G_{i,r+1}(x)$ from $G_{i,1}(x), G_{i,2}(x), \ldots, G_{i,r}(x)$ for some $r = r_k$, for the infinitely many $k \in \mathcal{K}$. This contradiction to the assumption that G passes all next-bit tests will finish the proof of the theorem.

Since A is able to distinguish between the uniform distribution and the distribution induced by G, we get for $k \in \mathcal{K}$ that

$$\frac{1}{P(k)} < \text{prob}(A(i,z) = 1 : i \leftarrow K(1^k), z \xleftarrow{p_{i,l(k)}} \{0,1\}^{l(k)})$$

$$- \text{prob}(A(i,z) = 1 : i \leftarrow K(1^k), z \xleftarrow{p_{i,0}} \{0,1\}^{l(k)})$$

$$= \sum_{r=0}^{l(k)-1} (\text{prob}(A(i,z) = 1 : i \leftarrow K(1^k), z \xleftarrow{p_{i,r+1}} \{0,1\}^{l(k)})$$

$$- \text{prob}(A(i,z) = 1 : i \leftarrow K(1^k), z \xleftarrow{p_{i,r}} \{0,1\}^{l(k)})).$$

We conclude that for $k \in \mathcal{K}$, there is some r_k, $0 \le r_k < l(k)$, with

$$\frac{1}{P(k)l(k)} < \text{prob}(A(i,z) = 1 : i \leftarrow K(1^k), z \overset{p_{i,r_k+1}}{\Leftarrow} \{0,1\}^{l(k)})$$

$$- \text{prob}(A(i,z) = 1 : i \leftarrow K(1^k), z \overset{p_{i,r_k}}{\Leftarrow} \{0,1\}^{l(k)}).$$

This means that $A(i,z)$ yields 1 for

$$z = (G_{i,1}(x), G_{i,2}(x), \ldots, G_{i,r_k}(x), b, b_{r_k+2}, \ldots, b_{l(k)})$$

with higher probability if b is equal to $G_{i,r_k+1}(x)$ and not a truly random bit.

On inputs $i \in I_k, z_1 \ldots z_r$ ($0 \le r < l(k)$), algorithm $\tilde{A}$ is defined as follows:

1. Choose truly random bits $b, b_{r+2}, \ldots, b_{l(k)}$, and set

$$z := (z_1, \ldots, z_r, b, b_{r+2}, \ldots, b_{l(k)}).$$

2. Let

$$\tilde{A}(i, z_1 \ldots z_r) := \begin{cases} b & \text{if } A(i,z) = 1, \\ 1 - b & \text{if } A(i,z) = 0. \end{cases}$$

Applying Lemma B.13, we get

$$\text{prob}(\tilde{A}(i, G_{i,1}(x) \ldots G_{i,r_k}(x)) = G_{i,r_k+1}(x) : i \leftarrow K(1^k), x \overset{u}{\leftarrow} X_i)$$

$$= \frac{1}{2} + \text{prob}(\tilde{A}(i, G_{i,1}(x) \ldots G_{i,r_k}(x)) = b \mid$$

$$G_{i,r_k+1}(x) = b : i \leftarrow K(1^k), x \overset{u}{\leftarrow} X_i)$$

$$- \text{prob}(\tilde{A}(i, G_{i,1}(x) \ldots G_{i,r_k}(x)) = b : i \leftarrow K(1^k), x \overset{u}{\leftarrow} X_i)$$

$$= \frac{1}{2} + \text{prob}(A(i,z) = 1 : i \leftarrow K(1^k), z \overset{p_{i,r_k+1}}{\Leftarrow} \{0,1\}^{l(k)})$$

$$- \text{prob}(A(i,z) = 1 : i \leftarrow K(1^k), z \overset{p_{i,r_k}}{\Leftarrow} \{0,1\}^{l(k)})$$

$$> \frac{1}{2} + \frac{1}{P(k)l(k)},$$

for the infinitely many $k \in \mathcal{K}$. This is the desired contradiction and completes the proof of Yao's Theorem. $\qquad \square$

Exercises

1. Let $I = (I_k)_{k \in \mathbb{N}}$ be a key set with security parameter k, and let $G = (G_i)_{i \in I}$ be a computationally perfect pseudorandom bit generator with polynomial stretch function l. Let $\pi = (\pi_i)_{i \in I}$ be a family of permutations, where π_i is a permutation of $\{0,1\}^{l(k)}$ for $i \in I_k$. Assume that π can be computed by a polynomial algorithm Π, i.e., $\pi_i(y) = \Pi(i,y)$. Let $\pi \circ G$ be the composition of π and G: $x \mapsto \pi_i(G_i(x))$.

   Show that $\pi \circ G$ is also a computationally perfect pseudorandom bit generator.

2. Give an example of a computationally perfect pseudorandom bit generator $G = (G_i)_{i \in I}$ and a family of permutations π, such that $\pi \circ G$ is not computationally perfect.
   (According to Exercise 1, π cannot be computable in polynomial time.)

3. Let $G = (G_i)_{i \in I}$ be a pseudorandom bit generator with polynomial stretch function l and key generator K.
   Show that G is computationally perfect if and only if next bits in the past cannot be predicted; i.e., for every probabilistic polynomial algorithm $A(i, z_{r+1} \ldots z_{l(k)})$ which, given $i \in I_k$, outputs a bit ("the next bit in the past") from $l(k) - r$ input bits z_j, there is a $k_0 \in \mathbb{N}$ such that for all $k \geq k_0$ and all $1 \leq r \leq l(k)$

$$\mathrm{prob}(G_{i,r}(x) = A(i, G_{i,r+1}(x) \ldots G_{i,l(k)}(x)) : i \leftarrow K(1^k), x \stackrel{u}{\leftarrow} X_i)$$
$$\leq \frac{1}{2} + \frac{1}{P(k)}.$$

4. Let Q be a positive polynomial, and let $G = (G_i)_{i \in I}$ be a computationally perfect pseudorandom bit generator with

$$G_i : \{0,1\}^{Q(k)} \longrightarrow \{0,1\}^{Q(k)+1} \quad (i \in I_k),$$

   i.e., G extends the binary length of the seeds by 1. Recursively, define the pseudorandom bit generators G^l by

$$G^1 := G, \quad G_i^l(x) := (G_{i,1}(x), G_i^{l-1}(G_{i,2}(x), \ldots, G_{i,Q(k)+1}(x))).$$

   As before, we denote by $G_{i,j}^l(x)$ the j-th bit of $G_i^l(x)$. Let l vary with the security parameter k, i.e., $l = l(k)$, and assume that $l : \mathbb{N} \longrightarrow \mathbb{N}$ is a polynomial function.
   Show that G^l is computationally perfect.

5. Prove the following stronger version of Yao's Theorem (Theorem 8.7).
   Let $I = (I_k)_{k \in \mathbb{N}}$ be a key set with security parameter k, and let $G = (G_i : X_i \longrightarrow \{0,1\}^{l(k)})_{i \in I}$ be a pseudorandom bit generator with polynomial stretch function l and key generator K.
   Let $f = (f_i : X_i \longrightarrow Y_i)_{i \in I}$ be a Monte-Carlo computable family of maps. Then, the following statements are equivalent:
   a. For every probabilistic polynomial algorithm $A(i, y, z)$ and every positive polynomial P, there is a $k_0 \in \mathbb{N}$ such that for all $k \geq k_0$ and all $0 \leq r < l(k)$

$$\mathrm{prob}(G_{i,r+1}(x) = A(i, f_i(x), G_{i,1}(x) \ldots G_{i,r}(x)) :$$
$$i \leftarrow K(1^k), x \stackrel{u}{\leftarrow} X_i)$$
$$\leq \frac{1}{2} + \frac{1}{P(k)}.$$

b. For every probabilistic polynomial algorithm $A(i, y, z)$ and every positive polynomial P, there is a $k_0 \in \mathbb{N}$ such that for all $k \geq k_0$

$$| \operatorname{prob}(A(i, f_i(x), z) = 1 : i \leftarrow K(1^k), x \xleftarrow{u} X_i, z \xleftarrow{u} \{0,1\}^{l(k)})$$
$$- \operatorname{prob}(A(i, f_i(x), G_i(x)) = 1 : i \leftarrow K(1^k), x \xleftarrow{u} X_i)|$$
$$\leq \frac{1}{P(k)}.$$

In this exercise a setting is modeled in which some information $f_i(x)$ about the seed x is known to the adversary.

6. Let $f = (f_i : D_i \longrightarrow R_i)_{i \in I}$ be a family of one-way functions with key generator K, and let $B = (B_i : D_i \longrightarrow \{0,1\}^{l(k)})_{i \in I}$ be a family of l-bit predicates which is computable by a Monte Carlo algorithm (l a polynomial function). Let $B_i = (B_{i,1}, B_{i,2}, \ldots, B_{i,l(k)})$. We call B an l-bit *hard-core predicate* for f (or *simultaneously secure bits* of f), if for every probabilistic polynomial algorithm $A(i, y, z_1, \ldots, z_l)$ and every positive polynomial P, there is a $k_0 \in \mathbb{N}$ such that for all $k \geq k_0$

$$| \operatorname{prob}(A(i, f_i(x), B_{i,1}(x), \ldots, B_{i,l(k)}(x)) = 1 : i \leftarrow K(1^k), x \xleftarrow{u} D_i)$$
$$- \operatorname{prob}(A(i, f_i(x), z) = 1 : i \leftarrow K(1^k), x \xleftarrow{u} D_i, z \xleftarrow{u} \{0,1\}^{l(k)})|$$
$$\leq \frac{1}{P(k)}.$$

For $l = 1$, the definition is equivalent to our previous Definition 6.15 of hard-core bits (see Exercise 7 in Chapter 6).

Now assume that B is an l-bit hard-core predicate, and let $C = (C_i : \{0,1\}^{l(k)} \longrightarrow \{0,1\})_{i \in I}$ be a Monte-Carlo computable family of predicates with $\operatorname{prob}(C_i(x) = 0 : x \xleftarrow{u} \{0,1\}^{l(k)}) = 1/2$ for all $i \in I$. Show that the composition $C \circ B, x \in D_i \mapsto C_i(B_i(x))$, is a hard-core predicate for f.

7. Let $f = (f_i : D_i \longrightarrow R_i)_{i \in I}$ be a family of one-way functions with key generator K, and let $B = (B_i : D_i \longrightarrow \{0,1\}^{l(k)})_{i \in I}$ be a family of l-bit predicates for f. Let $B_i = (B_{i,1}, B_{i,2}, \ldots, B_{i,l(k)})$. Assume that knowing $f_i(x)$ and $B_{i,1}(x), \ldots, B_{i,j-1}(x)$ does not help in the computation of $B_{i,j}(x)$. More precisely, assume that for every probabilistic polynomial algorithm $A(i, y, z)$ and every positive polynomial P, there is a $k_0 \in \mathbb{N}$ such that for all $1 \leq j \leq l(k)$

$$\operatorname{prob}(A(i, f_i(x), B_{i,1}(x) \ldots B_{i,j-1}(x)) = B_{i,j}(x) : i \leftarrow K(1^k), x \xleftarrow{u} X_i)$$
$$\leq \frac{1}{2} + \frac{1}{P(k)}.$$

(In particular, the $(B_{i,j})_{i \in I}$ are hard-core predicates for f.) Show that the bits $B_{i,1}, \ldots, B_{i,l}$ are simultaneously secure bits for f.

Examples:

The $\lfloor \log_2(|n|) \rfloor$ least-significant bits are simultaneously secure for the RSA (and the Square) one-way function (Exercise 12 in Chapter 7). If p is a prime, $p - 1 = 2^t a$ and a is odd, then the bits at the positions $t, t + 1, \ldots, t + \lfloor \log_2(|p|) \rfloor$ (counted from the right, starting with 0) are simultaneously secure for the discrete exponential one-way function (Exercise 5 in Chapter 7).

8. Let $f = (f_i : D_i \longrightarrow D_i)_{i \in I}$ be a family of one-way permutations with key generator K, and let $B = (B_i)_{i \in I}$ be an l-bit hard-core predicate for f. Let G be the following pseudorandom bit generator with stretch function lQ (Q a positive polynomial):

$$G := \left(G_i : D_i \longrightarrow \{0, 1\}^{l(k)Q(k)} \right)_{k \in \mathbb{N}, i \in I_k}$$

$$x \in D_i \longmapsto (B_i(x), B_i(f_i(x)), B_i(f_i^2(x)), \ldots, B_i(f_i^{Q(k)-1}(x))).$$

Prove a statement that is analogous to Theorem 8.4. In particular, prove that G is computationally perfect.

Example:

Taking the Square one-way permutation $x \mapsto x^2$ ($x \in QR_n$, with n a product of distinct primes) and the $\lfloor \log_2(|n|) \rfloor$ least-significant bits, we get the generalized Blum-Blum-Shub generator. It is used in the Blum-Goldwasser probabilistic encryption scheme (see Chapter 9).

9. Provably Secure Encryption

This chapter deals with provable security. It is desirable that mathematical proofs show that a given cryptosystem resists certain types of attacks. The security of cryptographic schemes and randomness are closely related. An encryption method provides secrecy only if the ciphertexts appear sufficiently random to the adversary. Therefore, probabilistic encryption algorithms are required. The pioneering work of Shannon on provable security, based on his information theory, is discussed in Section 9.1. For example, we prove that Vernam's one-time pad is a perfectly secret encryption. Shannon's notion of perfect secrecy may be interpreted in terms of probabilistic attacking algorithms that try to distinguish between two candidate plaintexts (Section 9.2). Unfortunately, Vernam's one-time pad is not practical in most situations. In Section 9.3, we give important examples of probabilistic encryption algorithms that are practical. One-way permutations with hard-core predicates yield computationally perfect pseudorandom bit generators (Chapter 8), and these can be used to define "public-key pseudorandom one-time pads", by analogy to Vernam's one-time pad: the plaintext bits are XORed with pseudorandom bits generated from a short, truly random (one-time) seed. More recent notions of provable security, which include the computational complexity of attacking algorithms, are considered in Section 9.4. The computational analogue of Shannon's perfect secrecy, ciphertext-indistinguishability, is defined. A typical security proof for probabilistic public-key encryption schemes is given. We show that the public-key one-time pads, introduced in Section 9.3, provide computationally perfect secrecy against passive eavesdroppers, who perform ciphertext-only or chosen-plaintext attacks. Encryption schemes that are secure against adaptively-chosen-ciphertext attacks, are considered in Section 9.5. The security proof for Boneh's SAEP is a typical proof in the random oracle model, the proof for Cramer-Shoup's public key encryption scheme is based solely on a standard number-theoretic assumption and the collision-resistance of the hash function used. Finally, a short introduction to some results of the "unconditional security approach" is given in Section 9.6. In this approach, the goal is to design practical cryptosystems which provably come close to perfect information-theoretic security, without relying on unproven assumptions about problems from computational number theory.

9.1 Classical Information-Theoretic Security

A deterministic public-key encryption algorithm E necessarily leaks information to an adversary. For example, recall the small-message-space attack on RSA (Section 3.3.3). An adversary intercepts a ciphertext c and knows that the transmitted message m is from a small set $\{m_1, \ldots, m_r\}$ of possible messages. Then he easily finds out m by computing the ciphertexts $E(m_1), \ldots, E(m_r)$ and comparing them with c. This example shows that randomness in encryption is necessary to ensure real secrecy. Learning the encryption $c = E(m)$ of a message m, an adversary should not be able to predict the ciphertext the next time when m is encrypted by E. This observation applies also to symmetric-key encryption schemes. Thus, to obtain a provably secure encryption scheme, we have to study randomized encryption algorithms.

Definition 9.1. An encryption algorithm E, which on input $m \in M$ outputs a ciphertext $c \in C$, is called a *randomized encryption* if E is a nondeterministic probabilistic algorithm.

The random behavior of a randomized encryption E is caused by its coin tosses. These coin tosses may be considered as the random choice of a one-time key (for each message to be encrypted a new random key is chosen, independently of the previous choices). Take, for example, Vernam's one-time pad which is the classical example of a randomized (and provably secure) cipher. We recall its definition.

Definition 9.2. Let $n \in \mathbb{N}$ and $M := C := \{0, 1\}^n$. The randomized encryption E which encrypts a message $m \in M$ by XORing it bitwise with a randomly and uniformly chosen bit sequence $k \xleftarrow{u} \{0, 1\}^n$ of the same length, $E(m) := m \oplus k$, is called *Vernam's one-time pad*.

As the name indicates, key k is used only once: each time a message m is encrypted, a new bit sequence is randomly chosen as the encryption key. This choice of the key is viewed as the coin tosses of a probabilistic algorithm. The security of a randomized encryption algorithm is related to the level of randomness caused by its coin tosses. More randomness means more security. Vernam's one-time pad includes a maximum of randomness and hence, provably, provides a maximum of security, as we will see below (Theorem 9.5).

The problem with Vernam's one-time pad is that truly random keys of the same length as the message have to be generated and securely transmitted to the recipient. This is rarely a practical operation (for an example, see Section 2.1). Later (in Section 9.4), we will see how to obtain practical, but still provably secure probabilistic encryption methods, by using high quality pseudorandom bit sequences as keys.

The classical notion of security of an encryption algorithm is based on Shannon's information theory and his famous papers [Shannon48] and

[Shannon49]. Appendix B.4 gives an introduction to information theory and its basic notions, such as entropy, uncertainty and mutual information.

We consider a randomized encryption algorithm E mapping plaintexts $m \in M$ to ciphertexts $c \in C$. We assume that the messages to be encrypted are generated according to some probability distribution, i.e., M is assumed to be a probability space. The distribution on M and the algorithm E induce probability distributions on $M \times C$ and C (see Section 5.1). As usual, the probability space induced on $M \times C$ is denoted by MC and, for $m \in M$ and $c \in C$, $\mathrm{prob}(c \,|\, m)$ denotes the probability that c is the ciphertext if the plaintext is m. Analogously, $\mathrm{prob}(m \,|\, c)$ is the probability that m is the plaintext if c is the ciphertext.[1]

Without loss of generality, we assume that $\mathrm{prob}(m) > 0$ for all $m \in M$ and that $\mathrm{prob}(c) > 0$ for all $c \in C$.

Definition 9.3 *(Shannon).* The encryption E is *perfectly secret* if C and M are independent, i.e., the distribution of MC is the product of the distributions on M and C:

$$\mathrm{prob}(m, c) = \mathrm{prob}(m) \cdot \mathrm{prob}(c), \text{ for all } m \in M, c \in C.$$

Perfect secrecy can be characterized in different ways.

Proposition 9.4. *The following statements are equivalent:*

1. *E is perfectly secret.*
2. *The mutual information $I(M; C) = 0$.*
3. *$\mathrm{prob}(m \,|\, c) = \mathrm{prob}(m)$, for all $m \in M$ and $c \in C$.*
4. *$\mathrm{prob}(c \,|\, m) = \mathrm{prob}(c)$, for all $m \in M$ and $c \in C$.*
5. *$\mathrm{prob}(c \,|\, m) = \mathrm{prob}(c \,|\, m')$, for all $m, m' \in M$ and $c \in C$.*
6. *$\mathrm{prob}(E(m) = c) = \mathrm{prob}(c)$, for all $m \in M$ and $c \in C$.*
7. *$\mathrm{prob}(E(m) = c) = \mathrm{prob}(E(m') = c)$, for all $m, m' \in M$ and $c \in C$;*
   i.e., the distribution of $E(m)$ does not depend on m.

Proof. All statements of Proposition 9.4 are contained in, or immediately follow from Proposition B.32. For the latter two statements, observe that

$$\mathrm{prob}(c \,|\, m) = \mathrm{prob}(E(m) = c),$$

by the definition of $\mathrm{prob}(E(m) = c)$ (see Chapter 5).   □

Remarks:

1. The probabilities in statement 7 only depend on the coin tosses of E. This means, in particular, that the perfect secrecy of an encryption algorithm E does not depend on the distribution of the plaintexts.

---
[1] The notation is introduced on p. 328 in Appendix B.1.

2. Let Eve be an attacker trying to discover information about the plaintexts from the ciphertexts that she is able to intercept. Assume that Eve is well informed and knows the distribution of the plaintexts. Then perfect secrecy means that her uncertainty about the plaintext (as precisely defined in information theory, see Appendix B.4, Definition B.27) is the same whether or not she observes the ciphertext c: learning the ciphertext does not increase her information about the plaintext m. Thus, perfect secrecy really means unconditional security against ciphertext-only attacks.

   A perfectly secret randomized encryption E also withstands the other types of attacks, such as the known-plaintext attacks and adaptively-chosen-plaintext/ciphertext attacks discussed in Section 1.3. Namely, the security is guaranteed by the randomness caused by the coin tosses of E. Encrypting, say, r messages, means applying E r times. The coin tosses within one of these executions of E are independent of the coin tosses in the other executions (in Vernam's one-time pad, this corresponds to the fact that an individual key is chosen independently for each message). Knowing details about previous encryptions does not help the adversary. Each encryption is a new and independent random experiment and, hence, the probabilities $\text{prob}(c|m)$ are the same, whether we take them conditional on other plaintext-ciphertext pairs (m', c') or not. Note that additional knowledge of the adversary is included by conditioning the probabilities on this knowledge.

3. The mutual information is a typical measure defined in information theory (see Definition B.30). It measures the average amount of information Eve obtains about the plaintext m when learning the ciphertext c.

Vernam's one-time pad is a perfectly secret encryption. More generally, we prove the following theorem.

Theorem 9.5 *(Shannon). Let $M := C := K := \{0,1\}^n$, and let E be a one-time pad, which encrypts $m := (m_1, \ldots, m_n) \in M$ by XORing it with a random key string $k := (k_1, \ldots, k_n) \in K$, chosen independently from m:*

$$E(m) := m \oplus k := (m_1 \oplus k_1, \ldots, m_n \oplus k_n).$$

Then E is perfectly secret if and only if K is uniformly distributed.

Proof. We have

$$\text{prob}_{MC}(m, c) = \text{prob}_{MK}(m, m \oplus c) = \text{prob}_M(m) \cdot \text{prob}_K(m \oplus c).$$

If M and C are independent, then

$$\text{prob}_M(m) \cdot \text{prob}_C(c) = \text{prob}_{MC}(m, c) = \text{prob}_M(m) \cdot \text{prob}_K(m \oplus c).$$

Hence

$$\operatorname{prob}_K(m \oplus c) = \operatorname{prob}_C(c), \text{ for all } m \in M.$$

This means that $\operatorname{prob}_K(k)$ is the same for all $k \in K$. Thus, K is uniformly distributed. Conversely, if K is uniformly distributed, then

$$\operatorname{prob}_C(c) = \sum_{m \in M} \operatorname{prob}_{MK}(m, m \oplus c) = \sum_{m \in M} \operatorname{prob}_M(m) \cdot \operatorname{prob}_K(m \oplus c)$$

$$= \sum_{m \in M} \operatorname{prob}_M(m) \cdot \frac{1}{2^n}$$

$$= \frac{1}{2^n}.$$

Hence, C is also distributed uniformly, and we obtain:

$$\operatorname{prob}_{MC}(m, c) = \operatorname{prob}_{MK}(m, m \oplus c)$$

$$= \operatorname{prob}_M(m) \cdot \operatorname{prob}_K(m \oplus c) = \operatorname{prob}_M(m) \cdot \frac{1}{2^n}$$

$$= \operatorname{prob}_M(m) \cdot \operatorname{prob}_C(c).$$

Thus, M and C are independent. $\qquad \square$

Remarks:

1. Note that we do not consider the one-time pad as a cipher for plaintexts of varying length: we have to assume that all plaintexts have the same length n. Otherwise some information, namely the length of the plaintext, leaks to adversary Eve, and the encryption could not be perfectly secret.
2. There is a high price to pay for the perfect secrecy of Vernam's one-time pad. For each message to be encrypted, of length n, n independent random bits have to be chosen for the key. One might hope to find a more sophisticated, perfectly secret encryption method requiring less randomness. Unfortunately, this hope is destroyed by the following result which was proven by Shannon ([Shannon49]).

Theorem 9.6. *Let E be a randomized encryption algorithm with the deterministic extension $E_D : M \times K \longrightarrow C$. Each time a message $m \in M$ is encrypted, a one-time key k is chosen randomly from K (according to some probability distribution on K), independently from the choice of m. Assume that the plaintext m can be recovered from the ciphertext c and the one-time key k (no other information is necessary for decryption). Then, if E is perfectly secret, the uncertainty of the keys cannot be smaller than the uncertainty of the messages:*

$$H(K) \geq H(M).$$

Remark. The uncertainty of a probability space M (see Definition B.27) is maximal and equal to $log_2(|M|)$ if the distribution of M is uniform (Proposition B.28). Hence, if $M = \{0, 1\}^n$ as in Theorem 9.5, then the entropy of any key set K – yielding a perfectly secret encryption – is at least n. Thus, the random choice of $k \in K$ requires the choice of at least n truly random bits.

Note at this point that the perfect secrecy of an encryption does not depend on the distribution of the plaintexts (Proposition 9.4). Therefore, we may assume that M is uniformly distributed and, as a consequence, that $H(M) = n$.

Proof. The plaintext m can be recovered from the ciphertext c and the one-time key k. This means that there is no uncertainty about the plaintext if both the ciphertext and the key are known, i.e., the conditional entropy $H(M|KC) = 0$ (see Definition B.30). Perfect secrecy means $I(M;C) = 0$ (Proposition 9.4), or equivalently, $H(C) = H(C|M)$ (Proposition B.32). Since M and K are assumed to be independent, $I(K;M) = I(M;K) = 0$ (Proposition B.32). We compute by use of Proposition B.31 and Definition B.33 the following:

$$
\begin{aligned}
H(K) - H(M) &= I(K;M) + H(K|M) - I(M;K) - H(M|K) \\
&= H(K|M) - H(M|K) \\
&= I(K;C|M) + H(K|CM) - I(M;C|K) - H(M|KC) \\
&= I(K;C|M) + H(K|CM) - I(M;C|K) \\
&\geq I(K;C|M) - I(M;C|K) \\
&= H(C|M) - H(C|KM) - H(C|K) + H(C|KM) \\
&= H(C|M) - H(C|K) \\
&= H(C) - H(C) + I(K;C) = I(K;C) \\
&\geq 0.
\end{aligned}
$$

The proof of Theorem 9.6 is finished.   □

Remark. In Vernam's one-time pad it is not possible, without destroying perfect secrecy, to use the same randomly chosen key for the encryption of two messages. This immediately follows, for example, from Theorem 9.6. Namely, such a modified Vernam one-time pad may be described as a probabilistic algorithm from $M \times M$ to $C \times C$, with the deterministic extension

$$ M \times M \times K \longrightarrow C \times C, \quad (m, m', k) \mapsto (m \oplus k, m' \oplus k), $$

where $M = K = C = \{0,1\}^n$. Assuming the uniform distribution on M, we have

$$ H(K) = n < H(M \times M) = 2n. $$

9.2 Perfect Secrecy and Probabilistic Attacks

We model the behavior of an adversary Eve by probabilistic algorithms, and show the relation between the failure of such algorithms and perfect secrecy.

In Section 9.3, we will slightly modify this model by restricting the computing power of the adversary to polynomial resources.

As in Section 9.1, let E be a randomized encryption algorithm that maps plaintexts $m \in M$ to ciphertexts $c \in C$ and is used by Alice to encrypt her messages. As before, Alice chooses the messages $m \in M$ according to some probability distribution. The distribution on M and the algorithm E induce probability distributions on $M \times C$ and C. $\text{prob}(m, c)$ is the probability that m is the chosen message and that the probabilistic encryption of m yields c.

We first consider a probabilistic algorithm A which on input $c \in C$ outputs a plaintext $m \in M$. Algorithm A models an adversary Eve performing a ciphertext-only attack and trying to decrypt ciphertexts. Recall that the coin tosses of a probabilistic algorithm are independent of any other random events in the given setting (see Chapter 5). Thus, the coin tosses of A are independent of the choice of the message and the coin tosses of E. This is a reasonable model, because sender Alice, generating and encrypting messages, and adversary Eve operate independently. We have

$$\text{prob}(m, c, A(c) = m) = \text{prob}(m, c) \cdot \text{prob}(A(c) = m),$$

for $m \in M$ and $c \in C$ (see Chapter 5). $\text{prob}(A(c) = m)$ is the conditional probability that $A(c)$ yields m, assuming that m and c are fixed. It is determined by the coin tosses of A. The *probability of success* of A is given by

$$\text{prob}_{\text{success}}(A) := \sum_{m,c} \text{prob}(m, c) \cdot \text{prob}(A(c) = m)$$

$$= \sum_{m,c} \text{prob}(m) \cdot \text{prob}(E(m) = c) \cdot \text{prob}(A(c) = m)$$

$$= \text{prob}(A(c) = m : m \leftarrow M, c \leftarrow E(m)).$$

Proposition 9.7. *If E is perfectly secret, then for every probabilistic algorithm A which on input $c \in C$ outputs a plaintext $m \in M$*

$$\text{prob}_{\text{success}}(A) \leq \max_{m \in M} \text{prob}(m).$$

Proof.

$$\text{prob}_{\text{success}}(A) = \sum_{m,c} \text{prob}(m, c) \cdot \text{prob}(A(c) = m)$$

$$= \sum_{c} \text{prob}(c) \cdot \sum_{m} \text{prob}(m \mid c) \cdot \text{prob}(A(c) = m)$$

$$= \sum_{c} \text{prob}(c) \cdot \sum_{m} \text{prob}(m) \cdot \text{prob}(A(c) = m) \qquad \text{(by Proposition 9.4)}$$

$$\leq \max_{m \in M} \text{prob}(m) \cdot \sum_c \text{prob}(c) \cdot \sum_m \text{prob}(A(c) = m)$$

$$= \max_{m \in M} \text{prob}(m),$$

and the proposition follows.   □

Remarks:

1. In Proposition 9.7, as in the whole of Section 9.2, we do not assume any limits for the resources of the algorithms. The running time and the memory requirements may be exponential.
2. Proposition 9.7 says that for a perfectly secret encryption, selecting a plaintext with maximal probability from M, without looking at the ciphertext, is optimum under all attacks that try to derive the plaintext from the ciphertext. If M is uniformly distributed, then randomly selecting a plaintext is an optimal strategy.

Perfect secrecy may also be described in terms of distinguishing algorithms.

Definition 9.8. A *distinguishing algorithm* for E is a probabilistic algorithm A which on inputs $m_0, m_1 \in M$ and $c \in C$ outputs an $m \in \{m_0, m_1\}$.

Remark. A distinguishing algorithm A models an adversary Eve, who, given a ciphertext c and two plaintext candidates m_0 and m_1, tries to find out which one of the both is the correct plaintext, i.e., which one is encrypted as c. Again, recall that the coin tosses of A are independent of a random choice of the messages and the coin tosses of the encryption algorithm (see Chapter 5). Thus, the adversary Eve and the sender Alice, generating and encrypting messages, are modeled as working independently.

Proposition 9.9. E *is perfectly secret if and only if for every probabilistic distinguishing algorithm* A *and all* $m_0, m_1 \in M$,

$$\text{prob}(A(m_0, m_1, c) = m_0 : c \leftarrow E(m_0))$$
$$= \text{prob}(A(m_0, m_1, c) = m_0 : c \leftarrow E(m_1)).$$

Proof. E is perfectly secret if and only if the distribution of $E(m)$ does not depend on m (Proposition 9.4). Thus, the equality obviously holds if E is perfectly secret.

Conversely, assume that E is not perfectly secret. There are no limits for the running time of our algorithms. Then there is an algorithm P which starts with a description of the encryption algorithm E and analyzes the paths and coin tosses of E and, in this way, computes the probabilities $\text{prob}(c|m)$:

$$P(c, m) := \text{prob}(c|m), \text{ for all } c \in C, m \in M.$$

We define the following distinguishing algorithm:

$$A(m_0, m_1, c) := \begin{cases} m_0 \text{ if } P(c, m_0) > P(c, m_1), \\ m_1 \text{ otherwise.} \end{cases}$$

Since E is not perfectly secret, there are $m_0, m_1 \in M$ and $c_0 \in C$, such that $P(c_0, m_0) = \mathrm{prob}(c_0 \,|\, m_0) > P(c_0, m_1) = \mathrm{prob}(c_0 \,|\, m_1)$ (Proposition 9.4). Let

$$C_0 := \{c \in C \mid \mathrm{prob}(c \,|\, m_0) > \mathrm{prob}(c \,|\, m_1)\} \text{ and}$$

$$C_1 := \{c \in C \mid \mathrm{prob}(c \,|\, m_0) \leq \mathrm{prob}(c \,|\, m_1)\}.$$

Then $A(m_0, m_1, c) = m_0$ for $c \in C_0$, and $A(m_0, m_1, c) = m_1$ for $c \in C_1$. We compute

$$\begin{aligned}
\mathrm{prob}(A(m_0, m_1, c) &= m_0 : c \leftarrow E(m_0)) \\
&\quad - \mathrm{prob}(A(m_0, m_1, c) = m_0 : c \leftarrow E(m_1)) \\
&= \sum_{c \in C} \mathrm{prob}(c \,|\, m_0) \cdot \mathrm{prob}(A(m_0, m_1, c) = m_0) \\
&\quad - \sum_{c \in C} \mathrm{prob}(c \,|\, m_1) \cdot \mathrm{prob}(A(m_0, m_1, c) = m_0) \\
&= \sum_{c \in C_0} \mathrm{prob}(c \,|\, m_0) - \mathrm{prob}(c \,|\, m_1) \\
&\geq \mathrm{prob}(c_0 \,|\, m_0) - \mathrm{prob}(c_0 \,|\, m_1) \\
&> 0,
\end{aligned}$$

and see that a violation of perfect secrecy causes a violation of the equality condition. The proof of the proposition is finished. $\qquad\square$

Proposition 9.10. *E is perfectly secret if and only if for every probabilistic distinguishing algorithm A and all $m_0, m_1 \in M$, with $m_0 \neq m_1$,*

$$\mathrm{prob}(A(m_0, m_1, c) = m : m \xleftarrow{u} \{m_0, m_1\}, c \leftarrow E(m)) = \frac{1}{2}.$$

Proof.

$$\begin{aligned}
\mathrm{prob}(A(m_0, m_1, c) &= m : m \xleftarrow{u} \{m_0, m_1\}, c \leftarrow E(m)) \\
&= \frac{1}{2} \cdot \mathrm{prob}(A(m_0, m_1, c) = m_0 : c \leftarrow E(m_0)) \\
&\quad + \frac{1}{2} \cdot \mathrm{prob}(A(m_0, m_1, c) = m_1 : c \leftarrow E(m_1)) \\
&= \frac{1}{2} + \frac{1}{2} \cdot (\mathrm{prob}(A(m_0, m_1, c) = m_0 : c \leftarrow E(m_0)) \\
&\quad - \mathrm{prob}(A(m_0, m_1, c) = m_0 : c \leftarrow E(m_1)),
\end{aligned}$$

and the proposition follows from Proposition 9.9. $\qquad\square$

Remark. Proposition 9.10 characterizes a perfectly secret encryption scheme in terms of a passive eavesdropper A, who performs a ciphertext-only attack. But, as we observed before, the statement would remain true, if we model an (adaptively-)chosen-plaintext/ciphertext attacker by algorithm A (see the remark after Proposition 9.4).

9.3 Public-Key One-Time Pads

Vernam's one-time pad is provably secure (Section 9.1) and thus appears to be a very attractive encryption method. However, there is the problem that truly random keys of the same length as the message have to be generated and securely transmitted to the recipient. The idea now is to use high quality ("cryptographically secure") pseudorandom bit sequences as keys and to obtain in this way practical, but still provably secure randomized encryption methods.

Definition 9.11. Let $I = (I_k)_{k \in \mathbb{N}}$ be a key set with security parameter k, and let $G = (G_i)_{i \in I}$, $G_i : X_i \longrightarrow \{0,1\}^{l(k)}$ $(i \in I_k)$, be a pseudorandom bit generator with polynomial stretch function l and key generator K (see Definition 8.2).

The probabilistic polynomial encryption algorithm $E(i, m)$ which, given (a public key) $i \in I_k$, encrypts a message $m \in \{0,1\}^{l(k)}$ by bitwise XORing it with the pseudorandom sequence $G_i(x)$, generated by G_i from a randomly and uniformly chosen seed $x \in X_i$,

$$E(i, m) := m \oplus G_i(x), \quad x \stackrel{u}{\leftarrow} X_i,$$

is called the *pseudorandom one-time pad induced by G*. Keys i are assumed to be generated by K.

Example. Let $f = (f_i : D_i \longrightarrow D_i)_{i \in I}$ be a family of one-way permutations with hard-core predicate $B = (B_i : D_i \longrightarrow \{0,1\})_{i \in I}$, and let Q be a polynomial. f, B and Q induce a pseudorandom bit generator $G(f, B, Q)$ with stretch function Q (see Definition 8.3), and hence a pseudorandom one-time pad.

We will see in Section 9.4 that computationally perfect pseudorandom generators (Definition 8.2), such as the $G(f, B, Q)$s, lead to provably secure encryption schemes. Nevertheless, one important problem remains unsolved: how to transmit the secret one-time key – the randomly chosen seed x – to the recipient of the message?

If G is induced by a family of trapdoor permutations with hard-core predicate, there is an easy answer. We send x hidden by the one-way function together with the encrypted message. Knowing the trapdoor, the recipient is able to determine x.

Definition 9.12. Let $I = (I_k)_{k \in \mathbb{N}}$ be a key set with security parameter k, and let Q be a positive polynomial. Let $f = (f_i : D_i \longrightarrow D_i)_{i \in I}$ be a family of trapdoor permutations with hard-core predicate B and key generator K. Let $G(f, B, Q)$ be the induced pseudorandom bit generator with stretch function Q. For every recipient of messages, a public key $i \in I_k$ (and the associated trapdoor information) is generated by K.

The probabilistic polynomial encryption algorithm $E(i, m)$ which encrypts a message $m \in \{0, 1\}^{Q(k)}$ as

$$E(i, m) := (m \oplus G(f, B, Q)_i(x), f_i^{Q(k)}(x)),$$

with x chosen randomly and uniformly from D_i (for each message m), is called the *public-key one-time pad* induced by f, B and Q.

Remarks:

1. Recall that we get $G(f, B, Q)_i(x)$ by repeatedly applying f_i to x and taking the hard-core bits $B_i(f_i^j(x))$ of the sequence

$$x, f_i(x), f_i^2(x), f_i^3(x), \ldots, f_i^{Q(k)-1}(x)$$

   (see Definition 8.3). In order to encrypt the seed x, we then apply f_i once more. Note that we cannot take $f_i^j(x)$ with $j < Q(k)$ as the encryption of x, because this would reveal bits from the sequence $G(f, B, Q)_i(x)$.
2. Since f_i is a permutation of D_i and the recipient Bob knows the trapdoor, he has an efficient algorithm for f_i^{-1}. He is able to compute the sequence

$$x, f_i(x), f_i^2(x), f_i^3(x), \ldots, f_i^{Q(k)-1}(x)$$

   from $f_i^{Q(k)}(x)$, by repeatedly applying f_i^{-1}. In this way, he can easily decrypt the ciphertext.
3. In the public-key one-time pad, the basic pseudorandom one-time pad is augmented by an asymmetric (i.e., public-key) way of transmitting the one-time symmetric encryption key x.
4. Like the basic pseudorandom one-time pad, the augmented version is provably secure against passive attacks (see Theorem 9.16). Supplying the encrypted key $f_i^{Q(k)}(x)$ does not diminish the secrecy of the encryption scheme.
5. Pseudorandom one-time pads and public-key one-time pads are straightforward analogies to the classical probabilistic encryption method, Vernam's one-time pad. We will see in Section 9.4 that more recent notions of secrecy, such as indistinguishability (introduced in [GolMic84]), are also analogous to the classical notion in Shannon's work. The statements on the secrecy of pseudorandom and public-key one-time pads (see Section 9.4) are analogous to the classical results by Shannon.

6. The notion of *probabilistic public-key encryption*, whose security may be rigorously proven in a complexity theoretic model, was suggested by Goldwasser and Micali ([GolMic84]). They introduced the hard-core predicates of trapdoor functions (or, more generally, trapdoor predicates) as the basic building blocks of such schemes. The implementation of probabilistic public-key encryption, given in [GolMic84] and known as *Goldwasser-Micali probabilistic encryption* (see Exercise 7), is based on the quadratic residuosity assumption (Definition 6.11). During encryption, messages are expanded by a factor proportional to the security parameter k. Thus, this implementation is quite wasteful in space and bandwidth and is therefore not really practical. The public-key one-time pads, introduced in [BluGol85] and [BluBluShu86], avoid this large message expansion. They are the *efficient implementations of (asymmetric) probabilistic encryption*.

9.4 Passive Eavesdroppers

We study the security of public-key encryption schemes against passive eavesdroppers, who perform ciphertext-only attacks. In a public-key encryption scheme, a ciphertext-only attacker (as everybody) can encrypt messages of his choice at any time by using the publicly known key. Therefore, security against ciphertext-only attacks in a public-key encryption scheme also includes security against adaptively-chosen-plaintext attacks.

The stronger notion of security against adaptively-chosen-ciphertext attacks is considered in the subsequent Section 9.5.

Throughout this section we consider a probabilistic polynomial encryption algorithm $E(i, m)$, such as the pseudorandom or public-key one-time pads defined in Section 9.3. Here, $I = (I_k)_{k \in \mathbb{N}}$ is a key set with security parameter k and, for every $i \in I$, E maps plaintexts $m \in M_i$ to ciphertexts $c := E(i, m) \in C_i$. The keys i are generated by a probabilistic polynomial algorithm K and are assumed to be public. In our examples, the encryption E is derived from a family $f = (f_i)_{i \in I}$ of one-way permutations, and the index i is the public key of the recipient.

We define distinguishing algorithms A for E completely analogous to Definition 9.8. Now the computational resources of the adversary, modeled by A, are limited. A is required to be polynomial.

Definition 9.13. A *probabilistic polynomial distinguishing algorithm* for E is a probabilistic polynomial algorithm $A(i, m_0, m_1, c)$ which on inputs $i \in I, m_0, m_1 \in M_i$ and $c \in C_i$ outputs an $m \in \{m_0, m_1\}$.

Below we show that pseudorandom one-time pads induced by computationally perfect pseudorandom bit generators have computationally perfect secrecy. This result is analogous to the classical result by Shannon that Vernam's one-time pad is perfectly secret. Using truly random bit sequences as the key in the one-time pad, the probability of success of an attack with unlimited resources – which tries to distinguish between two candidate plaintexts – is equal to $1/2$; so there is no use in observing the ciphertext. Using computationally perfect pseudorandom bit sequences, the probability of success of an attack with polynomial resources is, at most, negligibly more than $1/2$.

Definition 9.14. The encryption E is called *ciphertext-indistinguishable* or (for short) *indistinguishable*, if for every probabilistic polynomial distinguishing algorithm $A(i, m_0, m_1, c)$ and every probabilistic polynomial sampling algorithm S, which on input $i \in I$ yields $S(i) = \{m_0, m_1\} \subset M_i$, and every positive polynomial $P \in \mathbb{Z}[X]$, there is a $k_0 \in \mathbb{N}$ such that for all $k \geq k_0$:

$$\mathrm{prob}(A(i, m_0, m_1, c) = m : i \leftarrow K(1^k), \{m_0, m_1\} \leftarrow S(i),$$
$$m \xleftarrow{u} \{m_0, m_1\}, c \leftarrow E(i, m)) \leq \frac{1}{2} + \frac{1}{P(k)}.$$

Remarks:

1. The definition is a definition in the "public-key model": the keys i are public and hence available to the distinguishing algorithms. It can be adapted to a private-key setting. See the analogous remark after the definition of pseudorandom generators (Definition 8.2).
2. Algorithm A models a passive adversary, who performs a ciphertext-only attack. But, everybody knows the public key i and can encrypt messages of his choice at any time. This implies that the adversary algorithm A may include the encryption of messages of its choice. We see that ciphertext-indistinguishability, as defined here, means security against adaptively-chosen-plaintext attacks. Chosen-ciphertext attacks are considered in Section 9.5.
3. The output of the sampling algorithm S is a subset $\{m_0, m_1\}$ with two members, and therefore $m_0 \neq m_1$.
   If the message spaces M_i are unrestricted, like $\{0, 1\}^*$, then it is usually required that the two candidate plaintexts m_0, m_1, generated by S, have the same bit length. This additional requirement is reasonable. Typically, the length of the plaintext and the length of the ciphertext are closely related. Hence, information about the length of the plaintexts necessarily leaks to an adversary, and plaintexts of different length can easily be distinguished. For the same reason, we considered Vernam's one-time pad as a cipher for plaintexts of a fixed length in Section 9.1 (see the remark after Theorem 9.5).
   In this section, we consider only schemes where all plaintexts are of the same bit length.
4. In view of Proposition 9.10, the definition is the computational analogy to the notion of perfect secrecy, as defined by Shannon.

Non-perfect secrecy means that some algorithm A is able to distinguish between distinct plaintexts m_0 and m_1 (given the ciphertext) with a probability $> 1/2$. The running time of A may be exponential. If an encryption scheme is not ciphertext-indistinguishable, then an algorithm with polynomial running time is able to distinguish between distinct plaintexts m_0 and m_1 (given the ciphertext) with a probability significantly larger than $1/2$. In addition, the plaintexts m_0 and m_1 can be found in polynomial time by some probabilistic algorithm S. This additional requirement is adequate. A secrecy problem can only exist for messages which can be generated in practice by using a probabilistic polynomial algorithm. The message generation is modeled uniformly by S for all keys i.[2]

5. The notion of ciphertext-indistinguishability was introduced by Gold-wasser and Micali ([GolMic84]). They call it *polynomial security* or *polynomial-time indistinguishability*. Ciphertext-indistinguishable encryption schemes are also called *schemes with indistinguishable encryptions*. Another notion of security was introduced in [GolMic84]. An encryption scheme is called *semantically secure*, if it has the following property: Whatever a passive adversary Eve is able to compute about the plaintext in polynomial time given the ciphertext, she is also able to compute in polynomial time without the ciphertext. The messages to be encrypted are assumed to be generated by a probabilistic polynomial algorithm. Semantic security is equivalent to indistinguishability ([GolMic84]; [MicRacSlo88]; [WatShiIma03]; [Goldreich04]).

6. Recall that the execution of a probabilistic algorithm is an independent random experiment (see Chapter 5). Thus, the coin tosses of the distinguishing algorithm A are independent of the coin tosses of the sampling algorithm S and the coin tosses of the encryption algorithm E. This reflects the fact that sender and adversary operate independently.

7. The probability in our present definition is also taken over the random generation of a key i, with a given security parameter k. Even for very large k, there may be insecure keys i such that A is able to distinguish successfully between two plaintext candidates. However, when randomly generating keys by the key generator, the probability of obtaining an insecure one is negligibly small (see Proposition 6.17 for a precise statement).

8. Ciphertext-indistinguishable encryption algorithms are necessarily randomized encryptions. When encrypting a plaintext m twice, the probability that we get the same ciphertext must be negligibly small. Otherwise it would be easy to distinguish between two messages m_0 and m_1 by comparing the ciphertext with encryptions of m_0 and m_1.

---

[2] By applying a (more general) non-uniform model of computation (non-uniform polynomial-time algorithms instead of probabilistic polynomial algorithms, see, for example, [Goldreich99], [Goldreich04]), one can dispense with the sampling algorithm S.

Theorem 9.15. *Let E be the pseudorandom one-time pad induced by a computationally perfect pseudorandom bit generator G. Then E is ciphertext-indistinguishable.*

Proof. The proof runs in exactly the same way as the proof of Theorem 9.16, yielding a contradiction to the assumption that G is computationally perfect. See below for the details. $\qquad\qquad\qquad\qquad\qquad\qquad\qquad\qquad\qquad\qquad\Box$

General pseudorandom one-time pads leave open how the secret encryption key – the randomly chosen seed – is securely transmitted to the receiver of the message. Public-key one-time pads provide an answer. The key is encrypted by the underlying one-way permutation and becomes part of the encrypted message (see Definition 9.12). The indistinguishability is preserved.

Theorem 9.16. *Let E be the public-key one-time pad induced by a family $f = (f_i : D_i \longrightarrow D_i)_{i \in I}$ of trapdoor permutations with hard-core predicate B. Then E is ciphertext-indistinguishable.*

Remark. Theorem 9.16 states that public-key cryptography provides variants of the one-time pad which are provably secure and practical. XORing the plaintext with a pseudorandom bit sequence generated from a short random seed by a trapdoor permutation with hard-core predicate (e.g. use the Blum-Blum-Shub generator or the RSA generator, see Section 8.1,) yields an encryption with indistinguishability. Given the ciphertext, an adversary is provably not able to distinguish between two plaintexts. In addition, it is possible in this public-key one-time pad to securely transmit the key string (more precisely, the seed of the pseudorandom sequence) to the recipient, simply by encrypting it by means of the one-way function.

Of course, the security proof for a public-key one-time pad, such as the RSA- or Blum-Blum-Shub-based one-time pad, is conditional. It depends on the validity of basic unproven (though widely believed) assumptions, such as the RSA assumption (Definition 6.7), or the factoring assumption (Definition 6.9).

Computing the pseudorandom bit sequences using a one-way permutation requires complex computations, such as exponentiation and modular reductions. Thus, the classical private-key symmetric encryption methods, like the DES (see Chapter 2) or stream ciphers, using shift registers to generate pseudorandom sequences (see, e.g., [MenOorVan96], Chapter 6), are much more efficient than public-key one-time pads, and hence are better suited for large amounts of data.

However, notice that the one-way function of the Blum-Blum-Shub generator (see Chapter 8) is a quite simple one. Quadratic residues $x \bmod n$ are squared: $x \mapsto x^2 \bmod n$. A public-key one-time pad, whose efficiency is comparable to standard RSA encryption, can be implemented based on this generator.

Namely, suppose $n = pq$, with distinct primes $p, q \equiv 3 \bmod 4$ of binary length k. Messages m of length l are to be encrypted. In order to encrypt m, we randomly choose an element from $\mathbb{Z}_n^*$ and square it to get a random x in QR_n. This requires $O(k^2)$ steps. To get the pseudorandom sequence and to encrypt the random seed x we have to compute l squares modulo n, which comes out to $O(k^2 l)$ steps. XORing requires $O(l)$ steps. Thus, encryption is finished in $O(k^2 l)$ steps.

To decrypt, we first compute the seed x from x^{2^l} by drawing the square root l times in QR_n. We can do this by drawing the square roots modulo p and modulo q, and applying the Chinese Remainder Theorem (see Proposition A.62).

Assume, in addition, that we have even chosen $p, q \equiv 7 \bmod 8$. Then $p + 1/8$ is in $\mathbb{N}$, and for every quadratic residue $a \in \mathrm{QR}_p$ we get a square root b which again is an element of QR_p by setting

$$b = a^{(p+1)/4} = \left(a^{(p+1)/8} \right)^2.$$

Here, note that
$$b^2 = a^{(p+1)/2} = a \cdot a^{(p-1)/2} = a,$$

since $a^{(p-1)/2} = \left(\frac{a}{p} \right) = 1$ for quadratic residues $a \in \mathrm{QR}_p$ (Proposition A.52). Thus, we can derive $x \bmod p$ from $y = x^{2^l}$ by

$$x \bmod p = y^u \bmod p, \text{ with } u = \left(\frac{p+1}{4} \right)^l \bmod (p-1).$$

The exponent u can be reduced modulo $p - 1$, since $a^{p-1} = 1$ for all $a \in \mathbb{Z}_p^*$. We assume that the message length l is fixed. Then the exponent u can be computed in advance, and we see that figuring out $x \bmod p$ (or $x \bmod q$) requires at most k squarings applying the repeated squaring algorithm (Algorithm A.26). Thus, it can be done in $O(k^3)$ steps.

Reducing $x^{2^l} \bmod p$ (and $x^{2^l} \bmod q$) at the beginning and applying the Chinese Remainder Theorem at the end requires at most $O(k^2)$ steps. Summarizing, we see that computing x requires $O(k^3)$ steps. Now, completing the decryption essentially means performing an encryption whose cost is $O(k^2 l)$, as we saw above. Hence, the complete decryption procedure takes $O(k^3 + k^2 l)$ steps. If $l = O(k)$, this is equal to $O(k^3)$ and thus of the same order as the running time of an RSA encryption.

The efficiency of the Blum-Blum-Shub-based public-key one-time pad (as well as that of the RSA-based one) can be increased even further, by modifying the generation of the pseudorandom bit sequence. Instead of taking only the least-significant bit of $x^{2^j} \bmod n$, you may take the $\lfloor \log_2(|n|) \rfloor$ least-significant bits after each squaring. These bits form a $\lfloor \log_2(|n|) \rfloor$-bit hard-core predicate of the modular squaring function and are simultaneously secure (see Exercise 7 in Chapter 8). The resulting public-key one-time pad is called *Blum-Goldwasser probabilistic encryption* ([BluGol85]). It is also ciphertext-indistinguishable (see Exercise 8 in Chapter 8).

Our considerations are not only valid asymptotically, as the O notation might suggest. Take for example $k = 512$ and $|n| = 1024$, and encrypt 1024-bit messages m. In the $x^2 \bmod n$ public-key one-time pad, always use the $\log_2(|n|) = 10$ least-significant bits. To encrypt a message m, about 100 modular squarings of 1024-bit numbers are necessary. To decrypt a ciphertext, we first determine the seed x by at most $1024 = 512 + 512$ modular squarings and multiplications of 512-bit numbers, and then compute the plaintext using about 100 modular squarings, as in encryption. Encrypting and decrypting messages $m \in \mathbb{Z}_n$ by RSA requires up to 1024 modular squarings and multiplications of 1024-bit (encryption) or 512-bit (decryption) numbers, with the actual number depending on the size of the encryption and decryption exponents. In the estimates, we did not count the (few) operations associated with applying the Chinese Remainder Theorem during decryption.

Proof (of Theorem 9.16). Let K be the key generator of f and $G := G(f, B, Q)$ be the pseudorandom bit generator (Chapter 8). Recall that

$$E(i, m) = (m \oplus G_i(x), f_i^{Q(k)}(x)),$$

where $i \in I = (I_k)_{k \in \mathbb{N}}$, $m \in \{0, 1\}^{Q(k)}$ (for $i \in I_k$) and the seed x is randomly and uniformly chosen from D_i.

The image of the uniform distribution on D_i under $f_i^{Q(k)}$ is again the uniform distribution on D_i, because f_i is bijective.

You can obtain a proof of Theorem 9.15 simply by omitting "$\times D_i$", $y, f_i^{Q(k)}(x)$ everywhere (and by replacing $Q(k)$ by $l(k)$). Our proof yields a contradiction to Theorem 8.4. In the proof of Theorem 9.15, you get the completely analogous contradiction to the assumption that the pseudorandom bit generator G is computationally perfect.

Now assume that there is a probabilistic polynomial distinguishing algorithm $A(i, m_0, m_1, c, y)$, with inputs $i \in I, m_0, m_1 \in M_i, c \in \{0, 1\}^{Q(k)}$ (if $i \in I_k$), $y \in D_i$, a probabilistic polynomial sampling algorithm $S(i)$ and a positive polynomial P, such that

$\text{prob}(A(i, m_0, m_1, c, y) = m :$
$$i \leftarrow K(1^k), \{m_0, m_1\} \leftarrow S(i), m \xleftarrow{u} \{m_0, m_1\}, (c, y) \leftarrow E(i, m))$$
$$= \text{prob}(A(i, m_0, m_1, m \oplus G_i(x), f_i^{Q(k)}(x)) = m :$$
$$i \leftarrow K(1^k), \{m_0, m_1\} \leftarrow S(i), m \xleftarrow{u} \{m_0, m_1\}, x \xleftarrow{u} D_i)$$
$$> \frac{1}{2} + \frac{1}{P(k)},$$

for infinitely many k. We define a probabilistic polynomial statistical test $\tilde{A} = \tilde{A}(i, z, y)$, with inputs $i \in I, z \in \{0, 1\}^{Q(k)}$ (if $i \in I_k$) and $y \in D_i$ and output in $\{0, 1\}$:

1. Apply $S(i)$ and get $\{m_0, m_1\} := S(i)$.

2. Randomly choose m in $\{m_0, m_1\} : m \xleftarrow{u} \{m_0, m_1\}$.
3. Let

$$\tilde{A}(i, z, y) := \begin{cases} 1 \text{ if } A(i, m_0, m_1, m \oplus z, y) = m, \\ 0 \text{ otherwise.} \end{cases}$$

The statistical test $\tilde{A}$ will be able to distinguish between the pseudorandom sequences produced by G and truly random, uniformly distributed sequences, thus yielding the desired contradiction. We have to compute the probability

$$\text{prob}(\tilde{A}(i, z, y) = 1 : i \leftarrow K(1^k), (z, y) \leftarrow \{0, 1\}^{Q(k)} \times D_i)$$

for both the uniform distribution on $\{0, 1\}^{Q(k)} \times D_i$ and the distribution induced by $(G, f^{Q(k)})$. More precisely, we have to compare the probabilities

$$p_{k, G} := \text{prob}(\tilde{A}(i, G_i(x), f_i^{Q(k)}(x)) = 1 : i \leftarrow K(1^k), x \xleftarrow{u} D_i) \quad \text{and}$$

$$p_{k, \text{uni}} := \text{prob}(\tilde{A}(i, z, y) = 1 : i \leftarrow K(1^k), z \xleftarrow{u} \{0, 1\}^{Q(k)}, y \xleftarrow{u} D_i).$$

Our goal is to prove that

$$p_{k, G} - p_{k, \text{uni}} > \frac{1}{P(k)},$$

for infinitely many k. This contradicts Theorem 8.4 and finishes the proof. From the definition of $\tilde{A}$ we get

$$\text{prob}(\tilde{A}(i, z, y) = 1 : i \leftarrow K(1^k), (z, y) \leftarrow \{0, 1\}^{Q(k)} \times D_i)$$
$$= \text{prob}(A(i, m_0, m_1, m \oplus z, y) = m : i \leftarrow K(1^k),$$
$$(z, y) \leftarrow \{0, 1\}^{Q(k)} \times D_i, \{m_0, m_1\} \leftarrow S(i), m \xleftarrow{u} \{m_0, m_1\}).$$

Since the random choice of (z, y) and the random choice of m in the probabilistically computed pair $S(i)$ are independent, we may switch them in the probability and obtain

$$\text{prob}(\tilde{A}(i, z, y) = 1 : i \leftarrow K(1^k), (z, y) \leftarrow \{0, 1\}^{Q(k)} \times D_i)$$
$$= \text{prob}(A(i, m_0, m_1, m \oplus z, y) = m : i \leftarrow K(1^k),$$
$$\{m_0, m_1\} \leftarrow S(i), m \xleftarrow{u} \{m_0, m_1\}, (z, y) \leftarrow \{0, 1\}^{Q(k)} \times D_i).$$

Now consider $p_{k, G}$:

$$p_{k, G} = \text{prob}(A(i, m_0, m_1, m \oplus G_i(x), f_i^{Q(k)}(x)) = m :$$
$$i \leftarrow K(1^k), \{m_0, m_1\} \leftarrow S(i), m \xleftarrow{u} \{m_0, m_1\}, x \xleftarrow{u} D_i).$$

We assumed that this probability is $> 1/2 + 1/P(k)$, for infinitely many k. The following computation shows that $p_{k, \text{uni}} = 1/2$ for all k, thus completing the proof. We have

$$p_{k,\text{uni}} = \text{prob}(A(i, m_0, m_1, m \oplus z, y) = m : i \leftarrow K(1^k), \{m_0, m_1\} \leftarrow S(i),$$
$$m \xleftarrow{u} \{m_0, m_1\}, z \xleftarrow{u} \{0, 1\}^{Q(k)}, y \xleftarrow{u} D_i)$$
$$= \sum_i \text{prob}(K(1^k) = i) \cdot \sum_{m_0, m_1} \text{prob}(S(i) = \{m_0, m_1\}) \cdot p_{i, m_0, m_1},$$

with

$$p_{i, m_0, m_1} = \text{prob}(A(i, m_0, m_1, m \oplus z, y) = m :$$
$$m \xleftarrow{u} \{m_0, m_1\}, z \xleftarrow{u} \{0, 1\}^{Q(k)}, y \xleftarrow{u} D_i).$$

Vernam's one-time pad is perfectly secret (Theorem 9.5). Thus, the latter probability is equal to $1/2$ by Proposition 9.10. Note that the additional input $y \xleftarrow{u} D_i$ of A, not appearing in Proposition 9.10, can be viewed as additional coin tosses of A. So, we finally get the desired equation

$$p_{k,\text{uni}} = \sum_i \text{prob}(K(1^k) = i) \sum_{m_0, m_1} \text{prob}(S(i) = \{m_0, m_1\}) \cdot \frac{1}{2} = \frac{1}{2},$$

and the proof of the theorem is finished. $\square$

9.5 Chosen-Ciphertext Attacks

In the preceding section, we studied encryption schemes which are ciphertext-indistinguishable and provide security against a passive eavesdropper, who performs a ciphertext-only or an adaptively-chosen-plaintext attack. The schemes may still be insecure against an attacker, who manages to get temporary access to the decryption device and who executes a chosen-ciphertext or an adaptively-chosen-ciphertext attack (see Section 1.3 for a classification of attacks). Given a ciphertext c, such an attacker Eve tries to get information about the plaintext. In the course of her attack, Eve can get decryptions of ciphertexts c' from the decryption device, with the only restriction that $c' \neq c$. She has temporary access to a "decryption oracle".

Consider, for example, the efficient implementation of Goldwasser-Micali's probabilistic encryption, which we called public-key one-time pad and which we studied in the preceding sections. A public-key one-time pad encrypts a message m by XORing it bitwise with a pseudorandom key stream $G(x)$, where x is a secret random seed. The pseudorandom bit generator G is induced by a family $f = (f_i : D_i \longrightarrow D_i)_{i \in I}$ of trapdoor permutations with hard-core predicate B. The public-key-encrypted seed $f^l(x)$ (where l is the bit length of the messages m) is transmitted together with the encrypted message $m \oplus G(x)$ (see Section 9.4 above).

These encryption schemes are provably secure against passive eavesdroppers (Theorem 9.16), but they are insecure against a chosen-ciphertext attacker Eve. Eve submits ciphertexts (c, y) for decryption, where y is any element in the domain of f and $c = c_1c_2 \ldots c_l$ is any bit string of length l. If Eve only obtains the last bit m_l of the plaintext $m = m_1m_2 \ldots m_l$ from the decryption device, then she immediately derives the hard-core bit $B(f^{-1}(y))$ of $f^{-1}(y)$, since $B(f^{-1}(y)) = c_l \oplus m_l$. Therefore, Eve has an oracle that provides her with the hard-core bit $B(f^{-1}(y))$ for every y. Now assume that f is the RSA function modulo a composite $n = pq$ or the Rabin function, which squares quadratic residues modulo $n = pq$, as in Blum-Blum-Shub-encryption, and B is the least significant bit. The hard-core-bit oracle enables Eve to compute the inverse of the RSA or Rabin function modulo n by using an efficient algorithm, which calls the oracle as a subroutine. We constructed these algorithms in Sections 7.2 and 7.3. Then, of course, Eve can also compute the seed x from $f^l(x)$ and derive the plaintext $m = c \oplus G(x)$ for every ciphertext c.

We have to worry about adaptively-chosen-ciphertext attacks. One can imagine scenarios where Bob, the owner of the secret decryption key, might think that decryption requests are reasonable – for example, if an incomplete configuration or control of privileges enables attacker Eve from time to time to get access to the decryption device. If a system is secure against chosen-ciphertext attacks, then it also resists partial chosen-ciphertext attacks. In such an attack, adversary Eve does not get the full plaintext in response to her decryption requests, but only some partial information. Partial-chosen-ciphertext attacks are a real danger in practice. We just discussed a partial-chosen-ciphertext attack against public-key one-time pads. In Section 3.3.3, we described Bleichenbacher's 1-Million-Chosen-Ciphertext Attack against PKCS#1(v1.5)-based schemes, which is a practical example of a partial-chosen-ciphertext attack.

Therefore, it is desirable to have encryption schemes which are provably secure against adaptively-chosen-ciphertext attacks. We give two examples of such schemes. The security proof of the first scheme, Boneh's SAEP, relies on the random oracle model, which we described in Section 3.4.5, and the factoring assumption (Definition 6.9). The security proof of the second scheme, Cramer-Shoup's public key encryption scheme, is based solely on a standard number-theoretic assumption of the hardness of a computational problem and on a standard hash function assumption (collision-resistance). The stronger random oracle model is not needed.

We start with a definition of the security notion. The definition includes a precise description of the attack model. As before, we strive for ciphertext-indistinguishability. This notion can be extended to cover adaptively-chosen-ciphertext attacks ([NaoYun90]; [RacSim91]).

Definition 9.17. Let E be a public-key encryption scheme.

1. An *adaptively-chosen-ciphertext attack algorithm* A against E is a probabilistic polynomial algorithm that interacts with its environment, called the challenger, as follows:

a. **Setup:** The challenger C randomly generates a public-secret key pair (pk, sk) for E and calls A with the public key pk as input. The secret key sk is kept secret.

b. **Phase I:** The adversary A issues a sequence of decryption requests for various ciphertexts c'. The challenger responds with the decryption of the valid ciphertexts c'.

c. **Challenge:** At some point, algorithm A outputs two distinct messages m_0, m_1. The challenger selects a message $m \in \{m_0, m_1\}$ at random and responds with the "challenge" ciphertext c, which is an encryption $E(pk, m)$ of m.

d. **Phase II:** The adversary A continues to request the decryption of ciphertexts c', with the only constraint that $c' \neq c$. The challenger decrypts c', if c' is valid, and sends the plaintext to A. Finally, A terminates and outputs $m' \in \{m_0, m_1\}$.

The attacker A is successful, if $m' = m$.

We also call A, more precisely, an *adaptively-chosen-ciphertext distinguishing algorithm*.

2. E is *ciphertext-indistinguishable against adaptively-chosen-ciphertext attacks*, if for every adaptively-chosen-ciphertext distinguishing algorithm A, the probability of success is $\leq 1/2 + \varepsilon$, with ε negligible.

We also call such an encryption scheme E *indistinguishability-secure*, or *secure*, for short, *against adaptively-chosen-ciphertext attacks*.

Remarks:

1. In the real attack, the challenger is Bob, the legitimate owner of a public-secret key pair (pk, sk). He is attacked by the adversary A.

2. If the message space is unrestricted, like $\{0, 1\}^*$, then it is usually required that the two candidate plaintexts m_0, m_1 have the same bit length. This additional requirement is reasonable. Typically, the length of the plaintext and the length of the ciphertext are closely related. Hence, information about the length of the plaintexts necessarily leaks to an adversary, and plaintexts of different length can easily be distinguished (also see the remark on Vernam's one-time pad after Theorem 9.5).

   If the message space is a number-theoretic or geometric domain, as in the typical public-key encryption scheme, then, usually, all messages have the same bit length. Take, for example, $\mathbb{Z}_n$. The messages $m \in \mathbb{Z}_n$ are all encoded as bit strings of length $\lfloor \log_2(n) \rfloor + 1$; they are padded out with the appropriate number of leading zeros, if necessary.

3. In the literature, adaptively-chosen-ciphertext attacks are sometimes denoted by the acronym CCA2, and sometimes they are simply called chosen-ciphertext attacks. Non-adaptive chosen-ciphertext attacks, which can request the decryption of ciphertexts only in phase I of the attack, are often called *lunchtime attacks* or *midnight attacks* and denoted by the acronym CCA1.

4. The notion of semantic security (see the remarks after Definition 9.14) can also be carried over to the adaptively-chosen-ciphertext setting. It means that whatever an adversary Eve is able to compute about the plaintext m in polynomial time given the ciphertext c, she is also able to compute in polynomial time without the ciphertext, even if Eve gets the decryption of any adaptively chosen ciphertexts $c' \neq c$. As shown recently, semantic security and indistinguishability are also equivalent in the case of adaptively-chosen-ciphertext attacks ([WatShiIma03]; [Goldreich04]).

9.5.1 A Security Proof in the Random Oracle Model

Boneh's Simplified OAEP – SAEP. As an example, we study Boneh's Simple-OAEP encryption, or SAEP for short ([Boneh01]). In the encryption scheme, a collision-resistant hash function h is used. The security proof for SAEP is a proof in the random oracle model. Basically, this means that h is assumed to be a truly random function (see page 66 for a more precise description of the random oracle model).

In contrast to Bellare's OAEP, which we studied in Section 3.3.4, SAEP applies Rabin encryption (Section 3.6.1) and not the RSA function. Compared to OAEP, the padding scheme is considerably simplified. It requires only one cryptographic hash function. The slightly more complex padding scheme SAEP+ is provably secure and can be applied to the RSA and the Rabin function (see [Boneh01]). As OAEP, it requires an additional hash function G. We do not discuss SAEP+ here.

Key Generation. Let $k \in \mathbb{N}$ be an even security parameter (e.g. $k = 1024$). Bob generates a $(k+2)$-bit modulus $n = pq$, with $2^{k+1} < n < 2^{k+1} + 2^k$ (i.e., the two most significant bits of n are 10), where p and q are $(k/2 + 1)$-bit primes, with $p \equiv q \equiv 3 \bmod 4$. The primes p, q are chosen randomly. The public key is n, the private key is (p, q).

The security parameter is split into 3 parts, $k = l + s_0 + s_1$, with $l \leq k/4$ and $l + s_0 \leq k/2$. In practice, typical values for these parameters are $k = 1024, l = 256, s_0 = 128, s_1 = 640$. The constraints on the lengths of the security parameters are necessary for the security proof (see Theorem 9.19 below).

We make use of a collision-resistant hash function

$$h : \{0,1\}^{s_1} \longrightarrow \{0,1\}^{l+s_0}.$$

Notation. As usual, we denote by 0^r (or 1^r) the constant bit string $000 \ldots 0$ (or $111 \ldots 1$) of length r ($r \in \mathbb{N}$). As always, let $\|$ denote the concatenation of strings and $\oplus$ be the bitwise XOR operator.

Encryption. To encrypt an l-bit message $m \in \{0,1\}^l$ for Bob, Alice proceeds in the following steps:

1. She chooses a random bit string $r \in \{0,1\}^{s_1}$.

2. She appends s_0 0-bits to m to obtain the $(l + s_0)$-bit string $x := m \| 0^{s_0}$.
3. She sets $y = (x \oplus h(r)) \| r$.
4. She views the k-bit string y as a k-bit integer and applies the Rabin trapdoor function modulo n to obtain the ciphertext c:

$$c := y^2 \bmod n.$$

Note that $y < 2^k < n/2$.

Remarks:

1. At a first glance, the length l of the plaintexts might appear small (recall that a typical value is $l = 256$). But usually we encrypt only short data by using a public-key method, for example, session keys for a symmetric cipher, such as Triple-DES or AES, and for this purpose 256 bits are really sufficient.
2. All ciphertexts are quadratic residues modulo n. But not every quadratic residue appears as a ciphertext. Let c be a quadratic residue modulo n. Then c is the encryption of some plaintext in $\{0,1\}^l$, if there is a square root y of c modulo n, such that
   a. $y < 2^k$, i.e., we may consider y as a bit string of length k, and
   b. the s_0 least-significant bits of $v \oplus h(r)$ are all 0, where $y = v \| r, v \in \{0,1\}^{l+s_0}, r \in \{0,1\}^{s_1}$.
   In this case, the plaintext m, whose encryption is c, consists of the l most-significant bits of $v \oplus h(r)$, i.e., $v \oplus h(r) = m \| 0^{s_0}$, and we call c a *valid ciphertext* (or a *valid encryption of m*) with respect to y.

The security of SAEP is based on the factoring assumption: it is practically infeasible to compute the prime factors p and q of n (see Section 6.5 for a precise statement of the assumption). Decrypting ciphertexts requires drawing square roots. Without knowing the prime factors of n, it is infeasible to compute square roots modulo n. The ability to compute modular square roots is equivalent to the ability to factorize the modulus n (Proposition A.62, Lemma A.63). Bob can compute square roots modulo n because he knows the secret factors p and q.

We recall from Proposition A.62 some basics on computing square roots.

Let $c \in \mathbb{Z}_n$ be a quadratic residue, $c \neq 0$. If c is prime to n, then c has 4 distinct square roots modulo n. If c is not prime to n, i.e., if c is a multiple of p or q, then c has only 2 square roots modulo n. In the latter case, the factors p, q of n can be easily derived by computing $\gcd(c, n)$ with the Euclidean algorithm. The probability for this case is negligibly small, if c is a random quadratic residue.

If $y^2 \bmod n = c$, then also $(n - y)^2 \bmod n = c$. Hence, exactly two of the 4 roots (or one of the two roots) are $< n/2$.

The residue $[0] \in \mathbb{Z}_n$ has the only square root $[0]$.

If both primes p and q are $\equiv 3 \bmod 4$, as here in SAEP, and if the factors p and q are known, then the square roots of c can be easily and efficiently computed as follows:

1. $p + 1/4$ and $q + 1/4$ are integers, and the powers $z_p = c^{(p+1)/4} \bmod p$ of $c \bmod p$ and $z_q = c^{(q+1)/4} \bmod q$ of $c \bmod q$ are square roots of c modulo p and modulo q (see Proposition A.60). Recall that $z_p^2 = c^{(p+1)/2} = c \cdot c^{(p-1)/2}$ and $c^{(p-1)/2} \equiv 1 \bmod p$, if $c \bmod p \neq 0$ (see Euler's criterion, Proposition A.52). If $c \bmod p \neq 0$, then $\pm z_p$ are the two distinct square roots of $c \bmod p$. If $c \bmod p = 0$, then there is the single root $z_p = 0$ of c modulo p.

2. To get the square roots of c modulo n, we map $(\pm z_p, \pm z_q)$ to $\mathbb{Z}_n$ by applying the inverse Chinese remainder map. If c is prime to n, then we obtain 4 distinct roots. If $z_p = 0$ or $z_q = 0$, we get 2 distinct roots, and if both z_p and z_q are 0, then there is the only root 0 (see Proposition A.62).

Decryption. Bob decrypts a ciphertext c by using his secret key (p, q) as follows:

1. He computes the square roots of c modulo n, as just described. In this computation, he tests that $z_p^2 \equiv c \bmod p$ and $z_q^2 \equiv c \bmod q$. If either test fails, then c is not a quadratic residue and Bob rejects c.

2. Two of the four roots (or one of the two roots) are $> n/2$ and hence can be discarded. Bob is left with 2 square roots y_1, y_2 (let $y_1 = y_2$, if there is only one left). If neither of y_1, y_2 is $< 2^k$, then Bob rejects c.
From now on, we assume that $y_1 < 2^k$. If $y_2 \geq 2^k$, then Bob does not have to distinguish between two candidates, and thus he can simplify the following steps (omit the processing of y_2). So, assume now that both y_1 and y_2 are $< 2^k$. Then, we may view them as strings in $\{0,1\}^k$.

3. Bob writes $y_1 = v_1 \| r_1$ and $y_2 = v_2 \| r_2$, with $v_1, v_2 \in \{0,1\}^{l+s_0}$ and $r_1, r_2 \in \{0,1\}^{s_1}$, and computes $x_1 = v_1 \oplus h(r_1)$ and $x_2 = v_2 \oplus h(r_2)$.

4. He writes $x_1 = m_1 \| t_1$ and $x_2 = m_2 \| t_2$, with $m_1, m_2 \in \{0,1\}^l$ and $t_1, t_2 \in \{0,1\}^{s_0}$. If either none or both of t_1, t_2 are $00\ldots0$, then Bob rejects c. Otherwise, let $i \in \{1,2\}$ be the unique i with $t_i = 00\ldots0$. Then, m_i is the decryption of c.

Remark. It might happen that Bob rejects a valid encryption c of a message m, because c is valid with respect to both roots y_1 and y_2 and hence both t_1 and t_2 are 00...0. But the probability for this event is negligibly small, at least if h comes close to a random function.

Namely, assume $y_1 \neq y_2$ and c is a valid encryption of m with respect to y_1, i.e., $y_1 = ((m\|0^{s_0}) \oplus h(r_1))\|r_1$. If h is assumed to be a random function, then the probability that c is also a valid ciphertext with respect to y_2 is about $1/2^{s_0}$. There are 2 cases. If $r_1 \neq r_2$, then $h(r_2)$ is randomly generated independently from $h(r_1)$. Hence, the probability that the s_0 least-significant bits of $h(r_2)$ are just the s_0 least-significant bits of v_2 is $1/2^{s_0}$. If $r_1 = r_2$ and c is valid with respect to both roots, then the $(s_0 + s_1)$ least-significant bits of y_1 and y_2 coincide, i.e., $y_2 = y_1 + 2^{s_0+s_1}\delta$, with absolute value $|\delta| < 2^l$. Since $|\delta| < 2^l < 2^{k/2} < p, q$, we know that δ is a unit modulo n. From

$$y_2^2 = (y_1 + 2^{s_0+s_1}\delta)^2 = y_1^2 + 2^{s_0+s_1+1}\delta y_1 + 2^{2(s_0+s_1)}\delta^2 \equiv c \equiv y_1^2 \bmod n,$$

we conclude that $y_1 \equiv -2^{s_0+s_1-1}\delta \bmod n$, and the probability for that is $1/2^{s_0+s_1-1}$.

The rejection of valid ciphertexts can be completely avoided by a slight modification of the encryption algorithm. Alice repeats the random generation of r, until y has Jacobi symbol 1. Then Bob can always select the correct square root by taking the unique square root $< 2^k$ with Jacobi symbol 1. We described a similar approach in Section 3.6.1 on Rabin's encryption. However, this makes the encryption scheme less efficient, and it is not necessary.

The proof of security for SAEP is based on an important result due to Coppersmith ([Coppersmith97]).

Theorem 9.18. *Let n be an integer, and let $f(X) \in \mathbb{Z}_n[X]$ be a monic polynomial of degree d. Then there is an efficient algorithm which finds all $x \in \mathbb{Z}$ such that the absolute value $|x| < n^{1/d}$ and $f(x) \equiv 0 \bmod n$.*

For a proof, see [Coppersmith97]. The special case $f(X) = X^d - c, c \in \mathbb{Z}_n$, is easy. To get the solutions x with $|x| < n^{1/d}$, you can compute the ordinary d-th roots of c, because for $0 \leq x < n^{1/d}$ we have $x^d \bmod n = x^d$. To compute the ordinary d-th roots is easy, since $x \mapsto x^d$ (without taking residues) is strictly monotonic (take, for example, the simple Algorithm 3.4).

Theorem 9.19. *Assume that the hash function h in SAEP is a random oracle. Let $n = pq$ be a key for SAEP with security parameter $k = l + s_0 + s_1$ (i.e., $2^{k+1} < n < 2^{k+1} + 2^k$). Assume $l \leq k/4$ and $l + s_0 \leq k/2$. Let $A(n, l, s_0, s_1)$ be a probabilistic distinguishing algorithm with running time t that performs an adaptively-chosen-ciphertext attack against SAEP and has a probability of success $\geq 1/2 + \varepsilon$. Let q_d be the number of A's decryption queries, and let q_h be the number of A's queries of the random oracle h.*

Then there is a probabilistic algorithm B for factoring the modulus n with

$$\text{running time} \quad t + O(q_d q_h t_C + q_d t_C') \quad \text{and}$$

$$\text{probability of success} \geq \frac{1}{6} \cdot \varepsilon \cdot \left(1 - \frac{2q_d}{2^{s_0}} - \frac{2q_d}{2^{s_1}}\right).$$

Here, t_C (resp. t_C') is the running time of Coppersmith's algorithm for finding "small-size" roots of polynomials of degree 2 (resp. 4) modulo n (i.e., roots with absolute value $\leq n^{1/2}$ resp. $\leq n^{1/4}$).

Remarks:

1. Recall that typical values of the parameters are $k = 1024, l = 256, s_0 = 128, s_1 = 640$. The number q_d of decryption queries that an adversary can issue in practice should be limited by 2^{40}. We see that the fractions $2q_d/2^{s_0}$ and $2q_d/2^{s_1}$ are negligibly small.

2. Provided the factoring assumption is true, we conclude from Boneh's theorem that a probabilistic polynomial attacking algorithm like A, whose probability of success is $\geq 1/2 + 1/P(k)$, P a positive polynomial (i.e., A has a non-negligible "advantage"), cannot exist. Otherwise, B would factor moduli n in polynomial time. Thus, SAEP is indistinguishability-secure against adaptively-chosen-ciphertext attacks (in the random oracle model).

But Boneh's result is more precise: It gives a *tight reduction* from the problem of attacking SAEP to the problem of factoring. If we had a successful attacking algorithm A against SAEP, we could give precise estimates for the running time and the probability of successfully factoring n. Or, conversely, if we can state reasonable bounds for the running time and the probability of success in factoring numbers n of a given bit length, we can derive a concrete lower bound for the running time of algorithms attacking SAEP with a given probability of success.

Proof. We sketch the basic ideas of the proof. In particular, we illustrate how the random oracle assumption is applied. For more details we refer to [Boneh01].

It suffices to construct an algorithm S which on inputs n, l, s_0, s_1 and $c = \alpha^2 \bmod n$, for a randomly chosen α with $0 \leq \alpha < 2^k$, outputs a square root α', $0 \leq \alpha' < 2^k$, of c with probability $\geq \varepsilon \cdot (1 - 2q_d/2^{s_0} - 2q_d/2^{s_1})$. Namely, with probability $> 1/3$, a number c with $0 \leq c < n$ has two distinct square roots modulo n in $[0, 2^k[$ [3]. Hence $\alpha \neq \alpha'$ with probability $\geq 1/6$ and then n can be factored by computing $\gcd(n, \alpha - \alpha')$ (see Lemma A.63 and Section 3.6.1).

In the following we describe the algorithm S. The algorithm efficiently computes a square root $< 2^k$ of c, without knowing p, q, in two cases, which we study first.

1. Let $y = v \| r$, where v and r are bit strings of lengths $(l + s_0)$ and s_1. If y is a root of c, then v is a root of the quadratic polynomial $(2^{s_1} X + r)^2 - c$ modulo n, with $0 \leq v < 2^{l+s_0} \leq 2^{k/2} < n^{1/2}$.

   Thus, if S happens to know or guess correctly the s_1 lower significant bits r of a root y, then it efficiently finds y by applying the following algorithm

   $CompRoot_1(c, r)$:

   Compute the roots $v < 2^{k/2}$ of $(2^{s_1} X + r)^2 - c$ modulo n by using Coppersmith's algorithm. If such a v is found, return the root $y = v \| r$.

2. Let $y = m \| w$ be a square root modulo n of c, with $m \in \{0, 1\}^l$ and $w \in \{0, 1\}^{s_0 + s_1}$. Let $c' \in \mathbb{Z}_n$ be a further quadratic residue, $c' \neq c$, and assume that c' has a k-bit square root $y' = m' \| w$ modulo n, whose $(s_0 + s_1)$ lower significant bits w are the same as those of y. Then we

---

[3] See Fact 2 in [Boneh01]; n is chosen between 2^{k+1} and $2^{k+1} + 2^k$ to get this estimate.

may write $y' = y + 2^{s_0+s_1}\delta$, where $\delta = m' - m$ is an integer with absolute value $|\delta| < 2^l \le 2^{k/4} < n^{1/4}$. y is a common root of the polynomials $f(X) = X^2 - c$ and $g(X, \delta) = (X + 2^{s_0+s_1}\delta)^2 - c'$ modulo n. Therefore, δ is a root of the resultant $\mathcal{R}$ of $f(X)$ and $g(X, \Delta) = (X + 2^{s_0+s_1}\Delta)^2 - c'$. The resultant $\mathcal{R}$ is a polynomial modulo n in Δ of degree 4 (see, for example, [Lang05], Chapter IV). Since $|\delta| < 2^l \le 2^{k/4} < n^{1/4}$, we can compute δ efficiently by using Coppersmith's algorithm for polynomials of degree 4. The greatest common divisor $\gcd(f(X), g(X, \delta))$ is $X - y$. Thus, if S happens to get such an element c', then it efficiently finds a square root y of c by applying the following algorithm $CompRoot_2(c, c')$:
Compute the roots δ with absolute value $|\delta| < 2^l$ of the resultant $\mathcal{R}$ modulo n by using Coppersmith's algorithm. Compute the greatest common divisor $X - y$ of $X^2 - c$ and $(X + 2^{s_0+s_1}\delta)^2 - c'$ modulo n. If such a δ (and then y) is found, return the root y.

Algorithm S interacts with the attacking algorithm A. In the real attack, A interacts with Bob, the legitimate owner of the public-secret key pair $(n, (p, q))$, to obtain decryptions of ciphertexts of its choice, and with the random oracle h to get hash values $h(m)$ (in practice, interacting typically means to communicate with another computer program). Now S is constructed to replace both Bob and the random oracle h in the attack. It "simulates" Bob and h.

Each time, when A issues a query, S has a chance to compute a square root of c, and of course S terminates when a root is found.

S has no problems answering the hash queries. Since h is a random oracle, S can assign a randomly generated $v \in \{0, 1\}^{l+s_0}$ as hash value $h(r)$. The only restriction is: If the hash value of m is queried more than once, then always the same value has to be provided. Therefore, S has to store the list $\mathcal{H}$ of hash value pairs $(r, h(r))$ that it has given to A.

The structure of S is the following.

S calls A with the public key n and the security parameters l, s_0, s_1 as input. Then it waits for the queries of A.

1. If A queries the hash value for r, then
   a. if $(r, h(r))$ is on the list $\mathcal{H}$ of previous responses to hash queries, then S again sends $h(r)$ as hash value to A;
   b. else S applies algorithm $CompRoot_1(c, r)$; if S finds a root of c modulo n in this way, then it returns the root and terminates;
   c. else S picks a random bit string v of length $(l + s_0)$, puts (r, v) on its list $\mathcal{H}$ of hash values and sends $v = h(r)$ as hash value to A.
2. if A queries the decryption of a ciphertext c', then
   a. S applies algorithm $CompRoot_2(c, c')$; if S finds a root of c modulo n in this way, it returns the root and terminates;
   b. else S applies, for each $(r', h(r'))$ on the list $\mathcal{H}$ of previous hash values, algorithm $CompRoot_1(c', r')$; if S finds a square root $v' \| r'$ of c' in this

way, then it computes $w' = v' \oplus h(r')$; if the s_0 lower significant bits of w' are all 0, i.e., $w' = m' \| 0^{s_0}$, then S sends m' as plaintext to A, else S rejects c' as an "invalid ciphertext";

 c. else S could not find a square root of c' in the previous step b and rejects c' as an "invalid ciphertext";

3. if A produces the two candidate plaintexts $m_0, m_1 \in \{0,1\}^l$, then S sends c as encryption of m_0 or m_1 to A.

If A terminates, then S also terminates (if it has not terminated before).

To analyze algorithm S, let y_1 and y_2 be the two square roots of c with $0 \le y_1, y_2 < 2^k$ (take $y_2 = y_1$, if there is only one). The goal of S is to compute y_1 or y_2. We decompose $y_i = v_i \| r_i$, with $r_i \in \{0,1\}^{s_1}$.

We consider several cases.

1. If A happens to query the hash value for r_1 or r_2, then S successfully finds one of the roots y_1, y_2 in step 1b.
2. If A happens to query the decryption of a ciphertext c', and if c' has a k-bit square root y' modulo n, whose $(s_0 + s_1)$ lower significant bits are the same as those of y_1 or y_2, then S successfully finds one of the roots y_1, y_2 in step 2a.

We observed above that S can easily play the role of the random oracle h and answer hash queries. Sometimes, S can answer decryption requests correctly.

3. If A queries the decryption of a ciphertext c' that is valid with respect to its square root y' modulo n, and if A has previously asked for the hash value $h(r')$ of the s_1 rightmost bits r' of y' ($c' = y'^2 \bmod n; y' = v' \| r', r' \in \{0,1\}^{s_1}$), then S responds in step 2b to A with the correct plaintext.

But S does not know the secret key, and so S cannot perfectly simulate Bob. We now study the two cases where, from A's point of view, the behavior of S might appear different from Bob's behavior.

In the real attack, Bob sends a valid encryption of m_0 or m_1 as challenge ciphertext to A. The choice of the number c, which S sends as challenge ciphertext, has nothing to do with m_0 or m_1. Therefore, we have to study the next case 4.

4. The number c, which S presents as challenge ciphertext at the end of phase I, is, from A's point of view, not a valid encryption of m_0 or m_1 with respect to y_i, where $i = 1$ or $i = 2$. This means that $v_i \oplus h(r_i) \ne m_b \| 0^{s_0}$ or, equivalently, $h(r_i) \ne (m_b \| 0^{s_0}) \oplus v_i$ for $b = 0, 1$.
   This can only happen either
   - if A has asked for $h(r_i)$ in phase I, and S has responded with a non-appropriate hash value for r_i, or

- if A has asked in phase I for the decryption of a ciphertext c' with
  $c' = y'^2 \bmod n, y' = v' \| r', r' \in \{0,1\}^{s_1}, v' \in \{0,1\}^{l+s_0}$ and $r' = r_i$, i.e.,
  the s_0 rightmost bits of y' and y_i are equal[4]; in this case the answer of
  S might have put some restriction on $h(r_i)$.

Otherwise, the hash value $h(r_i)$, which is randomly generated by the
oracle, is independent of A's point of view at the end of phase I. This
implies that, from A's point of view, $h(r_i) = (m_b \| 0^{s_0}) \oplus v_i$ has the same
probability as $h(r_i) = w$, for any other $(l + s_0)$-bit string w. Therefore c
is a valid encryption of m_0 or m_1 with respect to y_i.

That A has asked for $h(r_i)$ before can be excluded, because then S has
already successfully terminated in step 1b.

The number c is generated by squaring a randomly chosen α, and c is not
known to A in phase I. Therefore, the random choice of c is independent
from $A's$ decryption queries in phase I. Hence, the probability that a
particular decryption query in phase I involves r_1 or r_2 is $2 \cdot 1/2^{s_1}$. There
are at most q_d decryption queries. Thus, the probability of case 4 is
$\leq q_d \cdot 2/2^{s_1}$.

5. Attacker A asks for the decryption of a ciphertext c' and S rejects it (in
   step 2b or step 3), whereas Bob, by using his secret key, accepts c' and
   provides A with the plaintext m'. Then, we are not in case 2, because,
   in case 2, S successfully terminates in step 2a before giving an answer to
   A.

   We assume that c is a valid ciphertext for m_0 or m_1 with respect to both
   square roots y_1 and y_2, i.e., that we are not in case 4. This means that
   for $i = 1$ and $i = 2$, we have $v \oplus h(r_i) = m_b \| 0^{s_0}$ for $b = 0$ or $b = 1$.
   Decrypting c', Bob finds:

   $$c' = y'^2 \bmod n, y' = v' \| r', v' \in \{0,1\}^{l+s_0}, r' \in \{0,1\}^{s_1}, v' \oplus h(r') = m' \| 0^{s_0}.$$

   If $r' = r_i$ for $i = 1$ or $i = 2$, then the s_0 rightmost bits of v' and v are
   the same – they are equal to the s_0 rightmost bits of $h(r') = h(r_i)$. Then
   the $(s_0 + s_1)$ rightmost bits of y' and y_i coincide and we are in case 2,
   which we have excluded. Hence $r' \neq r_1$ and $r' \neq r_2$.

   The hash value $h(r')$ has not been queried before, because otherwise we
   would be in case 3, and S would have accepted c' and responded with
   the plaintext m'.

   Hence, the hash value $h(r')$ is a random value which is independent from
   preceding hash queries and our assumption that c is a valid encryption
   of m_0 or m_1 (which constrains $h(r_1)$ and $h(r_2)$, see above). But then the
   probability that the s_0 rightmost bits of $h(r') \oplus v'$ are all 0 (which is
   necessary for Bob to accept) is $1/2^{s_0}$. Since there are at most 2 square
   roots y', and A queries the decryption of q_d ciphertexts, the probability
   that case 5 occurs is $\leq q_d \cdot 2/2^{s_0}$.

---

[4] Note that S successfully terminates in step 2a only if there are $(s_0 + s_1)$ common
rightmost bits.

Now assume that neither case 4 nor case 5 occurs. Following Boneh, we call this assumption $\mathcal{GoodSim}$.

Under this assumption $\mathcal{GoodSim}$, S behaves exactly like Bob and the random oracle h. Thus, A operates in the same probabilistic setting, and it therefore has the same probability of success $1/2 + \varepsilon$ as in the original attack.

If, in addition, cases 1 and 2 do not occur, then the randomly generated c (and r_1, r_2), and the hash values $h(r_1)$ and $h(r_2)$, which are generated by the random oracle, are independent of all hash and decryption queries issued by A and the responses given by S. Therefore A has not collected any information about $h(r_1)$ and $h(r_2)$. This means that the ciphertext c hides the plaintext m_b perfectly in the information-theoretic sense – the encryption includes a bitwise XOR with a truly random bit string $h(r_i)$. We have seen in Section 9.2 that then A's probability of correctly distinguishing between m_0 and m_1 is $1/2$.

Therefore, the advantage ε in A's probability of success necessarily results from the cases 1, 2 (we still assume $\mathcal{GoodSim}$). In the cases 1 and 2, the algorithm S successfully computes a square root of c. Hence the probability of success $\mathrm{prob}_{success}(S | \mathcal{GoodSim})$ of algorithm S assuming $\mathcal{GoodSim}$ is $\geq \varepsilon$. The probability of the cases 4 and 5 is $\leq 2q_d/2^{s_1} + 2q_d/2^{s_0}$ and hence $\mathrm{prob}(\mathcal{GoodSim}) \geq 1 - 2q_d/2^{s_1} - 2q_d/2^{s_0}$, and we conclude that the probability of success of S is

$$\geq \mathrm{prob}(\mathcal{GoodSim}) \cdot \mathrm{prob}_{success}(S | \mathcal{GoodSim}) \geq \varepsilon \cdot \left(1 - \frac{2q_d}{2^{s_1}} - \frac{2q_d}{2^{s_0}}\right).$$

The running time of S is essentially the running time of A plus the running times of the Coppersmith algorithms. The algorithm for quadratic polynomials is called after each of the q_h hash queries (step 1b). After each of the q_d decryption queries, it may be called for all of the $\leq q_h$ known hash pairs $(r, h(r))$ (step 2b). The algorithm for polynomials of degree 4 is called after each of the q_d decryption queries (step 2a). Thus we can estimate the running time of S by

$$t + O(q_d q_h t_C + q_d t'_C),$$

and we see that the algorithm S gives a tight reduction from the problem of attacking SAEP to the problem of factoring.   □

Remark. True random functions can not be implemented in practice. Therefore, a proof in the random oracle model – treating hash functions as equivalent to random functions – can never be a complete proof of security for a cryptographic scheme. But, intuitively, the random oracle model seems to be reasonable. In practice, a well-designed hash function should never have any features that distinguish it from a random function and that an attacker could exploit (see Section 3.4.4).

In recent years, doubts about the random oracle model have been expressed. Examples of cryptographic schemes were constructed which are provably secure in the random oracle model, but are insecure in any real-world implementation, where the random oracle is replaced by a real hash function ([CanGolHal98]; [GolTau03]; [GolTau03a]; [MauRenHol04]; [CanGolHal04]). However, the examples appear contrived and far from systems that would be designed in the real world. The confidence in the soundness of the random oracle assumption is still high among cryptographers, and the random oracle model is still considered a useful tool for validating cryptographic constructions. See, for example, [KobMen05] for a discussion of this point.

Nevertheless, it is desirable to have encryption schemes whose security can be proven solely under a standard assumption that some computational problem in number theory can not be efficiently solved. An example of such a scheme is given in the next section.

9.5.2 Security Under Standard Assumptions

The Cramer-Shoup Public Key Encryption Scheme. The security of the Cramer-Shoup public key encryption scheme ([CraSho98]) against adaptively-chosen-ciphertext attacks can be proven assuming that the decision Diffie-Hellman problem (Section 4.5.3) can not be solved efficiently and that the hash function used is collision-resistant. The random oracle model is not needed. In Chapter 10, we will give examples of signature schemes whose security can be proven solely under the assumed difficulty of computational problems (for example, Cramer-Shoup's signature scheme).

First, we recall the decision Diffie-Hellman problem (see Section 4.5.3).

Let p and q be large prime numbers, such that q is a divisor of $p-1$, and let G_q be the (unique) subgroup of order q of $\mathbb{Z}_p^*$. G_q is a cyclic group, and every element $g \in \mathbb{Z}_p^*$ of order q is a generator of G_q (see Lemma A.40).

Given $g_1, u_1 = g_1^x, g_2 = g_1^y, u_2$ with random elements $g_1, u_2 \in G_q$ and randomly chosen exponents $x, y \in \mathbb{Z}_q^*$, decide if $u_2 = g_1^{xy}$. This is equivalent to decide, for randomly (and independently) chosen elements $g_1, u_1, g_2, u_2 \in G_q$, if

$$\log_{g_2}(u_2) = \log_{g_1}(u_1).$$

If the equality holds, we say that (g_1, u_1, g_2, u_2) has the Diffie-Hellman property.

The *decision Diffie-Hellman assumption* says that no probabilistic polynomial algorithm exists to solve the decision Diffie-Hellman problem.

Notation. For pairs $u = (u_1, u_2)$, $x = (x_1, x_2)$ and a scalar value r we shortly write $u^x := u_1^{x_1} u_2^{x_2}$, $u^{rx} := u_1^{rx_1} u_2^{rx_2}$.

Key Generation. Bob randomly generates large prime numbers p and q, such that q is a divisor of $p-1$. He randomly chooses a pair $g = (g_1, g_2)$ of elements $g_1, g_2 \in G_q$.[5]

---

[5] Compare, for example, the key generation in the Digital Signature Standard, Section 3.5.3.

Then Bob randomly chooses three pairs of exponents $x = (x_1, x_2), y = (y_1, y_2), z = (z_1, z_2)$, with $x_1, x_2, y_1, y_2, z_1, z_2 \in \mathbb{Z}_q^*$ and computes modulo p:

$$d = g^x = g_1^{x_1} g_2^{x_2}, e = g^y = g_1^{y_1} g_2^{y_2}, f = g^z = g_1^{z_1} g_2^{z_2}.$$

Bob's public key is (p, q, g, d, e, f), his private key is (x, y, z).

For encryption, we need a collision-resistant hash function

$$h : \{0, 1\}^* \longrightarrow \mathbb{Z}_q^* = \{0, 1, \ldots q - 1\}.$$

h outputs bit strings of length $|q| = \lfloor \log_2(q) \rfloor + 1$. In practice, as in DSS, we might have $|q| = 160$ and $h = $ SHA-1, see Section 3.5.3.

Encryption. Alice can encrypt messages $m \in G_q$ for Bob, i.e., elements m of order q in $\mathbb{Z}_p^*$. To encrypt a message m for Bob, Alice chooses a random $r \in \mathbb{Z}_q^*$ and computes the ciphertext

$$c = (u_1, u_2, w, v) := (g_1^r, g_2^r, f^r \cdot m, v), \text{ with } v := d^r e^{r \cdot h(u_1, u_2, w)}).$$

Note that the Diffie-Hellman property holds for (g_1, u_1, g_2, u_2).

Decryption. Bob decrypts a ciphertext $c = (u_1, u_2, w, v)$ by using his private key (x, y, z) as follows:

1. Bob checks the verification code v. He checks if $u^{x + y \cdot h(u_1, u_2, w)} = v$. If this equation does not hold, he rejects c.
2. He recovers the plaintext by computing $w \cdot u^{-z}$.

Remarks:

1. We follow here the description of Cramer-Shoup's encryption scheme in [KobMen05]. The actual scheme in [CraSho98] is the special case with $z_2 = 0$. Cramer and Shoup prove the security of the scheme using a slightly weaker assumption on the hash function. They assume that h is a member of a universal one-way family of hash functions (as defined in Section 10.4).
2. If $c = (u_1, u_2, w, v)$ is the correct encryption of a message m, then the verification code v passes the check in step 1,

$$u^{x + y \cdot h(u_1, u_2, w)} = (g^r)^x \cdot (g^r)^{y \cdot h(u_1, u_2, w)}$$

$$= (g^x)^r \cdot (g^y)^{r \cdot h(u_1, u_2, w)} = d^r \cdot e^{r \cdot h(u_1, u_2, w)} = v,$$

and the plaintext m is recovered in step 2,

$$f^r = (g^z)^r = (g^r)^z = u^z, \text{ hence } w \cdot u^{-z} = m \cdot f^r \cdot f^{-r} = m.$$

3. The private key $(x, y, z) = ((x_1, x_2), (y_1, y_2), (z_1, z_2))$ is an element in the $\mathbb{Z}_q$-vectorspace $\mathbb{Z}_q^6$. Publishing the public key (d, e, f) means to publish the following conditions on (x, y, z):

$$(1)\ d = g^x, \quad (2)\ e = g^y, \quad (3)\ f = g^z.$$

These are linear equations for $x_1, x_2, y_1, y_2, z_1, z_2$. They can also be written as (with $\lambda := \log_{g_1}(g_2)$)

$$(1)\ \log_{g_1}(d) = x_1 + \lambda \cdot x_2$$
$$(2)\ \log_{g_1}(e) = \qquad\qquad y_1 + \lambda \cdot y_2$$
$$(3)\ \log_{g_1}(f) = \qquad\qquad\qquad z_1 + \lambda \cdot z_2.$$

The equations define lines L_x, L_y, L_z in the plane $\mathbb{Z}_q^2$. For a given public key (d, e, f), the private key element x (resp. y, z) is a (uniformly distributed) random element of L_x (resp. L_y, L_z); each of the elements in L_x (resp. L_y, L_z) has the same probability $1/q$ of being x (resp. y, z).

4. Let $c := (u_1, u_2, w, v)$ be a ciphertext-tuple. Then c is accepted for decryption only if the verification condition (4) $v = u^{x + y \cdot \alpha}$, with $\alpha = h(u_1, u_2, w)$, holds. The verification condition is a linear equation for x and y:

$$(4)\ \log_{g_1}(v) = \log_{g_1}(u_1)x_1 + \lambda \log_{g_2}(u_2)x_2 + \alpha \log_{g_1}(u_1)y_1 + \alpha \lambda \log_{g_2}(u_2)y_2.$$

Equations $(1), (2)$ and (4) are linearly independent, if and only if $\log_{g_1}(u_1) \neq \log_{g_2}(u_2)$.
The ciphertext-tuple c is the correct encryption of a message, if and only if c satisfies the verification condition and $\log_{g_1}(u_1) = \log_{g_2}(u_2)$, i.e., (g_1, u_1, g_2, u_2) has the Diffie-Hellman property. In this case, we call c a *valid ciphertext*.
Now assume that $\log_{g_1}(u_1) \neq \log_{g_2}(u_2)$. Then, the probability that c passes the check of the verification code and is not rejected in the decryption step 1 is $1/q$ and hence negligibly small. Namely, $(1), (2)$ and (4) are linearly independent. This implies that (x, y) is an element of the line in $\mathbb{Z}_q^4$ which is defined by $(1), (2), (4)$. The probability for that is $1/q$, since (x, y) is a random element of the 2-dimensional space $L_x \times L_y$.

Theorem 9.20. *Let the decision Diffie-Hellman assumption be true and let h be a collision-resistant hash function. Then the probability of success of any attacking algorithm A, which on input of a random public key (p, q, g, d, e, f) executes an adaptively-chosen-ciphertext attack and tries to distinguish between two plaintexts, is $\leq 1/2 + \varepsilon$, with ε negligible.*

Proof. We discuss the essential ideas of the proof. For more details, we refer to [CraSho98].

The proof runs by contradiction. Assume that there is an algorithm A which on input of a randomly generated public key successfully distinguishes between ciphertexts in an adaptively-chosen-ciphertext attack, with a probability of success of $1/2 + \varepsilon$, ε non-negligible.

Then we can construct a probabilistic polynomial algorithm S which answers the decision Diffie-Hellman problem with a probability close to 1, in contradiction to the decision Diffie-Hellman assumption. The algorithm S successfully finds out whether a random 4-tuple (g_1, u_1, g_2, u_2) has the Diffie-Hellman property or not.

As in the proof of Theorem 9.19, algorithm S interacts with the attacking algorithm A. In the real attack, A interacts with Bob, the legitimate owner of the secret key, to obtain decryptions of ciphertexts of its choice (in practice, interacting typically means to communicate with another computer program). Now S is constructed to replace Bob in the attack. S "simulates" Bob.

On input of $(p, q, g_1, u_1, g_2, u_2)$ the algorithm S repeatedly generates a random private key (x, y, z) and computes the corresponding public key $d = g^x, e = g^y, f = g^z$. Then S calls A with the public key, and A executes its attack. The difference to the real attack is that A communicates with S instead of Bob. At some point (end of phase I of the attack), A outputs two plaintexts m_0, m_1. S randomly selects a bit $b \in \{0, 1\}$, sets

$$w := u^z m_b, \alpha := h(u_1, u_2, w), v := u^{x + \alpha \cdot y}$$

and sends $c := (u_1, u_2, w, v)$ to A, as an encryption of m_b. The verification code v is correct by construction, even if $\log_{g_1} u_1 \neq \log_{g_2} u_2$ and c is not a valid encryption of m_b.

At any time in phase I or phase II, A can request the decryption of a ciphertext $c' = (u_1', u_2', w', v')$, $c' \neq c$. If c' satisfies the verification condition, then equation (4) holds for c', i.e., we have, with $\alpha' = h(u_1', u_2', w')$:

$$(5) \log_{g_1}(v') = \log_{g_1}(u_1')x_1 + \lambda \log_{g_2}(u_2')x_2 + \alpha' \log_{g_1}(u_1')y_1 + \alpha' \lambda \log_{g_2}(u_2')y_2.$$

We observe: If $\log_{g_1}(u_1) \neq \log_{g_2}(u_2)$ and $\log_{g_1}(u_1') \neq \log_{g_2}(u_2')$, then the equations $(1), (2), (4), (5)$ are linearly independent, if and only if $\alpha' \neq \alpha$.

If c' satisfies the verification condition, then, up to negligible probability, c' is a valid ciphertext (see Remark 4 above) and S answers with the correct plaintext. S can easily check the verification condition and decrypt the ciphertext, because it knows the secret key. Here, S behaves exactly like Bob in the real attack.

If (g_1, u_1, g_2, u_2) has the Diffie-Hellman property, then the ciphertext $c := (u_1, u_2, w, v)$, which S presents to A as an encryption of m_b, is a valid ciphertext, and the probability distribution of c is the same, as if Bob produced the ciphertext. Hence, in this case, A operates in the same setting as in the real attack.

If (g_1, u_1, g_2, u_2) does not have the Diffie-Hellman property, then A operates in a setting which does not occur in the real attack, since Bob only produces valid ciphertexts. We do not know how attacker A responds to the modified setting. For example, A could fail to terminate in its expected running time or it could output an error message or it could produce the usual output and answer, whether m_0 or m_1 is encrypted. But fortunately the concrete behavior of A is not relevant in this case: if (g_1, u_1, g_2, u_2) does not have the Diffie-Hellman property, then it is guaranteed that up to a negligible probability, m_b is perfectly hidden in $w = u^z m_b$ to A and A can not generate any advantage from knowing w in distinguishing between m_0 and m_1.

To understand this, assume that (g_1, u_1, g_2, u_2) does not have the Diffie-Hellman property. Then the probability that attacker A does get some information about the private key z by asking for the decryption of ciphertexts c' of its choice, $c' \neq c$, is negligibly small.

Namely, let A ask for the decryption of a ciphertext $c' = (u_1', u_2', w', v')$. Let $\alpha' = h(u_1', u_2', w')$.

Since we assume that $\log_{g_1}(u_1) \neq \log_{g_2}(u_2)$, equations $(1), (2), (4)$ are linearly independent and define a line L. From A's point of view, the private key $(x, y) = (x_1, x_2, y_1, y_2)$ is a random point on the line L.

We have to distinguish between several cases.

1. $\alpha' = \alpha$. Since h is collision-resistant[6], the probability that A can generate a triple $(u_1', u_2', w') \neq (u_1, u_2, w)$, with $h(u_1', u_2', w') = \alpha' = \alpha = h(u_1, u_2, w)$, is negligibly small. Hence the case $\alpha' = \alpha$ can happen, up to a negligible probability, only if $(u_1', u_2', w') = (u_1, u_2, w)$. But then $v' \neq v$, because $c' \neq c$, and hence $v' \neq (u')^{x + \alpha' \cdot y} = u^{x + \alpha \cdot y} = v$, and c' is rejected.

2. Now assume that $\alpha' \neq \alpha$ and (g_1, u_1', g_2, u_2') does not have the Diffie-Hellman property. If c' is not rejected, then the following equation (5) holds (it is equation (4) stated for c'):

$(5)\ \log_{g_1}(v')$

$$= \log_{g_1}(u_1')x_1 + \lambda \log_{g_2}(u_2')x_2 + \alpha' \log_{g_1}(u_1')y_1 + \alpha' \lambda \log_{g_2}(u_2')y_2.$$

Since $\alpha' \neq \alpha$, equation (5) puts, in addition to $(1), (2), (4)$, another linearly independent condition on (x, y). Hence, at most one point $(\tilde{x}, \tilde{y})$ satisfies these equations. Since (x, y) is a randomly chosen element on the line L, the probability that $(x, y) = (\tilde{x}, \tilde{y})$ is $\leq 1/q$. Hence, up to a negligible probability, c' is rejected.

3. The remaining case is that $\alpha' \neq \alpha$ and (g_1, u_1', g_2, u_2') has the Diffie-Hellman property. Either the check of the verification code v' fails and c' is rejected or equation (5) holds. In the latter case, S decrypts c' and provides A with the plaintext m'. Thus, what A learns from the

---

[6] Second-pre-image resistance would suffice.

decryption is the equation (6) $m' = w' \cdot u'^{-z}$, which can be reformulated as

$$(6) \quad \log_{g_1}(w' \cdot m'^{-1}) = \log_{g_1}(u'_1)z_1 + \lambda \log_{g_2}(u'_2)z_2$$
$$= \log_{g_1}(u'_1)(z_1 + \lambda z_2).$$

But this equation is linearly dependent on (3), so it does not give A any additional information on z. A learns from $m' = w' \cdot u'^{-z}$ the same information that it already knew from the public-secret-key equation (3) $f = e^z$: the private key z is a random element on the line L_z.

Summarizing, if (g_1, u_1, g_2, u_2) does not have the Diffie-Hellman property, then up to a negligible probability, from attacker A's point of view, the secret key z is a random point on a line in $\mathbb{Z}_q^2$. This means that from A's point of view, m_b is perfectly (in the information-theoretic sense) hidden in $w = f^z m_b$ by a random element of G_q. We learnt in Section 9.2 that therefore A's probability of distinguishing successfully between m_0 and m_1 is exactly $1/2$. Hence, the non-negligible advantage ε of A results solely from the case that (g_1, u_1, g_2, u_2) has the Diffie-Hellman property.

In order to decide whether (g_1, u_1, g_2, u_2) has the Diffie-Hellman property, S repeatedly randomly generates private key elements (x, y, z) and runs the attack with A. If A correctly determines which of the messages m_b was encrypted by S in significantly more than half of the iterations, then S can be almost certain that (g_1, u_1, g_2, u_2) has the Diffie-Hellman property. Otherwise, S is almost certain that it does not.   □

9.6 Unconditional Security of Cryptosystems

The security of many currently used cryptosystems, in particular that of all public-key cryptosystems, is based on the hardness of an underlying computational problem, such as factoring integers or computing discrete logarithms. Security proofs for these systems show that the ability of an adversary to perform a successful attack contradicts the assumed difficulty of the computational problem. A typical security proof was given in Section 9.4. We proved that public-key one-time pads induced by one-way permutations with a hard-core predicate are computationally secret. Thus, the security of the encryption scheme is reduced to the one-way feature of function families, such as the RSA or modular squaring families, and the one-way feature of these families is, in turn, based on the assumed hardness of inverting modular exponentiation or factoring a large integer (see Chapter 6). The security proof is conditional, and there is some risk that in the future, the underlying condition will turn out to be false.

On the other hand, Shannon's information-theoretic model of security provides unconditional security. The perfect secrecy of Vernam's one-time pad is not dependent on the hardness of a computational problem, or limits of the computational power of an adversary.

Although perfect secrecy is not reachable in most practical situations, there are various promising attempts to design practical cryptosystems whose security is not based on assumptions and which provably come close to perfect information-theoretic security.

One approach is quantum cryptography, introduced by Bennett and Brassard. Two parties agree on a secret key by transmitting polarized photons over a fiber-optic channel. The secrecy of the key is based on the uncertainty principle of quantum mechanics ([BraCre96]).

In other approaches, the unconditional security of cryptosystems is based on the fact that communication channels are noisy (and hence, an eavesdropper never gets all the information), or on the limited storage capacity of an adversary (see, e.g., [Maurer99] for an overview on information-theoretic cryptography).

9.6.1 The Bounded Storage Model

We will give a short introduction to encryption schemes designed by Maurer et al., whose unconditional security is guaranteed by a limit on the total amount of storage capacity available to an adversary. Most of the encryption schemes studied in this approach are similar to the one proposed in [Maurer92], which we are going to describe now.

Alice wants to transmit messages $m \in M := \{0,1\}^n$ to Bob. She uses a one-time pad for encryption, i.e., she XORs the message m bitwise with a one-time key k. As usual, we have a probability distribution on M which, together with the probabilistic choice of the keys, yields a probability distribution on the set C of ciphertexts. Without loss of generality, we assume that $\text{prob}(m) > 0$ for all $m \in M$, and $\text{prob}(c) > 0$ for all $c \in C$. The eavesdropper Eve is a passive attacker: observing the ciphertext $c \in C \subseteq \{0,1\}^n$, she tries to get information about the plaintext.

Alice and Bob extract the key k for the encryption of a message m from a publicly accessible "table of random bits". Security is achieved if Eve has access only to some part of the table. This requires some clever realization of the public table of random bits. A possibly realistic scenario (the "satellite scenario") is that the random bits are broadcast by some radio source (e.g. a satellite or a natural deep-space radio source) with a sufficiently high data rate (see the example below) and Eve can store only parts of it.

So, we assume that there is a public source broadcasting truly (uniformly distributed) random bits to Alice, Bob and Eve at a high speed. The communication channels are assumed to be error free.

Alice and Bob select their key bits from the bit stream over some time period T, according to some private strategy not known to Eve. The ciphertext is transmitted later, after the end of T. Due to her limited storage resources, Eve can store only a small part of the bits broadcast during T.

To extract the one-time key k for a message $m \in \{0,1\}^n$, Alice and Bob synchronize on the source and listen to the broadcast bits over the time period T. Let R (called the randomizer) be the random bits transmitted during T, and let $(r_{i,j} \mid 1 \leq i \leq l, 0 \leq j \leq t-1)$ be these bits arranged as elements of a matrix with l rows and t columns. Thus, R contains $|R| = lt$ bits. Typically, l is small (about 50) and t is huge, even when compared with the length n of the plaintext message m.

Alice and Bob have agreed on a (short) private key $(s_1, \ldots, s_l)$ in advance, with s_i randomly chosen in $\{0, \ldots, t-1\}$ (with respect to the uniform distribution), and take as key $k := k_0 \ldots k_{n-1}$, with

$$k_j := r_{1,(s_1+j) \bmod t} \oplus r_{2,(s_2+j) \bmod t} \cdots \oplus r_{l,(s_l+j) \bmod t}.$$

In other words, Alice and Bob select from each row i the bit string b_i of length n, starting at the randomly chosen "seed" s_i (jumping back to the beginning if the end of the row is reached) and then get their key k by XORing these strings b_i.

Attacker Eve also listens to the random source during T and stores some of the bits, hoping that this will help her to extract information when the encrypted message is transmitted after the end of T. Due to her limited storage space, she stores only q of the bits $r_{i,j}$. The bits are selected by some probabilistic algorithm.[7] For each of these q bits, she knows the value and possibly the position in the random bit stream R. Eve's knowledge about the randomizer bits is summarized by the random variable S. You may also consider S as a probabilistic algorithm, returning q positions and bit values. As usual in the study of a one-time pad encryption scheme, Eve may know the distribution on M. We assume that she has no further a priori knowledge about the messages $m \in M$ actually sent to Bob by Alice. The following theorem is proven in [Maurer92].

Theorem 9.21. *There exists an event $\mathcal{E}$, such that for all probabilistic strategies S for storing q bits of the randomizer R*

$$I(M; CS \mid \mathcal{E}) = 0 \text{ and } \mathrm{prob}(\mathcal{E}) \geq 1 - n\delta^l,$$

where $\delta := q/|R|$ is the fraction of randomizer bits stored by Eve.

Proof. We sketch the basic idea and refer the interested reader to [Maurer92]. Let $(s_1, \ldots, s_l)$ be the private key of Alice and Bob, and $k := k_0 \ldots k_{n-1}$ be the key extracted from R by Alice and Bob. Then bit k_j is derived from R by XORing the bits $r_{1,(s_1+j) \bmod t}, r_{2,(s_2+j) \bmod t}, \ldots, r_{l,(s_l+j) \bmod t}$.

If Eve's storage strategy missed only one of these bits, then the resulting bit k_j appears truly random to her, despite her knowledge S of the randomizer. The probability of the event $\mathcal{F}$ that Eve stores with her strategy all the bits $r_{1,(s_1+j) \bmod t}, r_{2,(s_2+j) \bmod t}, \ldots, r_{l,(s_l+j) \bmod t}$, for at least one j, $0 \leq j \leq n-1$, is very small – it turns out to be $\leq n\delta^l$.

---

[7] Note that there are no restrictions on the computing power of Eve.

The "security event" $\mathcal{E}$ is defined as the complement of $\mathcal{F}$. If $\mathcal{E}$ occurs, then, from Eve's point of view, the key extracted from R by Alice and Bob is truly random, and we have the situation of Vernam's one-time pad, with a mutual information which equals 0 (see Theorem 9.5).   □

Remarks:

1. The mutual information in Theorem 9.21 is conditional on an event $\mathcal{E}$. This means that all the entropies involved are computed with the conditional probabilities assuming $\mathcal{E}$ (see the final remark in Appendix B.4).
2. In [Maurer92] a stronger version is proven. It also includes the case that Eve has some a priori knowledge V of the plaintext messages, where V is jointly distributed with M. Then the mutual information $I(M, CS \,|\, V, \mathcal{E})$ between M and CS, conditioned over V and assuming $\mathcal{E}$, is 0. Conditioning over V means that the mutual information does not include the amount of information about M resulting from the knowledge of V (see Proposition B.36).
3. Adversary Eve cannot gain an advantage from learning the secret key $(s_1, \dots, s_l)$ after the broadcast of the randomizer R. Therefore, Theorem 9.23 is a statement on *everlasting* security.
4. The model of attack applied here is somewhat restricted. In the first phase, while listening to the random source, the eavesdropper Eve does not exploit her full computing power; she simply stores some of the transmitted bits and does not use the bit stream as input for computations at that time. The general model of attack, where Eve may compute and store arbitrary bits of information about the randomizer, is considered below.

Example. This example is derived from a similar example in [CachMau97]. A satellite broadcasting random bits at a rate of 16 Gbit/s is used for one day to provide a randomizer table R with about $1.5 \cdot 10^{15}$ bits. Let R be arranged in $l := 100$ rows and $t := 1.5 \cdot 10^{13}$ columns. Let the plaintexts be 6 MB, i.e., $n \approx 5 \cdot 10^7$ bits. Alice and Bob have to agree on a private key $(s_1, \dots, s_l)$ of $100 \cdot \log_2(1.5 \cdot 10^{13}) \approx 4380$ bits. The storage capacity of the adversary Eve is assumed to be 100 TB, which equals about $8.8 \cdot 10^{14}$ bits. Then $\delta \approx 0.587$ and

$$\text{prob(not } \mathcal{E}) \le 5 \cdot 10^7 \cdot 0.587^{100} \approx 3.7 \cdot 10^{-16} < 10^{-15}.$$

Thus, the probability that Eve gets any additional information about the plaintext by observing the ciphertext and applying an optimal storage strategy is less than 10^{-15}.

Theorem 9.21 may also be interpreted in terms of distinguishing algorithms (see Proposition 9.10). We denote by E the probabilistic encryption algorithm which encrypts m as $c := m \oplus k$, with k randomly chosen as above.

Theorem 9.22. *For every probabilistic storage strategy S storing a fraction δ of all randomizer bits, and every probabilistic distinguishing algorithm $A(m_0, m_1, c, s)$ and all $m_0, m_1 \in M$, with $m_0 \neq m_1$,*

$$\mathrm{prob}(A(m_0, m_1, c, s) = m : m \xleftarrow{u} \{m_0, m_1\}, c \leftarrow E(m), s \leftarrow S) \leq \frac{1}{2} + n\delta^l.$$

Remark. This statement is equivalent to

$$| \, \mathrm{prob}(A(m_0, m_1, c, s) = m_0 : c \leftarrow E(m_0), s \leftarrow S)$$
$$- \mathrm{prob}(A(m_0, m_1, c, s) = m_0 : c \leftarrow E(m_1), s \leftarrow S) \, | \leq n\delta^l,$$

as the same computation as in the proof of Proposition 9.10 shows.

Proof. From Theorem 9.21 and Proposition B.32, we conclude that

$$\mathrm{prob}(c, s \, | \, m_0, \mathcal{E}) = \mathrm{prob}(c, s \, | \, m_1, \mathcal{E})$$

for all $m_0, m_1 \in M$, $c \in C$ and $s \in S$. Computing with probabilities conditional on $\mathcal{E}$, we get

$$\mathrm{prob}(A(m_0, m_1, c, s) = m_0 \, | \, \mathcal{E} : c \leftarrow E(m_0), s \leftarrow S)$$
$$= \sum_{c,s} \mathrm{prob}(c, s \, | \, m_0, \mathcal{E}) \cdot \mathrm{prob}(A(m_0, m_1, c, s) = m_0 \, | \, \mathcal{E})$$
$$= \sum_{c,s} \mathrm{prob}(c, s \, | \, m_1, \mathcal{E}) \cdot \mathrm{prob}(A(m_0, m_1, c, s) = m_0 \, | \, \mathcal{E})$$
$$= \mathrm{prob}(A(m_0, m_1, c, s) = m_0 \, | \, \mathcal{E} : c \leftarrow E(m_1), s \leftarrow S).$$

Using Lemma B.9 we get

$$| \, \mathrm{prob}(A(m_0, m_1, c, s) = m_0 : c \leftarrow E(m_0), s \leftarrow S)$$
$$- \mathrm{prob}(A(m_0, m_1, c, s) = m_0 : c \leftarrow E(m_1), s \leftarrow S) \, |$$
$$\leq \mathrm{prob}(\mathrm{not} \, \mathcal{E}) \leq n\delta^l,$$

as desired.   □

As we observed above, the model of attack just used is somewhat restricted, because the eavesdropper Eve does not use the bit stream as input for computations in the first phase – while listening to the random source. Security proofs for the general model of attack, where Eve may use her (unlimited) computing power at any time – without any restrictions – were given only recently in a series of papers ([AumRab99]; [AumDinRab02]; [DinRab02]; [DziMau02]; [Lu02]; [Vadhan03]).

The results obtained by Aumann and Rabin ([AumRab99]) were still restricted. The *randomness efficiency*, i.e., the ratio $\delta = q/|R|$ of Eve's storage capacity and the size of the randomizer, was very small.

Major progress on the bounded storage model was then achieved by Aumann, Ding and Rabin ([AumDinRab02]; [Ding01]; [DinRab02]) proving the security of schemes with a randomness efficiency of about 0.2.

Strong security results for the general model of attack were shown by Dziembowski and Maurer in [DziMau02] (an extended version is [DziMau04a]).

They can prove that keys k, whose length n is much longer than the length of the initial private key, can be securely generated with a randomness efficiency which may be arbitrarily close to 1. A randomness efficiency of 0.1 is possible for reasonable parameter sizes, which appear possible in practice (for example, with attacker Eve's storage capacity ≈ 125 TB, a derived key k of 1 GB for the one-time pad, an initial private key < 1 KB and a statistical distance $< 2^{-29}$ between the distribution of the derived key k and the uniform distribution, from Eve's point of view).

The schemes investigated in the cited papers are all very similar to the scheme from [Maurer92], which we explained here. Of course, security proofs for the general model of attack require more sophisticated methods of probability and information theory.

In all of the bounded-storage-model schemes that we referred to above one assumes that Alice and Bob share an initial secret key s, usually without considering how such a key s is obtained by Alice and Bob. A natural way would be to exchange the initial key by using a public-key key agreement protocol, for example, the Diffie-Hellman protocol (see Section 4.1.2). At first glance, this approach may appear useless, since the information-theoretic security against a computationally unbounded adversary Eve is lost – Eve could break the public-key protocol with her unlimited resources. However, if Eve is an attacker, who gains her infinite computing power (and then the initial secret key) only after the broadcast of the randomizer, then the security of the scheme might be preserved (see [DziMau04b] for a detailed discussion).

In [CachMau97], another approach is discussed. In a variant of their scheme, the key k for the one-time pad is generated within the bounded storage model, and Alice and Bob need not share an initial secret key s. The general model of attack is applied in [CachMau97] – adversary Eve may use her unlimited computing power at any time[8].

Some of the techniques used there are basic.[9] To illustrate these techniques, we give a short overview about parts of [CachMau97].

As before, there is some source for truly random bits. Alice, Bob and Eve receive these random bits over perfect channels without any errors. We are looking for bit sequences of length n to serve as keys in a one-time pad for encrypting messages $m \in \{0,1\}^n$. The random bit source generates N bits $R := (r_1, \ldots, r_N)$. The storage capacity q of Eve is smaller than N, so she is not able to store the whole randomizer. In contrast to the preceding model, she not only stores q bits of the randomizer, but also executes some probabilistic algorithm U while listening to the random source, to compute q bits of information from R (and store them in her memory). As before, we

---

[8] Unfortunately, the arising schemes are either impractical or attacker Eve's probability of success is non-negligible, see below.

[9] They are also applied, for example, in the noisy channel model, which we discuss in Section 9.6.2.

denote by $\delta := q/N$ the fraction of Eve's storage capacity with respect to the total number of randomizer bits.

In a first phase, called *advantage distillation*, Alice and Bob extract sufficiently many, say l, bits $S := (s_1, \ldots, s_l)$ from R, at randomly chosen positions $P := (p_1, \ldots, p_l)$:

$$s_1 := r_{p_1}, s_2 := r_{p_2}, \ldots, s_l := r_{p_l}.$$

It is necessary that the positions are chosen pairwise independently. The positions $p_1, \ldots, p_l$ are kept secret, until the broadcast of the randomizer is finished.

Alice and Bob can select the bits in two ways.

1. **Private key scenario:** Alice and Bob agree on the positions $p_1, \ldots, p_l$ in advance and share these positions as an initial secret key.
2. **Key agreement solely by public discussion:** Independently, Alice and Bob each select and store w bits of the randomizer R. The positions of the selected bits $t_1, \ldots, t_w$ and $u_1, \ldots, u_w$ are randomly selected, pairwise independent and uniformly distributed. When the broadcast of the randomizer is finished, Alice and Bob exchange the chosen positions over the public channel. Let $\{p_1, \ldots, p_l\} = \{t_1, \ldots, t_w\} \cap \{u_1, \ldots, u_w\}$. Then, Alice and Bob share the l randomizer bits at the positions $p_1, \ldots, p_l$. It is easy to see that the expected number l of common positions is $l = w^2/N$ (see, for example, Corollary B.17). Hence, on average, they have to select and store $\sqrt{lN}$ randomizer bits to obtain l common bits.

Since Eve can store at most q bits, her information about S is incomplete. For example, it can be proven that Eve knows at most a fraction δ of the l bits in S (in the information-theoretic sense). Thus, Alice and Bob have distilled an advantage. Let e be the integer part of Eve's uncertainty $H(S|\text{Eve's knowledge})$ about S. Then Eve lacks approximately e bits of information about S.

In a second phase, Alice and Bob apply a powerful technique, called *privacy amplification* or *entropy smoothing*, to extract f bits from S in such a way that Eve has almost no information about the resulting string $\tilde{S}$. Here, f is given by the Rényi entropy of order 2 (see below). Since this entropy is less than or equal to Shannon's entropy, we have $f \leq e$. Eve's uncertainty about $\tilde{S}$ is close to f, so from Eve's point of view, $\tilde{S}$ appears almost truly random. Thus, it can serve Alice and Bob as a provably secure key k in a one-time pad.

Privacy amplification is accomplished by randomly selecting a member from a so-called universal class of hash functions (see below). Alice randomly selects an element h from such a universal class $\mathcal{H}$ (with respect to the uniform distribution) and sends h to Bob via a public channel. Thus, Eve may even know $\mathcal{H}$ and h. Alice and Bob both apply h to S in order to obtain their key $k := h(S)$ for the one-time pad.

Let H and K be the random variables describing the probabilistic choice of the function h and the probabilistic choice of the key k. The following theorem is proven in [CachMau97].

Theorem 9.23. *Given a fixed storage capacity q of Eve and $\epsilon_1, \epsilon_2 > 0$, there exists a security event $\mathcal{E}$ such that*

$$\mathrm{prob}(\mathcal{E}) \geq 1 - \epsilon_1 \text{ and } I(K; H \,|\, U = u, P = p, \mathcal{E}) \leq \epsilon_2,$$

and hence in particular

$$I(K; UHP \,|\, \mathcal{E}) \leq \epsilon_2 \text{ and } I(K; UH \,|\, \mathcal{E}) \leq \epsilon_2,$$

provided the size N of the randomizer R and the number l of elements selected from R by S are sufficiently large.

Remarks:

1. Explicit formulae are derived in [CachMau97] which connect the bounds ϵ_1, ϵ_2, the size N of the randomizer, Eve's storage capacity q, the number l of chosen positions and the number f of derived key bits.
2. The third inequality follows from the second by Proposition B.35, and the fourth inequality from the third by Proposition B.36 (also observe the final remark in Appendix B.4).
3. $I(K; H \,|\, U = u, P = p, \mathcal{E}) \leq \epsilon_2$ means the following. Assume Eve has the specific knowledge $U = u$ about the randomizer and has learned the positions P of the bits selected from R by Alice and Bob after the broadcast of the randomizer R. Then the average amount of information (measured in bits in the information-theoretic sense) that Eve can derive about the key k from learning the hash function h is less than ϵ_2, provided the security event $\mathcal{E}$ occurs.

   Thus, in the private key scenario, the bound ϵ_2 also holds if the secret key shared by Alice and Bob (i.e., the positions of the selected randomizer bits) is later compromised. The security is everlasting.

As we mentioned above, a key step in the proof of Theorem 9.23 is privacy amplification to transform almost all the entropy of a bit string into a random bit string. For this purpose, it is not sufficient to work with the classical Shannon entropy as defined in Appendix B.4. Instead, it is necessary to use more general information measures: the *Rényi entropies* of order α ($0 \leq \alpha \leq \infty$, see [Rényi61]; [Rényi70]). Here, in particular, the Rényi entropy of order 2 – also called *collision entropy* – is needed.

Definition 9.24. Let S be a random variable with values in the finite set $\mathcal{S}$. The *collision probability* $\mathrm{prob}_c(S)$ of S is defined as

$$\mathrm{prob}_c(S) := \sum_{s \in \mathcal{S}} \mathrm{prob}(S = s)^2.$$

The *collision entropy* or *Rényi entropy (of order 2) of S* is

$$H_2(S) := -\log_2(\mathrm{prob}_c(S)) = -\log_2\left(\sum_{s \in S} \mathrm{prob}(S = s)^2\right).$$

$\mathrm{prob}_c(S)$ is the probability that two independent executions of S yield the same result. $H_2(S)$ measures the uncertainty that two independent executions of the random experiment S yield the same result.

The mathematical foundation of privacy amplification is the *Smoothing Entropy Theorem*. It states that almost all the collision entropy of a random variable S may be converted into uniform random bits by selecting a function h randomly from a universal class of hash functions and applying h to S (see, e.g., [Luby96], Lecture 8). Universal classes of hash functions were introduced by Carter and Wegman ([CarWeg79]; [WegCar81]).

Definition 9.25. A set $\mathcal{H}$ of functions $h : X \longrightarrow Y$ is called a *universal class of hash functions* if for all distinct $x_1, x_2 \in X$,

$$\mathrm{prob}(h(x_1) = h(x_2) : h \xleftarrow{u} \mathcal{H}) = \frac{1}{|Y|}.$$

$\mathcal{H}$ is called a *strongly universal class of hash functions* if for all distinct $x_1, x_2 \in X$ and all (not necessarily distinct) $y_1, y_2 \in Y$,

$$\mathrm{prob}(h(x_1) = y_1, h(x_2) = y_2 : h \xleftarrow{u} \mathcal{H}) = \frac{1}{|Y|^2}.$$

In particular, a strongly universal class is also universal. (Strongly) universal classes of hash functions behave like completely random functions with respect to collisions (or value pairs).

Example. A straightforward computation shows that the set of linear mappings $\{0,1\}^l \longrightarrow \{0,1\}^f$ is a strongly universal class of hash functions. There are smaller classes ([Stinson92]). For example, the set

$$\mathcal{H} := \{h_{a_0,a_1} : \mathbb{F}_{2^l} \longrightarrow \mathbb{F}_{2^f}, \; x \longmapsto \mathrm{msb}_f(a_0 \cdot x + a_1) \mid a_0, a_1 \in \mathbb{F}_{2^l}\}$$

is strongly universal ($l \geq f$), and the set

$$\mathcal{H} := \{h_a : \mathbb{F}_{2^l} \longrightarrow \mathbb{F}_{2^f}, \; x \longmapsto \mathrm{msb}_f(a \cdot x) \mid a \in \mathbb{F}_{2^l}\}$$

is universal. Here, we consider $\{0,1\}^m$ as equipped with the structure $\mathbb{F}_{2^m}$ of the Galois field with 2^m elements (see Appendix A.5), and msb_f denotes the f most-significant bits. See Exercise 8.

Remark. In the key generation scheme discussed, Alice and Bob select w (or l) bits at pairwise independent random positions from the N bits broadcast by the random source. They have to store the positions of these bits. At first glance, $w \cdot \log_2(N)$ bits are necessary to describe w positions. Since w is large, a huge number of bits have to be stored and transferred between Alice and Bob. Strongly universal classes of hash functions also provide a solution to this problem.

Assume $N = 2^m$, and consider $\{0,1\}^m$ as $\mathbb{F}_{2^m}$. Alice and Bob may work with the strongly universal class

$$\mathcal{H} = \{h : \mathbb{F}_{2^m} \longrightarrow \mathbb{F}_{2^m},\ x \longmapsto a_0 \cdot x + a_1 \mid a_0, a_1 \in \mathbb{F}_{2^m}\}$$

of hash functions. They fix pairwise different elements $x_1, \ldots, x_w \in \mathbb{F}_{2^m}$ in advance. $\mathcal{H}$ and the x_i may be known to Eve. Now, to select w positions randomly – uniformly distributed and pairwise independent – Alice or Bob randomly chooses some h from $\mathcal{H}$ (with respect to the uniform distribution) and applies h to the x_i. This yields w uniformly distributed and pairwise independent positions

$$h(x_1), h(x_2), \ldots, h(x_w).$$

Thus, the random choice of the w positions reduces to the random choice of an element h in $\mathcal{H}$, and this requires the random choice of $2m = 2\log_2(N)$ bits.

Example. Assume that Alice and Bob do not share an initial private key and the key is derived solely by a public discussion. Using the explicit formulae, you get the following example for the Cachin-Maurer scheme (see [CachMau97] for more details). A satellite broadcasting random bits at a rate of 40 Gbit/s is used for $2 \cdot 10^5$ seconds (about 2 days) to provide a randomizer R with about $N = 8.6 \cdot 10^{15}$ bits. The storage capacity of the adversary Eve is assumed to be $1/2$ PB, which equals about $4.5 \cdot 10^{15}$ bits. To get $l = 1.3 \cdot 10^7$ common positions and common random bits, Alice and Bob each have to select and store $w = \sqrt{lN} = 3.3 \cdot 10^{11}$ bits (or about 39 GB) from R. By privacy amplification, they get a key k of about 61 KB and Eve knows not more than 10^{-20} bits of k, provided that the security event $\mathcal{E}$ occurs. The probability of $\mathcal{E}$ is $\geq 1 - \varepsilon_1$ with $\varepsilon_1 = 10^{-3}$. Since l is of the order of $1/\varepsilon_1^2$, the probability that the security event $\mathcal{E}$ does not occur can not be reduced to significantly smaller values, without increasing the storage requirements for Alice to unreasonably high values. To choose the w positions of the randomizer bits which they store, Alice and Bob each randomly select a strongly universal hash function (see the preceding remark). So, to exchange these positions, Alice and Bob have to transmit a strongly universal hash function in each direction, which requires $2\log_2(N) \approx 106$ bits. For privacy amplification, either Alice or Bob chooses the random universal hash function and communicates it, which takes about l bits ≈ 1.5 MB. The large size of the hash functions may be substantially reduced by using "almost universal" hash functions.

Remarks:

1. Alice and Bob need – as the example demonstrates – a very large capacity for storing the positions and the values of the randomizer bits, and the size of this storage rapidly increases if the probability of the security event $\mathcal{E}$ is required to be significantly closer to 1 than 10^{-3}, which is certainly not negligible. To restrict adversary Eve's probability of success to negligible values would require unreasonably high storage capacities of Alice and Bob. This is also true for the private-key scenario.

2. If the key k is derived solely by a public discussion, then both Alice and Bob need storage on the order of $\sqrt{N}$, which is also on the order of $\sqrt{q}$ (recall that N is the size of the randomizer and q is the storage size of the attacker). It is shown in [DziMau04b] that these storage requirements can not be reduced. The Cachin-Maurer scheme is essentially optimal in terms of the ratio between the storage capacity of Alice and Bob and the storage capacity of adversary Eve. The practicality of schemes in the bounded storage model which do not rely on a shared initial secret key is therefore highly questionable.

9.6.2 The Noisy Channel Model

An introduction to the noisy channel model is, for example, given in the survey article [Wolf98]. As before, we use the "satellite scenario". Random bits are broadcast by some radio source. Alice and Bob receive these bits and generate a key from them by a public discussion. The eavesdropper, Eve, also receives the random bits and can listen to the communication channel between Alice and Bob. Again we assume that Eve is a passive adversary. There are other models including an active adversary (see, e.g., [Maurer97]; [MauWol97]). Though all communications are public, Eve gains hardly any information about the key. Thus, the generated key appears almost random to Eve and can be used in a provably secure one-time pad. The secrecy of the key is based on the fact that no information channel is error-free. The system also works in the case where Eve receives the random bits via a much better channel than Alice and Bob.

The key agreement works in three phases. As in Section 9.6.1, it starts with advantage distillation and ends with privacy amplification. There is an additional intermediate phase called *information reconciliation*.

During advantage distillation, Alice chooses a truly random key k, for example from the radio source. Before transmitting it to Bob, she uses a random bit string r to mask k and to make the transmission to Bob highly reliable (applying a suitable error detection code randomized by r). The random bit string r is taken from the radio source and commonly available to all participants. If sufficient redundancy and randomness is built in, which means that the random bit string r is sufficiently long, the error probability of the adversary is higher than the error probability of the legitimate recipient Bob. In this way, Alice and Bob gain an advantage over Eve.

When phase 1 is finished, the random string k held by Alice may still differ from the string k' received by Bob. Now, Alice and Bob start information reconciliation and interactively modify k and k', such that at the end, the probability that $k \neq k'$ is negligibly small. This must be performed without leaking too much information to the adversary Eve. Alice and Bob may, for example, try to detect differing positions in the string by comparing the parity bits of randomly chosen substrings.

After phase 2, the same random string k of size l is available to Alice and Bob with a very high probability, and they have an advantage over Eve. Eve's information about k, measured by Rényi entropies, is incomplete. Applying the privacy amplification techniques sketched in Section 9.6.1 Alice and Bob obtain their desired key.

Exercises

1. Let $n \in \mathbb{N}$. We consider the *affine cipher* modulo n. It is a symmetric encryption scheme. A key (a, b) consists of a unit $a \in \mathbb{Z}_n^*$ and an element $b \in \mathbb{Z}_n$. A message $m \in \mathbb{Z}_n$ is encrypted as $a \cdot m + b$.
   Is the affine cipher perfectly secret if we randomly (and uniformly) choose a key for each message m to be encrypted?

2. Let E be an encryption algorithm, which encrypts plaintexts $m \in M$ as ciphertexts $c \in C$, and let K denote the secret key used to decrypt ciphertexts.
   Show that an adversary's uncertainty about the secret key is at least as great as her uncertainty about the plaintext: $H(K \mid C) \geq H(M \mid C)$.

3. ElGamal's encryption (Section 3.5.1) is probabilistic. Is it computationally secret?

4. Consider Definition 9.14 of computationally secret encryptions. Show that an encryption algorithm $E(i, m)$ is computationally secret if and only if for every probabilistic polynomial distinguishing algorithm $A(i, m_0, m_1, c)$ and every probabilistic polynomial sampling algorithm S, which on input $i \in I$ yields $S(i) = \{m_0, m_1\} \subset M_i$, and every positive polynomial $P \in \mathbb{Z}[X]$, there is a $k_0 \in \mathbb{N}$ such that for all $k \geq k_0$,

$$\mathrm{prob}(A(i, m_0, m_1, c) = m_0 : i \leftarrow K(1^k), \{m_0, m_1\} \leftarrow S(i), c \leftarrow E(m_0))$$
$$- \mathrm{prob}(A(i, m_0, m_1, c) = m_0 : i \leftarrow K(1^k), \{m_0, m_1\} \leftarrow S(i), c \leftarrow E(m_1))$$
$$\leq \frac{1}{P(k)}.$$

5. Let $I_k := \{(n,e) \mid n = pq,\ p,q \text{ distinct primes}, |p| = |q| = k, e \in \mathbb{Z}^*_{\varphi(n)}\}$ and $(n,e) \xleftarrow{u} I_k$ be a randomly chosen RSA key. Encrypt messages $m \in \{0,1\}^r$, where $r \leq \log_2(|n|)$, in the following way. Pad m with leading random bits to get a padded plaintext $\overline{m}$ with $|\overline{m}| = |n|$. If $\overline{m} > n$, then repeat the padding of m until $\overline{m} < n$. Then encrypt m as $E(m) := c := \overline{m}^e \bmod n$.

   Prove that this scheme is computationally secret.

   Hint: use Exercise 7 in Chapter 8.

6. Let $I = (I_k)_{k \in \mathbb{N}}$ be a key set with security parameter, and let $f = (f_i : D_i \longrightarrow D_i)_{i \in I}$ be a family of one-way trapdoor permutations with hard-core predicate $B = (B_i : D_i \longrightarrow \{0,1\})_{i \in I}$ and key generator K. Consider the following probabilistic public-key encryption scheme ([GolMic84]; [GolBel01]): Let Q be a polynomial and $n := Q(k)$. A bit string $m := m_1 \ldots m_n$ is encrypted as a concatenation $c_1 \| \ldots \| c_n$, where $c_j := f_i(x_j)$ and x_j is a randomly selected element of D_i with $B_i(x_j) = m_j$.

   Describe the decryption procedure and show that the encryption scheme is computationally secret.

   Hints: Apply Exercise 4. Given a pair $\{m_0, m_1\}$ of plaintexts, construct a sequence of messages $\tilde{m}_1 := m_0, \tilde{m}_2, \ldots, \tilde{m}_n := m_1$, such that $\tilde{m}_{j+1}$ differs from $\tilde{m}_j$ in at most one bit. Then consider the sequence of distributions $c \leftarrow E(i, \tilde{m}_j)$ (also see the proof of Proposition 8.4).

7. We consider the Goldwasser-Micali probabilistic encryption scheme ([GolMic84]). Let $I_k := \{n \mid n = pq, p,q \text{ distinct primes }, |p| = |q| = k\}$ and $I := (I_k)_{k \in \mathbb{N}}$. As his public key, each user Bob randomly chooses an $n \xleftarrow{u} I_k$ (by first randomly choosing the secret primes p and q) and a quadratic non-residue $z \xleftarrow{u} QNR_n$ with Jacobi symbol $\left(\frac{z}{n}\right) = 1$ (he can do this easily, since he knows the primes p and q; see Appendix A.6). A bit string $m = m_1 \ldots m_n$ is encrypted as a concatenation $c_1 \| \ldots \| c_n$, where $c_j = x_j^2$ if $m_j = 1$, and $c_j = zx_j^2$ if $m_j = 0$, with a randomly chosen $x_j \xleftarrow{u} \mathbb{Z}^*_n$. In other words: A 1 bit is encrypted by a random quadratic residue, a 0 bit by a random non-residue.

   Describe the decryption procedure and show that the encryption scheme is computationally secret, provided the quadratic residuosity assumption (Definition 6.11) is true.

   Hint: The proof is similar to the proof of Exercise 6. Use Exercise 9 in Chapter 6.

8. Let $l \geq f$. Prove that:
   a. $\mathcal{H} := \{h_a : \mathbb{F}_{2^l} \longrightarrow \mathbb{F}_{2^f},\ x \longmapsto \mathrm{msb}_f(a \cdot x) \mid a \in \mathbb{F}_{2^l}\}$ is a universal class of hash functions.
   b. $\mathcal{H} := \{h_{a_0,a_1} : \mathbb{F}_{2^l} \longrightarrow \mathbb{F}_{2^f},\ x \longmapsto \mathrm{msb}_f(a_0 \cdot x + a_1) \mid a_0, a_1 \in \mathbb{F}_{2^l}\}$ is a strongly universal class of hash functions.

Here we consider $\{0,1\}^m$ as equipped with the structure $\mathbb{F}_{2^m}$ of the Galois field with 2^m elements. As before, msb_f denotes the f most-significant bits.

10. Provably Secure Digital Signatures

In previous sections, we discussed signature schemes (Full-Domain-Hash RSA signatures and PSS in Section 3.4.5; the Fiat-Shamir signature scheme in Section 4.2.5) that include a hash function h and whose security can be proven in the random oracle model. It is assumed that the hash function h is a random oracle, i.e., it behaves like a perfectly random function (see Sections 3.4.4 and 3.4.5). Perfectly random means that for all messages m, each of the k bits of the hash value $h(m)$ is determined by tossing a coin, or, equivalently, that the map $h : X \longrightarrow Y$ is randomly chosen from the set $\mathcal{F}(X,Y)$ of all functions from X to Y. In general, $\mathcal{F}(X,Y)$ is tremendously large. For example, if $X = \{0,1\}^n$ and $Y = \{0,1\}^k$, then $|\mathcal{F}(X,Y)| = 2^{k2^n}$. Thus, it is obvious that perfectly random oracles cannot be implemented.

Moreover, examples of cryptographic schemes were constructed that are provably secure in the random oracle model, but are insecure in any real-world implementation, where the random oracle is replaced by a real hash function. Although these examples are contrived, doubts on the random oracle model arose (see the remark on page 244 in Section 9.5).

Therefore, it is desirable to have signature schemes whose security can be proven solely under standard assumptions (like the RSA or the discrete logarithm assumption). Examples of such signature schemes are given in this chapter.

10.1 Attacks and Levels of Security

Digital signature schemes are public-key cryptosystems and are based on the one-way feature of a number-theoretical function. A *digital signature scheme*, for example the basic RSA signature scheme (Section 3.3.2) or ElGamal's signature scheme (Section 3.5.2), consists of the following:

1. A key generation algorithm K, which on input 1^k (k being the security parameter) produces a pair (pk, sk) consisting of a public key pk and a secret (private) key sk.
2. A signing algorithm $S(sk, m)$, which given the secret key sk of user Alice and a message m to be signed, generates Alice's signature σ for m.

3. A verification algorithm $V(pk, m, \sigma)$, which given Alice's public key pk, a message m and a signature σ, checks whether σ is a valid signature of Alice for m. Valid means that σ might be output by $S(sk, m)$, where sk is Alice's secret key.

Of course, all algorithms must be polynomial. The key generation algorithm K is always a probabilistic algorithm. In many cases, the signing algorithm is also probabilistic (see, e.g., ElGamal or PSS). The verification algorithm might be probabilistic, but in practice it usually is deterministic.

As with encryption schemes, there are different types of attacks on signature schemes. We may distinguish between (see [GolMicRiv88]):

1. *Key-only attack.* The adversary Eve only knows the public key of the signer Alice.
2. *Known-signature attack.* Eve knows the public key of Alice and has seen message-signature pairs produced by Alice.
3. *Chosen-message attack.* Eve may choose a list $(m_1, \ldots, m_t)$ of messages and ask Alice to sign these messages.
4. *Adaptively-chosen-message attack.* Eve can adaptively choose messages to be signed by Alice. She can choose some messages and gets the corresponding signatures. Then she can do cryptanalysis and, depending on the outcome of her analysis, she can choose the next message to be signed, and so on.

Adversary Eve's level of success may be described in increasing order as (see [GolMicRiv88]):

1. *Existential forgery.* Eve is able to forge the signature of at least one message, not necessarily the one of her choice.
2. *Selective forgery.* Eve succeeds in forging the signature of some messages of her choice.
3. *Universal forgery.* Although unable to find Alice's secret key, Eve is able to forge the signature of any message.
4. *Retrieval of secret keys.* Eve finds out Alice's secret key.

As we have seen before, signatures in the basic RSA, ElGamal and DSA schemes, without first applying a suitable hash function, can be easily existentially forged using a key-only attack (see Section 3). In the basic Rabin scheme, secret keys may be retrieved by a chosen-message attack (see Section 3.6.1). We may define the level of security of a signature scheme by the level of success of an adversary performing a certain type of attack. Different levels of security may be required in different applications.

In this chapter, we are interested in signature schemes which provide the maximum level of security. The adversary Eve cannot succeed in an existential forgery with a significant probability, even if she is able to perform an adaptively-chosen-message attack. As usual, the adversary is modeled as a probabilistic polynomial algorithm.

Definition 10.1. Let $\mathcal{D}$ be a digital signature scheme, with key generation algorithm K, signing algorithm S and verification algorithm V. An *existential forger* F for $\mathcal{D}$ is a probabilistic polynomial algorithm F that on input of a public key pk outputs a message-signature pair $(m, \sigma) := F(pk)$. F is successful on pk if σ is a valid signature of m with respect to pk, i.e., $V(pk, F(pk)) = $ accept. F performs an *adaptively-chosen-message attack* if, while computing $F(pk)$, F can repeatedly generate a message $\tilde{m}$ and then is supplied with a valid signature $\tilde{\sigma}$ for $\tilde{m}$.

Remarks. Let F be an existential forger performing an adaptively-chosen-message attack:

1. Let (pk, sk) be a key of security parameter k. Since the running time of $F(pk)$ is bounded by a polynomial in k (note that pk is generated in polynomial time from 1^k), the number of messages for which F requests a signature is bounded by $T(k)$, where T is a polynomial.
2. The definition leaves open who supplies F with the valid signatures. If F is used in an attack against the legitimate signer, then the signatures $\tilde{\sigma}$ are supplied by the signing algorithm S, $S(sk, \tilde{m}) = \tilde{\sigma}$, where sk is the private key associated with pk. In a typical security proof, the signatures are supplied by a "simulated signer", who is able to generate valid signatures without knowing the trapdoor information that is necessary to derive sk from pk. This sounds mysterious and impossible. Actually, for some time it was believed that a security proof for a signature scheme is not possible, because it would necessarily yield an algorithm for inverting the underlying one-way function. However, the security proof given by Goldwasser, Micali and Rivest for their GMR scheme (discussed in Section 10.3) proved the contrary. See [GolMicRiv88], Section 4: The paradox of proving signature schemes secure. The key idea for solving the paradox is that the simulated signer constructs signatures for keys whose form is a very specific one, whereas their probability distribution is the same as the distribution of the original keys (see, e.g., the proof of Theorem 10.12).
3. The signatures $\sigma_i, 1 \leq i \leq T(k)$, supplied to F are, besides pk, inputs to F. The messages $m_i, 1 \leq i \leq T(k)$, for which F requests signatures, are outputs of F. Let M_i be the random variable describing the i-th message m_i. Since F adaptively chooses the messages, message m_i may depend on the messages m_j and the signatures σ_j supplied to F for $m_j, 1 \leq j < i$. Thus M_i may be considered as a probabilistic algorithm with inputs pk and $(m_j, \sigma_j)_{1 \leq j < i}$.
   The *probability of success* of F for security parameter k is then computed as[1]

---

[1] Unless otherwise stated, we always mean F's probability of success when the signatures for the adaptively chosen messages are supplied by the legitimate signer S.

$$\text{prob}(V(pk, F(pk, (\sigma_i)_{1 \le i \le T(k)})) = \text{accept} : (pk, sk) \leftarrow K(1^k),$$
$$m_i \leftarrow M_i(pk, (m_j, \sigma_j)_{1 \le j < i}), \sigma_i \leftarrow S(sk, m_i), 1 \le i \le T(k)).$$

Definition 10.2. A digital signature scheme *is secure against adaptively-chosen-message attacks* if and only if for every existential forger F performing an adaptively-chosen-message attack and every positive polynomial P, there is a $k_0 \in \mathbb{N}$ such that for all security parameters $k \ge k_0$, the probability of success of F is $\le 1/P(k)$.

Remark. Fail-stop signature schemes provide an additional security feature. If a forger – even if he has unlimited computing power and can do an exponential amount of work – succeeds in generating a valid signature, then the legitimate signer Alice can prove with a high probability that the signature is forged. In particular, Alice can detect forgeries and then stop using the signing mechanism ("fail then stop"). The signature scheme is based on the assumed hardness of a computational problem, and the proof of forgery is performed by showing that this underlying assumption has been compromised. Fail-stop signature schemes were introduced by Waidner and Pfitzmann ([WaiPfi89]). We do not discuss fail-stop signatures here (see, e.g., [Pfitzmann96]; [MenOorVan96]; [Stinson95]; [BarPfi97]).

10.2 Claw-Free Pairs and Collision-Resistant Hash Functions

In many digital signature schemes, the message to be signed is first hashed with a collision-resistant hash function. Provably collision-resistant hash functions can be constructed from claw-free pairs of trapdoor permutations. In Section 10.3 we will discuss the GMR signature scheme introduced by Goldwasser, Micali and Rivest. It was the first signature scheme that was provably secure against adaptively-chosen-message attacks (without depending on the random oracle model), and it is based on claw-free pairs.

Definition 10.3. Let $f_0 : D \longrightarrow D$ and $f_1 : D \longrightarrow D$ be permutations of the same domain D. A pair (x, y) is called a *claw* of f_0 and f_1 if $f_0(x) = f_1(y)$.

Let $I = (I_k)_{k \in \mathbb{N}}$ be a key set with security parameter k. We consider families
$$f_0 = (f_{0,i} : D_i \longrightarrow D_i)_{i \in I}, f_1 = (f_{1,i} : D_i \longrightarrow D_i)_{i \in I}$$
of one-way permutations with common key generator K that are defined on the same domains.

Definition 10.4. (f_0, f_1) is called a *claw-free pair of one-way permutations* if it is infeasible to compute claws; i.e., for every probabilistic polynomial algorithm A which on input i outputs distinct elements $x, y \in D_i$, and for every positive polynomial P, there is a $k_0 \in \mathbb{N}$ such that for all $k \ge k_0$

$$\text{prob}(f_{0,i}(x) = f_{1,i}(y) : i \leftarrow K(1^k), \{x, y\} \leftarrow A(i)) \leq \frac{1}{P(k)} \, .$$

Claw-free pairs of one-way permutations exist, if, for example, factoring is hard.

Proposition 10.5. *Let* $I := \{n \mid n = pq, \ p, q \text{ primes}, p \equiv 3 \bmod 8, q \equiv 7 \bmod 8\}$. *If the factoring assumption (Definition 6.9) holds, then*

$$\text{CQ} := (f_n, g_n : \text{QR}_n \longrightarrow \text{QR}_n)_{n \in I},$$

where $f_n(x) := x^2$ *and* $g_n(x) := 4x^2$, *is a claw-free pair of one-way permutations (with trapdoor).*

Proof. Let $n = pq \in I$. Since both primes, p and q, are congruent 3 modulo 4, f_n is a permutation of QR_n (Proposition A.66). Four is a square and it is a unit in $\mathbb{Z}_n$. Thus, g_n is also a permutation of QR_n. Now, let $x, y \in \text{QR}_n$ with $x^2 = 4y^2 \bmod n$ be a claw. From $n \equiv 1 \bmod 4$ and $n \equiv -3 \bmod 8$, we conclude $\left(\frac{-1}{n}\right) = 1$ and $\left(\frac{2}{n}\right) = -1$ (Theorem A.57). We get $\left(\frac{\pm 2y}{n}\right) = \left(\frac{\pm 1}{n}\right) \cdot \left(\frac{2}{n}\right) \cdot \left(\frac{y}{n}\right) = 1 \cdot (-1) \cdot 1 = -1$, whereas $\left(\frac{x}{n}\right) = 1$ since x is a square. Thus $x \neq \pm 2y$, and the Euclidean algorithm yields a factorization of n. Namely, $0 = x^2 - 4y^2 = (x - 2y)(x + 2y) = 0$ and thus $\gcd(x^2 - 4y^2, n)$ is a non-trivial divisor of n. We see that an algorithm which, with some probability, finds claws of CQ yields an algorithm factoring n with the same probability, which is a contradiction to the factoring assumption. $\square$

Claw-free pairs of one-way permutations can be used to construct collision-resistant hash functions.

Definition 10.6. Let $I = (I_k)_{k \in \mathbb{N}}$ be a key set with security parameter k, and let K be a probabilistic polynomial sampling algorithm for I, which on input 1^k outputs a key $i \in I_k$. Let $k(i)$ be the security parameter of i (i.e., $k(i) = k$ for $i \in I_k$), and $g : \mathbb{N} \longrightarrow \mathbb{N}$ be a polynomial function. A family

$$\mathcal{H} = \left(h_i : \{0,1\}^* \longrightarrow \{0,1\}^{g(k(i))} \right)_{i \in I}$$

of hash functions is called a family of *collision-resistant* (or *collision-free*) hash functions (or a *collision-resistant hash function* for short) with key generator K, if:

1. The hash values $h_i(x)$ can be computed by a polynomial algorithm H with inputs $i \in I$ and $x \in \{0,1\}^*$.
2. It is computationally infeasible to find a collision; i.e., for every probabilistic polynomial algorithm A which on input $i \in I$ outputs messages $m_0, m_1 \in \{0,1\}^*, m_0 \neq m_1$, and for every positive polynomial P, there is a $k_0 \in \mathbb{N}$ such that

$$\text{prob}(h_i(m_0) = h_i(m_1) : i \leftarrow K(1^k), \{m_0, m_1\} \leftarrow A(i)) \leq \frac{1}{P(k)},$$

for all $k \geq k_0$.

Remark. Collision-resistant hash functions are one way (see Exercise 2).

Let (f_0, f_1) be a pair of one-way permutations, as above. For every $m \in \{0,1\}^*$, we may derive a family $f_m = (f_{m,i} : D_i \longrightarrow D_i)_{i \in I}$ as follows. For $m := m_1 \ldots m_l \in \{0,1\}^l$ and $x \in D_i$, let

$$f_{m,i}(x) := f_{m_1,i}(f_{m_2,i}(\ldots f_{m_l,i}(x)\ldots)).$$

If $m \in \{0,1\}^*$ is the concatenation $m := m_1 \| m_2$ of strings m_1 and m_2, then obviously $f_{m,i}(x) = f_{m_1,i}(f_{m_2,i}(x))$.

This family may now be used to construct a family $\mathcal{H} = (h_j)_{j \in J}$ of hash functions. Let $J_k := \{(i,x) \mid i \in I_k, x \in D_i\}$ and $J := \bigcup_{k \in \mathbb{N}} J_k$. We define

$$F_j(m) := f_{m,i}(x) \in D_i \text{ for } j = (i,x) \in J \text{ and } m \in \{0,1\}^*.$$

Our goal is a collision-resistant family of hash functions $\mathcal{H}$. To achieve this, we have to modify our construction a little and first encode our messages m in a prefix-free way. Let $[m]$ denote a prefix-free (binary) encoding of the messages m in $\{0,1\}^*$. Prefix-free means that no encoded $[m]$ appears as a prefix[2] in the encoding $[m']$ of an $m' \neq m$. For example, we might encode each 1 by 1, each 0 by 00 and terminate all encoded messages by 01.[3] We define

$$h_j(m) := F_j([m]), \text{ for } m \in \{0,1\}^* \text{ and } j \in J.$$

We will prove that $\mathcal{H}$ is a collision-resistant family of hash functions if the pair (f_0, f_1) is claw-free. Thus, we obtain the following proposition.

Proposition 10.7. *If claw-free pairs of one-way permutations exist, then collision-resistant hash functions also exist.*

Proof. Let $\mathcal{H}$ be the family of hash functions constructed above. Assume that $\mathcal{H}$ is not collision-resistant. This means that there is a positive polynomial P and a probabilistic polynomial algorithm A which on input $j = (i,x) \in J_k$ finds a collision $\{m, m'\}$ of h_j with non-negligible probability $\geq 1/P(k)$ (where P is a positive polynomial) for infinitely many k. Collision means that $f_{[m],i}(x) = f_{[m'],i}(x)$. Let $[m] = m_1 \ldots m_r$ and $[m'] = m'_1 \ldots m'_{r'}$ $(m_j, m'_{j'} \in \{0,1\})$, and let l be the smallest index u with $m_u \neq m'_u$. Such an index l exists, since $[m]$ is not a prefix of $[m']$, nor vice versa. We have $f_{m_l \ldots m_r,i}(x) = f_{m'_l \ldots m'_{r'},i}(x)$, since $f_{0,i}$ and $f_{1,i}$ are injective. Then $(f_{m_{l+1} \ldots m_r,i}(x), f_{m'_{l+1} \ldots m'_{r'},i}(x))$ is a claw of (f_0, f_1). The binary lengths r and r' of m and m' are bounded by a polynomial in k, since m and m' are computed by the polynomial algorithm A. Thus, the claw of (f_0, f_1) can be computed from the collision $\{m, m'\}$ in polynomial time. Hence, we can compute claws with non-negligible probability $\geq 1/P(k)$, for infinitely many k, which is a contradiction. □

---

[2] A string s is called a prefix of a string t if $t = s\|s'$ is the concatenation of s and another string s'.

[3] Efficient prefix-free encodings exist, such that $[m]$ has almost the same length as m (see, e.g., [BerPer85]).

Remarks:

1. The constructed hash functions are rather inefficient. Thus, in practice, custom-designed hash functions such as SHA are used, whose collision resistance cannot be proven rigorously (see Section 3.4).
2. Larger sets of pairwise claw-free one-way permutations may be used in the construction instead of one pair, for example sets with 2^s elements. Then s bits of the messages m are processed in one step. There are larger sets of pairwise claw-free one-way permutations that are based on the assumptions that factoring and the computation of discrete logarithms are infeasible (see [Damgård87]).
3. Another method of constructing provably collision-resistant hash functions is given in Exercise 8 in Chapter 3. It is based on the assumed infeasibility of computing discrete logarithms.

10.3 Authentication-Tree-Based Signatures

Again, we consider a claw-free pair (f_0, f_1) of one-way permutations, as above. In addition, we assume that f_0 and f_1 have trapdoors. Such a claw-free pair of trapdoor permutations and the induced functions f_m, as defined above, may be used to generate probabilistic signatures. Namely, Alice randomly chooses some $i \in I$ (with a sufficiently large security parameter k) and some $x \in D_i$, and publishes (i, x) as her public key. Then Alice, by using her trapdoor information, computes her signature $\sigma(i, x, m)$ for a message $m \in \{0,1\}^*$ as

$$\sigma(i, x, m) := f_{[m],i}^{-1}(x),$$

where $[m]$ denotes some (fixed) prefix-free encoding of m. Bob can verify Alice's signature σ by comparing $f_{[m],i}(\sigma)$ with x. Since $f_{[m],i}$ is one way and $m \mapsto f_{[m],i}(\sigma)$ is collision resistant, as we have just seen in the proof of Proposition 10.7, only Alice can compute the signature for m, and no one can use one signature σ for two different messages m and m'. Unfortunately, this scheme is a *one-time signature scheme*. This means that only one message can be signed by Alice with her key (i, x); otherwise the security is not guaranteed.[4] If two messages $m \neq m'$ were signed with the same reference value x, then a claw of f_0 and f_1 can be easily computed from $\sigma(i, x, m)$ and $\sigma(i, x, m')$ (see Exercise 5)[5], and this can be a severe security risk. If we use, for example, the claw-free pair of Proposition 10.5, then Alice's secret key (the factors of the modulus n) can be easily retrieved from the computed claw.

---

[4] More examples of one-time signature schemes may be found, e.g., in [MenOorVan96] and [Stinson95].
[5] It is not a contradiction to the assumed claw-freeness of the pair that the claw can be computed, because here the adversary is additionally supplied with two signatures which can only be computed by use of the secret trapdoor information.

In the *GMR signature scheme* ([GolMicRiv88]), Goldwasser, Micali and Rivest overcome this difficulty by using a new random reference value for each message m. Of course, it is not possible to publish all these reference values as a public key in advance, so the reference values are attached to the signatures. Then it is necessary to authenticate these reference values, and this is accomplished by a second claw-free pair of trapdoor permutations. The GMR scheme is based on two claw-free pairs

$$(f_0, f_1) = (f_{0,i}, f_{1,i} : D_i \longrightarrow D_i)_{i \in I}, (g_0, g_1) = (g_{0,j}, g_{1,j} : E_j \longrightarrow E_j)_{j \in J}$$

of trapdoor permutations, defined over $I = (I_k)_{k \in \mathbb{N}}$ and $J = (J_k)_{k \in \mathbb{N}}$. Each user Alice runs a key-generation algorithm $K(1^k)$ to randomly choose an $i \in I_k$ and an $j \in J_k$ (and generate the associated trapdoor information). Moreover, Alice generates a binary "authentication tree" of depth d. The nodes of this tree are randomly chosen elements in D_i. Then Alice publishes (i, j, r) as her public key, where r is the root of the authentication tree. The authentication tree has 2^d leaves $v_l, 1 \leq l \leq 2^d$. Alice can now sign up to 2^d messages. To sign the l-th message m, she takes the l-th leaf v_l and takes as the first part $\sigma_1(m)$ of the signature the previously defined probabilistic signature $f^{-1}_{[m],i}(v_l)$ with respect to the reference value v_l.[6] The second part $\sigma_2(m)$ of the signature authenticates v_l. It contains the elements $x_0 := r, x_1, \ldots, x_d := v_l$ on the path from the root r to the leaf v_l in the authentication tree and authentication values for each node $x_m, 1 \leq m \leq d$. The authentication value of x_m contains the parent x_{m-1} and both of its children c_0 and c_1 (one of them is x_m) and the signature $g^{-1}_{[c_0 \| c_1], j}(x_{m-1})$ of the concatenated children with respect to the reference value x_{m-1}, computed by the second claw-free pair. The children of a node are authenticated jointly. To verify Alice's signatures, Bob has to climb up the tree from the leaf in the obvious way. If he finally computes the correct root r, he accepts the signature.

Theorem 10.8. *The GMR signature scheme is secure against adaptively-chosen-message attacks.*

Proof. See [GolMicRiv88].   □

Remarks:

1. The full authentication tree and the authentication values for its nodes could be constructed in advance and stored. However, it is more efficient to develop it dynamically, as it is needed for signatures, and to store only the necessary information about its current state.

2. The size of a GMR signature is of order $O(kd)$ if the inputs of the claw-free pairs are of order $O(k)$ (as in the pair given in Proposition 10.5). In practice, this size is considerable if the desired number n of signatures and hence $d = \log_2(n)$ increases. For example, think only of 10000

---

[6] Here we simplify a little and omit the "bridge items".

signatures with a security parameter $k = 1024$. The size of the signatures and the number of computations of f – necessary to generate and to verify signatures – could be substantially reduced if authentication trees with much larger branching degrees could be used instead of the binary one, thus reducing the distance from a leaf to the root. Such signature schemes have been developed, for example, by Dwork and Naor ([DwoNao94]) and Cramer and Damgård ([CraDam96]). They are quite efficient and provably provide security against adaptively-chosen-message attacks if the RSA assumption 6.7 holds. For example, taking a 1024-bit RSA modulus, a branching degree of 1000 and a tree of depth 3 in the Cramer-Damgård scheme, Alice could sign up to 10^9 messages, with the size of each signature being less than 4000 bits.

10.4 A State-Free Signature Scheme

The signing algorithm in GMR or in other authentication-tree-based signature schemes is not state free: the signing algorithm has to store the current state of the authentication tree which depends on the already generated signatures, and the next signature depends on this state. In this section, we will describe a provably secure and quite efficient state-free digital signature scheme introduced by Cramer and Shoup ([CraSho2000]). The scheme is secure against adaptively-chosen-message attacks, provided the so-called *strong* RSA assumption (see below) holds. Another state-free signature scheme based on the strong RSA assumption, has been, for example, introduced in [GenHalRab99]. The security proof, which we will give below, shows the typical features of such a proof. It runs with a contradiction: a successful forging algorithm is used to construct an attacker A who successfully inverts the underlying one-way function. The main problem is that in a chosen-message attack, the forger F is only successful if he can request signatures from the legitimate signer. Now the legitimate signer, who uses his secret key, cannot be called during the execution of F, because A is only allowed to use publicly accessible information. Thus, a major problem is to substitute the legitimate signer by a simulation.

The moduli in the Cramer-Shoup signature scheme are defined with special types of primes.

Definition 10.9. A prime p is called a *Sophie Germain prime* if $2p + 1$ is also a prime.[7]

Remark. In the Cramer-Shoup signature scheme we have to assume that sufficiently many Sophie Germain primes exist. Otherwise there is no guarantee that keys can be generated in polynomial time. The security proof given below

---

[7] Sophie Germain (1776–1831) proved the first case of Fermat's Last Theorem for prime exponents p, for which $2p + 1$ is also prime ([Osen74]).

also relies on this assumption (see Lemma 10.11). There must be a positive polynomial P, such that the number of k-bit Sophie Germain primes p is $\geq 2^k/P(k)$. Today, there is no rigorous mathematical proof for this. It is not even known whether there are infinitely many Sophie Germain primes. On the other hand, it is conjectured and there are heuristic arguments and numerical evidence that the number of k-bit Sophie Germain primes is asymptotically equal to $c \cdot 2^k/k^2$, where c is an explicitly computable constant ([Koblitz88]; [BatHor62]; [BatHor65]). Thus, there is convincing evidence for the existence of sufficiently many Sophie Germain primes.

The Strong RSA Assumption. The security of the Cramer-Shoup signature scheme is based on the following *strong RSA assumption* introduced in [BarPfi97].

Let $I := \{n \in \mathbb{N} \mid n = pq,\ p \neq q \text{ prime numbers },\ |p| = |q|\}$ be the set of RSA moduli and $I_k := \{n \in I \mid n = pq, |p| = |q| = k\}$.

Definition 10.10 *(strong RSA assumption).* For every positive polynomial Q and every probabilistic polynomial algorithm A which on inputs $n \in I$ and $y \in \mathbb{Z}_n^*$ outputs an exponent $e > 1$ and an $x \in \mathbb{Z}_n^*$, there exists a $k_0 \in \mathbb{N}$ such that

$$\mathrm{prob}(x^e = y : n \xleftarrow{u} I_k, y \xleftarrow{u} \mathbb{Z}_n^*, (e, x) \leftarrow A(n, y)) \leq \frac{1}{Q(k)}$$

for $k \geq k_0$.

Remark. The strong RSA assumption implies the classical RSA assumption (Definition 6.7). In the classical RSA assumption, the attacking algorithm has to find an e-th root for $y \in \mathbb{Z}_n^*$, for a given e. Here the exponent is not given. The adversary is successful if, given some $y \in \mathbb{Z}_n^*$, she can find an exponent $e > 1$, such that she is able to extract the e-th root x of y. Today, the only known method for breaking either assumption is to solve the factorization problem.

Let
$I_{\mathrm{SG}} := \{n \in I \mid n = pq, p = 2\tilde{p} + 1, q = 2\tilde{q} + 1, \tilde{p}, \tilde{q} \text{ Sophie Germain primes}\}$
and $I_{\mathrm{SG},k} := I_{\mathrm{SG}} \cap I_k$.

Lemma 10.11. *Assume that there is a positive polynomial P, such that the number of k-bit Sophie Germain primes is $\geq 2^k/P(k)$. Then the strong RSA assumption implies that for every positive polynomial Q and every probabilistic polynomial algorithm A which on inputs $n \in I_{\mathrm{SG}}$ and $y \in \mathbb{Z}_n^*$ outputs an exponent $e > 1$ and an $x \in \mathbb{Z}_n^*$, there exists a $k_0 \in \mathbb{N}$ such that*

$$\mathrm{prob}(x^e = y : n \xleftarrow{u} I_{\mathrm{SG},k}, y \xleftarrow{u} \mathbb{Z}_n^*, (e, x) \leftarrow A(n, y)) \leq \frac{1}{Q(k)}$$

for $k \geq k_0$.

Proof. The distribution $n \xleftarrow{u} I_{SG,k}$ is polynomially bounded by the distribution $n \xleftarrow{u} I_k$ if the existence of sufficiently many Sophie Germain primes is assumed. Thus, we may replace $n \xleftarrow{u} I_k$ by $n \xleftarrow{u} I_{SG,k}$ in the strong RSA assumption (Proposition B.26). □

The Cramer-Shoup Signature Scheme. In the key generation and in the signing procedure, a probabilistic polynomial algorithm *GenPrime*(1^k) with the following properties is used:

1. On input 1^k, *GenPrime* outputs a k-bit prime.
2. If *GenPrime*(1^k) is executed $R(k)$ times (R a positive polynomial), then the probability that any two of the generated primes are equal is negligibly small; i.e., for every positive polynomial P there is a $k_0 \in \mathbb{N}$, such that for all $k \geq k_0$

$$\mathrm{prob}(e_{j_1} = e_{j_2} \text{ for some } j_1 \neq j_2 : e_j \leftarrow GenPrime(1^k), 1 \leq j \leq R(k))$$

$$\leq \frac{1}{P(k)}.$$

Such algorithms exist. For example, an algorithm that randomly and uniformly chooses primes of binary length k satisfies the requirements. Namely, the probability that $e_{j_1} = e_{j_2}$ (j_1 and j_2 fixed, $j_1 \neq j_2$) is about $k/2^k$ by the Prime Number Theorem (Theorem A.68), and there are $\binom{R(k)}{2} < R(k)^2/2$ subsets $\{j_1, j_2\}$ of $\{1, \ldots, R(k)\}$. There are suitable implementations of *GenPrime* which are much more efficient than the uniform sampling algorithm (see [CraSho2000]).

Let $N \in \mathbb{N}, N > 1$ be a constant. To set up a Cramer-Shoup signature scheme, we choose two security parameters k and l, with $k^{1/N} < l+1 < k-1$. Then we choose a collision-resistant hash function $h : \{0,1\}^* \longrightarrow \{0,1\}^l$. More precisely, by using $\mathcal{H}$'s key generator, we randomly select a hash function from $\mathcal{H}_l$, where $\mathcal{H}$ is a collision-resistant family of hash functions and $\mathcal{H}_l$ is the subset of functions with security parameter l (without loss of generality, we assume that the functions in $\mathcal{H}_l$ map to $\{0,1\}^l$). We proved in Section 10.2 that such collision-resistant families exist if the RSA assumption and, as a consequence, the factoring assumption hold. The output of h is considered as a number in $\{0, \ldots, 2^l - 1\}$. All users of the scheme generate their signatures by using the hash function h.[8]

Given k, l and h, each user Alice generates her public and secret key.

Key Generation.

1. Alice randomly chooses a modulus $n \xleftarrow{u} I_{SG,k}$, i.e., she randomly and uniformly chooses Sophie Germain primes $\tilde{p}$ and $\tilde{q}$ of length $k-1$ and sets $n := pq$, $p := 2\tilde{p} + 1$ and $q := 2\tilde{q} + 1$.

---

[8] In practice we may use, for example, $k = 512, l = 160$ and $h =$ SHA-1, which is believed to be collision resistant.

2. She chooses $g \xleftarrow{u} QR_n$ and $x \xleftarrow{u} QR_n$ at random and generates an $(l+1)$-bit prime $\tilde{e} := GenPrime(1^{l+1})$.

3. $(n, g, x, \tilde{e})$ is the public key; (p, q) is the secret key.

Remark. Using Sophie Germain primes ensures that the order $\frac{p-1}{2} \cdot \frac{q-1}{2} = \tilde{p} \cdot \tilde{q}$ of QR_n is a product of distinct primes. Thus it is a cyclic group; it is the cyclic subgroup of order $\tilde{p}\tilde{q}$ of $\mathbb{Z}_n^*$.

In the following, all computations are done in $\mathbb{Z}_n^*$ unless otherwise stated and, as usual, we identify $\mathbb{Z}_n = \{0, \dots, n-1\}$.

Signing. It is possible to sign arbitrary messages $m \in \{0,1\}^*$. To sign m, Alice generates an $(l+1)$-bit prime $e := GenPrime(1^{l+1})$ and randomly chooses $\tilde{y} \xleftarrow{u} QR_n$. She computes

$$\tilde{x} := \tilde{y}^{\tilde{e}} \cdot g^{-h(m)},$$

$$y := \left(x \cdot g^{h(\tilde{x})} \right)^{e^{-1}},$$

where e^{-1} is the inverse of e in $\mathbb{Z}_{\varphi(n)}^*$ (the powers are computed in $\mathbb{Z}_n^*$, which is of order $\varphi(n)$). The signature σ of m is $(e, y, \tilde{y})$.

Remarks:

1. Taking the e^{-1}-th power in the computation of y means computing the e-th root in $\mathbb{Z}_n^*$. Alice needs her secret key for this computation. Since $|e| = l + 1 < k - 1 = |\tilde{p}| = |\tilde{q}|$, the prime e does not divide $\varphi(n) = 4\tilde{p}\tilde{q}$. Hence, Alice can easily compute the inverse e^{-1} of e in $\mathbb{Z}_{\varphi(n)}^*$, by using her secret $\tilde{p}$ and $\tilde{q}$ (and the extended Euclidean algorithm, see Proposition A.16).

2. Signing is a probabilistic algorithm, because the prime e is generated probabilistically and a random quadratic residue $\tilde{y}$ is chosen. After these choices, the computation of the signature is deterministic. Therefore, we can describe the signature σ of m as the value of a mathematical function sign:

$$\sigma = \text{sign}(h, n, g, x, \tilde{e}, e, \tilde{y}, m).$$

To compute the function sign by an algorithm, Alice has to use her knowledge about the prime factors of n.

Verification. Recipient Bob verifies a signature $\sigma = (e, y, \tilde{y})$ of Alice for message m as follows:

1. First, he checks whether e is an odd $(l+1)$-bit number that is not divisible by $\tilde{e}$.

2. Then he computes

$$\tilde{x} := \tilde{y}^{\tilde{e}} \cdot g^{-h(m)}$$

and checks whether

$$x = y^e \cdot g^{-h(\tilde{x})}.$$

He accepts if both checks are affirmative; otherwise he rejects.

Remarks:

1. Note that the verification algorithm does not verify that e is a prime; it only checks whether e is odd. A primality test would considerably decrease the efficiency of verification, and the security of the scheme does not require it (as the security proof shows).
2. If Alice generates a signature $(e, y, \tilde{y})$ with $e = \tilde{e}$, then this signature is not accepted by the verification procedure. However, since both e and $\tilde{e}$ are generated by *GenPrime*, this happens only with negligible probability, and Alice could simply generate a new prime in this case.

Theorem 10.12. *If the strong RSA assumption holds, "many" Sophie Germain primes exist and $\mathcal{H}$ is collision resistant, then the Cramer-Shoup signature scheme is secure against adaptively-chosen-message attacks.*

Remark. There is a variant of the signature scheme which does not require the collision resistance of the hash function (see [CraSho2000]). The family $\mathcal{H}$ is only assumed to be a *universal one-way family of hash functions* ([NaoYun89]; [BelRog97]). The universal one-way property is weaker than full collision resistance: if an adversary Eve first chooses a message m and then a random key i is chosen, it should be infeasible for Eve to find $m' \neq m$ with $h_i(m) = h_i(m')$. Note that the size of the key can grow with the length of m.

In the proof of the theorem we need the following technical lemma.

Lemma 10.13. *There is a deterministic polynomial algorithm that for all k, given $n \in I_{\mathrm{SG},k}$, an odd natural number e with $|e| < k - 1$, a number f and elements $u, v \in \mathbb{Z}_n^*$ with $u^e = v^f$ as inputs, computes the r-th root $v^{r^{-1}} \in \mathbb{Z}_n^*$ of v for $r := e/d$ and $d := \gcd(e, f)$.*

Proof. e and hence r and d are prime to $\varphi(n)$, since $\varphi(n) = 4\tilde{p}\tilde{q}$, with Sophie Germain primes $\tilde{p}$ and $\tilde{q}$ of binary length $k - 1$, and e is an odd number with $|e| < k - 1$. Thus, the inverse elements r^{-1} of r and d^{-1} of d in $\mathbb{Z}_{\varphi(n)}^*$ exist and can be computed by the extended Euclidean algorithm (Proposition A.16). Let $s := f/d$. Since r is prime to s, the extended Euclidean algorithm (Algorithm A.5) computes integers m and m', with $sm + rm' = 1$. We have

$$u^r = \left(u^e\right)^{d^{-1}} = \left(v^f\right)^{d^{-1}} = v^s$$

and

$$(u^m \cdot v^{m'})^r = (v^s)^m \cdot (v^{m'})^r = v^{sm+rm'} = v.$$

By setting

$$v^{r^{-1}} = u^m \cdot v^{m'}$$

we obtain the r-th root of v. $\qquad\square$

Proof (of Theorem 10.12). The proof runs by contradiction.

Let $Forger(h, n, g, x, \tilde{e})$ be a probabilistic polynomial forging algorithm which adaptively requests the signatures for t messages, where $t = R(k)$ for some polynomial R, and then produces a valid forgery with non-negligible probability for infinitely many security parameters (k, l). By non-negligible, we mean that the probability is $> 1/Q(k)$ for some positive polynomial Q.

We will define an attacking algorithm A that on inputs $n \in I_{SG}$ and $z \in \mathbb{Z}_n^*$ successfully computes an r-th root modulo n of z, without knowing the prime factors of n (contradicting the strong RSA assumption, by Lemma 10.11).

On inputs $n \in I_{SG,k}$ and $z \in \mathbb{Z}_n^*$, A works as follows:

1. Randomly and uniformly select the second security parameter l and randomly choose a hash function $h \in \mathcal{H}_l$ (by using $\mathcal{H}$'s key generator).
2. In a clever way, generate the missing elements g, x and $\tilde{e}$ of a public key $(n, g, x, \tilde{e})$.
3. Interact with *Forger* to obtain a forged signature (m, σ) for the public key $(n, g, x, \tilde{e})$.

   Since the prime factors of n are not known in this setting, *Forger* cannot get the signatures he requests from the original signing algorithm. Instead, he obtains them from A. Since g, x and $\tilde{e}$ were chosen in a clever way, A is able to supply *Forger* with valid signatures without knowing the prime factors of n.
4. By use of the forged signature (m, σ), compute an r-th root modulo n of z for some $r > 1$.

A simulates the legitimate signer in step 3. We therefore also say that *Forger* runs against a simulated signer. Simulating the signer is the core of the proof. To ensure that *Forger* yields a valid signature with a non-negligible probability, the probabilistic setting where *Forger* operates must be identical (or at least very close) to the setting where *Forger* runs against the legitimate signer. This means, in particular, that the keys generated in step 2 must be distributed as are the keys in the original signature scheme, and the signatures supplied to *Forger* in step 3 must be distributed as if they were generated by the legitimate signer.

We denote by $m_i, 1 \leq i \leq t$, the messages for which signatures are requested by *Forger*, and by $\sigma_i = (e_i, y_i, \tilde{y}_i)$ the corresponding signatures supplied to *Forger*. Let (m, σ) be the output of *Forger*, i.e., m is a message $\neq m_i, 1 \leq i \leq t$, and $\sigma = (e, y, \tilde{y})$ is the forged signature of m. Let $\tilde{x}_i := \tilde{y}_i^{\tilde{e}} \cdot g^{-h(m_i)}$ and $\tilde{x} := \tilde{y}^{\tilde{e}} \cdot g^{-h(m)}$.

We distinguish three (overlapping) types of forgery:

1. Type 1. For some i, $1 \leq i \leq t$, e_i divides e and $\tilde{x} = \tilde{x}_i$.
2. Type 2. For some i, $1 \leq i \leq t$, e_i divides e and $\tilde{x} \neq \tilde{x}_i$.
3. Type 3. For all i, $1 \leq i \leq t$, e_i does not divide e.

Here, note that the number e in the forged signature can be non-prime (the verification procedure does not test whether e is a prime). The numbers e_i are primes (see below).

We may define a forging algorithm $Forger_1$ which yields the output of $Forger$ if it is of type 1, and otherwise returns some message and signature not satisfying the verification condition. Analogously, we define $Forger_2$ and $Forger_3$. Then the valid forgeries of $Forger_1$ are of type 1, those of $Forger_2$ are of type 2 and those of $Forger_3$ of type 3. If $Forger$ succeeds with non-negligible probability, then at least one of the three "single-type forgers" succeeds with non-negligible probability. Replacing $Forger$ by this algorithm, we may assume from now on that $Forger$ generates valid forgeries of one type.

Case 1: $Forger$ is of type 1. To generate the public key in step 2, A proceeds as follows:

1. Generate $(l+1)$-bit primes $e_i := GenPrime(1^{l+1})$, $1 \leq i \leq t$. Set

$$g := z^{2 \prod_{1 \leq i \leq t} e_i}.$$

2. Randomly choose $w \xleftarrow{u} \mathbb{Z}_n^*$ and set

$$x := w^{2 \prod_{1 \leq i \leq t} e_i}.$$

3. Set $\tilde{e} := GenPrime(1^{l+1})$.

To generate the signature for the i-th message m_i, A randomly chooses $\tilde{y}_i \xleftarrow{u} QR_n$ and computes

$$\tilde{x}_i := \tilde{y}_i^{\tilde{e}} \cdot g^{-h(m_i)} \quad \text{and} \quad y_i := \left(x \cdot g^{h(\tilde{x}_i)} \right)^{e_i^{-1}}.$$

Though A does not know the prime factors of n, she can easily compute the e_i-th root to get y_i, because she knows the e_i-th root of x and g by construction.

$Forger$ then outputs a forged signature $\sigma = (e, y, \tilde{y})$ for a message $m \notin \{m_1, \ldots, m_t\}$. If the forged signature does not pass the verification procedure, then the forger did not produce a valid signature. In this case, A returns a random exponent and a random element in $\mathbb{Z}_n^*$, and stops. Otherwise, $Forger$ yields a valid type-1 forgery. Thus, for some j, $1 \leq j \leq t$, we have $e_j \mid e$, $\tilde{x} = \tilde{x}_j$ and

$$\tilde{y}^{\tilde{e}} = \tilde{x} \cdot g^{h(m)}, \quad \tilde{y}_j^{\tilde{e}} = \tilde{x} \cdot g^{h(m_j)}.$$

Now A has to compute an r-th root of z. If $h(m) \neq h(m_j)$, which happens almost with certainty, since $\mathcal{H}$ is collision resistant, we get, by dividing the two equations, the equation

$$\tilde{z}^{\tilde{e}} = g^a = z^{2a \prod_{1 \leq i \leq t} e_i},$$

with $0 < a < 2^l$. Here recall that h yields l-bit numbers and these are $< 2^l$. Then $a := h(m) - h(m_j)$ if $h(m) - h(m_j) > 0$; otherwise $a := h(m_j) - h(m)$. Since $\tilde{e}$ is an $(l+1)$-bit prime and thus $\geq 2^l$, $\tilde{e}$ does not divide a. Moreover, $\tilde{e}$ and the e_i were chosen by *GenPrime*. Thus, with a high probability (stated more precisely below), $\tilde{e} \neq e_i, 1 \leq i \leq t$. In this case, A can compute $z^{\tilde{e}^{-1}}$ using Lemma 10.13 (note that $\tilde{e}$ is an $(l+1)$-bit prime and $l+1 < k-1$). A returns the exponent $\tilde{e}$ and the $\tilde{e}$-th root $z^{\tilde{e}^{-1}}$.

We still have to compute the probability of success of A. Interacting with the legitimate signer, *Forger* is assumed to be successful. This means that there exists a positive polynomial P, an infinite subset $\mathcal{K} \subseteq \mathbb{N}$ and an $l(k)$ for each $k \in \mathcal{K}$, such that *Forger* produces a valid signature with a probability $\geq 1/P(k)$ for all the pairs $(k, l(k)), k \in \mathcal{K}$, of security parameters. To state this more precisely, let $M_i(h, n, g, x, \tilde{e}, (m_j, \sigma_j)_{1 \leq j < i})$ be the random variable describing the choice of m_i by *Forger* ($1 \leq i \leq t$). The signatures σ_i are additional inputs to *Forger*. We have for $k \in \mathcal{K}$ and $l := l(k)$:

$$p_{\text{success},Forger}(k, l) = \text{prob}(Verify(h, n, g, x, \tilde{e}, m, \sigma) = \text{accept}:$$
$$h \leftarrow \mathcal{H}_l, n \xleftarrow{u} I_{\text{SG},k},$$
$$g \xleftarrow{u} \text{QR}_n, x \xleftarrow{u} \text{QR}_n, \tilde{e} \leftarrow GenPrime(1^{l+1}),$$
$$e_i \leftarrow GenPrime(1^{l+1}), \tilde{y}_i \xleftarrow{u} \text{QR}_n,$$
$$m_i \leftarrow M_i(h, n, g, x, \tilde{e}, (m_j, \sigma_j)_{1 \leq j < i}),$$
$$\sigma_i = \text{sign}(h, n, g, x, \tilde{e}, e_i, \tilde{y}_i, m_i), 1 \leq i \leq t,$$
$$(m, \sigma) \leftarrow Forger(h, n, g, x, \tilde{e}, (\sigma_i)_{1 \leq i \leq t}))$$
$$\geq \frac{1}{P(k)}.$$

Here recall that the signature σ_i of m_i can be derived deterministically from $h, n, g, x, \tilde{e}, e_i, \tilde{y}_i$ and m_i as the value of a mathematical function sign (see p. 276).

Let $\psi(z, (e_i)_{1 \leq i \leq t}) := g = z^{2 \prod_{1 \leq i \leq t} e_i}$ and $\chi(w, (e_i)_{1 \leq i \leq t}) := x = w^{2 \prod_{1 \leq i \leq t} e_i}$ be the specific g and x constructed by A. $A(n, z)$ succeeds in computing the root $z^{\tilde{e}^{-1}}$ if the *Forger* produces a valid signature, if $h(m) \neq h(m_i)$ and if $\tilde{e} \neq e_i, 1 \leq i \leq t$. All other steps in A are deterministic. Therefore, we may compute the probability of success of A (for security parameter k) as follows:

$$\text{prob}(v^r = z : n \xleftarrow{u} I_{\text{SG},k}, z \xleftarrow{u} \mathbb{Z}_n^*, (v, r) \leftarrow A(n, z))$$
$$\geq \text{prob}(Verify(h, n, \psi(z, (e_i)_i), \chi(w, (e_i)_i), \tilde{e}, m, \sigma) = \text{accept},$$
$$h(m) \neq h(m_i), \tilde{e} \neq e_i, 1 \leq i \leq t :$$
$$h \leftarrow \mathcal{H}_l, n \xleftarrow{u} I_{\text{SG},k}, z \xleftarrow{u} \mathbb{Z}_n^*, w \xleftarrow{u} \mathbb{Z}_n^*, \tilde{e} \leftarrow GenPrime(1^{l+1}),$$
$$e_i \leftarrow GenPrime(1^{l+1}), \tilde{y}_i \xleftarrow{u} \text{QR}_n,$$
$$m_i \leftarrow M_i(h, n, \psi(z, (e_i)_i), \chi(w, (e_i)_i), \tilde{e}, (m_j, \sigma_j)_{1 \leq j < i}),$$

$$\sigma_i = \text{sign}(h, n, \psi(z, (e_i)_i), \chi(w, (e_i)_i), \tilde{e}, e_i, \tilde{y}_i, m_i), 1 \leq i \leq t,$$
$$(m, \sigma) \leftarrow Forger(h, n, \psi(z, (e_i)_i), \chi(w, (e_i)_i), \tilde{e}, (\sigma_i)_{1 \leq i \leq t}))$$

$=: p_1$

Let Q be a positive polynomial. *GenPrime*, when called polynomially times, generates the same prime more than once only with negligible probability, and h is randomly chosen from a collision-resistant family $\mathcal{H}$ of hash functions. Thus, there is some k_0 such that both the probability that $\tilde{e} \neq e_i$ for $1 \leq i \leq t$ and the probability that $h(m) \neq h(m_i)$ for $1 \leq i \leq t$ are $\geq 1 - 1/Q(k)$, for $k \geq k_0$.[9] Hence, we get for $k \geq k_0$ that

$$p_1 \geq \text{prob}(Verify(h, n, \psi(z, (e_i)_i), \chi(w, (e_i)_i, \tilde{e}, m, \sigma) = \text{accept} :$$
$$h \leftarrow \mathcal{H}_l, n \stackrel{u}{\leftarrow} I_{SG,k}, z \stackrel{u}{\leftarrow} \mathbb{Z}_n^*, w \stackrel{u}{\leftarrow} \mathbb{Z}_n^*, \tilde{e} \leftarrow GenPrime(1^{l+1}),$$
$$e_i \leftarrow GenPrime(1^{l+1}), \tilde{y}_i \stackrel{u}{\leftarrow} \text{QR}_n,$$
$$m_i \leftarrow M_i(h, n, \psi(z, (e_i)_i), \chi(w, (e_i)_i), \tilde{e}, (m_j, \sigma_j)_{1 \leq j < i}),$$
$$\sigma_i = \text{sign}(h, n, \psi(z, (e_i)_i), \chi(w, (e_i)_i), \tilde{e}, e_i, \tilde{y}_i, m_i), 1 \leq i \leq t,$$
$$(m, \sigma) \leftarrow Forger(h, n, \psi(z, (e_i)_i), \chi(w, (e_i)_i), \tilde{e}, (\sigma_i)_{1 \leq i \leq t}))$$
$$\cdot \left(1 - \frac{1}{Q(k)}\right) \cdot \left(1 - \frac{1}{Q(k)}\right)$$

$=: p_2$

The first factor in p_2 is the probability that *Forger* successfully yields a valid signature when interacting with the simulated signer. This probability is equal to *Forger*'s probability of success, $p_{\text{success},Forger}(k, l)$, when he interacts with the legitimate signer. Namely, $\psi(z, (e_i)_i)$ and $\chi(w, (e_i)_i)$ are uniformly distributed quadratic residues, independent of the distribution of the e_i, since $z \stackrel{u}{\leftarrow} \mathbb{Z}_n^*$ and $w \stackrel{u}{\leftarrow} \mathbb{Z}_n^*$.[10] Thus, we may replace $\psi(z, (e_i)_i), z \stackrel{u}{\leftarrow} \mathbb{Z}_n^*$ and $\chi(w, (e_i)_i), w \stackrel{u}{\leftarrow} \mathbb{Z}_n^*$ by $g \stackrel{u}{\leftarrow} \text{QR}_n$ and $x \stackrel{u}{\leftarrow} \text{QR}_n$. We get

$$p_2 = p_{\text{success},Forger}(k, l) \cdot \left(1 - \frac{1}{Q(k)}\right)^2.$$

For $k \in \mathcal{K}$ and $l = l(k)$, we have $p_{\text{success},Forger}(k, l) \geq 1/P(k)$. The probability that A chooses $l(k)$ in her first step is $\geq 1/k$, and we finally obtain that

$$\text{prob}(v^r = z : n \stackrel{u}{\leftarrow} I_{SG,k}, z \stackrel{u}{\leftarrow} \mathbb{Z}_n^*, (v, r) \leftarrow A(n, z)) \geq \frac{1}{P(k)} \cdot \frac{1}{k} \cdot \left(1 - \frac{1}{Q(k)}\right)^2,$$

for the infinitely many $k \in \mathcal{K}$. This contradicts the strong RSA assumption. The proof of Theorem 10.12 is finished in the case where the forger is of type 1. The other cases are proven below. $\square$

---

[9] Note that the messages m and m_i are generated by a probabilistic polynomial algorithm, namely *Forger*.

[10] Here note that $\prod_i e_i$ and $\tilde{e}$ are prime to $\varphi(n) = 4\tilde{p}\tilde{q}$, because $\tilde{e}$ and the e_i are $(l+1)$-bit primes and $l + 1 < k - 1 = |\tilde{p}| = |\tilde{q}|$.

Remark. Before we study the next cases, let us have a look back at the proof just finished. The core of the proof is to simulate the legitimate signer and to supply valid signatures to the forger, without knowing the prime factors of the modulus n. We managed to do this by a clever choice of the second parts of the public keys. Here, the key point is that given a fixed first part of the public key, i.e., given a modulus n, the joint distribution of the second part $(g, x, \tilde{e})$ of the public key and the generated signatures is the same in the legitimate signer and in the simulation by A. This fact is often referred to as "the simulated signer perfectly simulates the legitimate signer".

Proof (of cases 2 and 3).

Case 2: *Forger* is of type 2. We may assume that the j with $e_j \mid e$ and $\tilde{x} \neq \tilde{x}_j$ is fixed. Namely, we may guess the correct j, i.e., we iterate over the polynomially many cases j and assume j fixed in each case (see Chapter 7, proof of Theorem 7.7, for an analogous argument).

To generate the missing elements g, x and $\tilde{e}$ of the public key, A proceeds as follows:

1. Generate $(l+1)$–bit primes $e_i := GenPrime(1^{l+1})$, $1 \leq i \leq t$. Choose a further prime $\tilde{e} := GenPrime(1^{l+1})$. Set

$$g := z^{2\tilde{e} \prod_{i \neq j} e_i}.$$

2. Randomly choose $w \xleftarrow{u} \mathbb{Z}_n^*$ and $u \xleftarrow{u} \mathbb{Z}_n^*$. Set

$$y_j := w^{2 \prod_{i \neq j} e_i} \text{ and } \tilde{x}_j := u^{2\tilde{e}}.$$

3. Let $x := y_j^{e_j} \cdot g^{-h(\tilde{x}_j)}$.

Then g and x are uniformly distributed quadratic residues, since z, w and u are uniformly distributed (and the exponents $\tilde{e}$ and e_i are prime to $\varphi(n)$, see the footnote on p. 281).

To generate the signature $(e_i, y_i, \tilde{y}_i)$ for the i-th message m_i, requested by *Forger*, A proceeds as follows:

1. If $i \neq j$, then A randomly chooses $\tilde{y}_i \xleftarrow{u} QR_n$ and computes

$$\tilde{x}_i := \tilde{y}_i^{\tilde{e}} \cdot g^{-h(m_i)} \text{ and } y_i := \left(x \cdot g^{h(\tilde{x}_i)}\right)^{e_i^{-1}}.$$

A can compute the e_i-th root, because the e_i-th roots of g and x are known to her by construction.

2. If $i = j$, then the value of y_j has already been computed above. Moreover, A can compute the correct value of $\tilde{y}_j = (\tilde{x}_j g^{h(m_j)})^{\tilde{e}^{-1}}$, because she knows the $\tilde{e}$-th root of $\tilde{x}_j$ and g. Note that $\tilde{y}_j$ is uniformly distributed, as required, since $\tilde{x}_j$ is uniformly distributed.

It is obvious from the construction that A generates signatures which satisfy the verification condition. *Forger* outputs a forged signature $\sigma = (e, y, \tilde{y})$ for a message $m \notin \{m_1, \ldots, m_t\}$. If the forged signature does not pass the verification procedure, A returns a random exponent and a random element in $\mathbb{Z}_n^*$, and stops. Otherwise, *Forger* yields a valid type-2 forgery, such that e_j divides e, i.e., $e = e_j \cdot f$ and $\tilde{y}^{\tilde{e}} \cdot g^{-h(m)} = \tilde{x} \neq \tilde{x}_j$. We have

$$y^e = \left(y^f\right)^{e_j} = x \cdot g^{h(\tilde{x})} \text{ and } y_j{}^{e_j} = x \cdot g^{h(\tilde{x}_j)}.$$

Now A has to compute an r-th root of z. If $h(\tilde{x}) \neq h(\tilde{x}_j)$, which happens almost with certainty, since $\mathcal{H}$ is collision resistant, we get, by dividing the two equations, the equation

$$\tilde{z}^{e_j} = g^a = z^{2a\tilde{e} \prod_{i \neq j} e_i},$$

with $0 < a < 2^l$. Since all $(l+1)$-bit primes are chosen by $GenPrime(1^{l+1})$, the probability that e_j is equal to $\tilde{e}$ or equal to an $e_i, i \neq j$, is negligibly small. If e_j is different from $\tilde{e}$ and the $e_i, i \neq j$, then we can compute $z^{e_j^{-1}}$ by Lemma 10.13. In this case, A returns the exponent e_j and the e_j-th root $z^{e_j^{-1}}$.

$A(n, z)$ succeeds in computing the root $z^{e_j^{-1}}$ if the *Forger* produces a valid signature, if $h(\tilde{x}) \neq h(\tilde{x}_j)$ and if e_j is different from $\tilde{e}$ and the $e_i, i \neq j$. As in case 1, it follows from the construction that A perfectly simulates the choice of the key together with the generation of signatures supplied to *Forger*. Thus, as we have seen in case 1, the *Forger*'s probability of success if he interacts with the simulated signer is the same as if he interacted with the legitimate signer. Computing the probabilities in a completely analogous way as in case 1, we derive that

$$\text{prob}(v^r = z : n \stackrel{u}{\leftarrow} I_{\text{SG},k}, z \stackrel{u}{\leftarrow} \mathbb{Z}_n^*, (v, r) \leftarrow A(n, z)) \geq \frac{1}{S(k)},$$

for some positive polynomial S and infinitely many k. This contradicts the strong RSA assumption (Lemma 10.11).

Case 3: *Forger* is of type 3. To complete the public key by g, x and $\tilde{e}$, A proceeds as follows:

1. Generate $(l+1)$-bit primes $\tilde{e}$ and $e_i, 1 \leq i \leq t$, by applying $GenPrime(1^{l+1})$. Set
$$g := z^{2\tilde{e} \prod_i e_i}.$$

2. Choose $a \stackrel{u}{\leftarrow} \{1, \ldots, n^2\}$ and set $x := g^a$.

A can easily generate valid signatures $(e_i, y_i, \tilde{y}_i)$ for messages $m_i, 1 \leq i \leq t$, requested by *Forger*. Namely, A chooses $\tilde{y}_i \stackrel{u}{\leftarrow} \text{QR}_n$ and computes $\tilde{x}_i = \tilde{y}_i^{e_i} \cdot g^{-h(m_i)}$ and $y_i = (x \cdot g^{h(\tilde{x}_i)})^{e_i^{-1}}$. The latter computation works, because due to the construction, the e_i-th roots of g and x can be immediately derived for $1 \leq i \leq t$.

Then *Forger* outputs a forged signature $\sigma = (e, y, \tilde{y})$ for a message $m \notin \{m_1, \ldots, m_t\}$. The signature is of type 3, i.e., e_i does not divide e, $1 \leq i \leq t$. If the signature is valid, we get the equation

$$y^e = x \cdot g^{h(\tilde{x})} = z^f, \quad \text{where } f = 2\tilde{e} \prod_i e_i \cdot (a + h(\tilde{x})),$$

with $\tilde{x} = \tilde{y}^{\tilde{e}} \cdot g^{-h(m)}$.

To compute a root of z, let $d = \gcd(e, f)$. Observe that e may be a non-prime, because the verification algorithm only tests whether e is odd. If $d < e$, i.e., e does not divide f, then $r := e/d > 1$, and we can compute the r-th root $z^{r^{-1}}$ by Lemma 10.13.

Thus, A succeeds in computing a root if *Forger* yields a valid forgery and if e does not divide f. To compute the probability that *Forger* yields a valid forgery, we have to consider the distribution of the keys (g, x), generated by A in step 2.

By the definition of n, QR_n is a cyclic group of order $\tilde{p}\tilde{q}$ with distinct Sophie Germain primes $\tilde{p}$ and $\tilde{q}$.

Note that $\tilde{e}$ and the e_i are $(l+1)$-bit primes, and $l+1 < k-1$. Thus, $\tilde{e}$ and none of the e_i are equal to $\tilde{p}$ or $\tilde{q}$, which are $(k-1)$-bit primes. Hence, $\tilde{e} \prod_i e_i$ is prime to $\varphi(n) = 4\tilde{p}\tilde{q}$. We conclude that g is uniformly distributed in QR_n, since z is uniformly chosen from $\mathbb{Z}_n^*$.

Let $a = b\tilde{p}\tilde{q} + c, 0 \leq c < \tilde{p}\tilde{q}$ (division with remainder). Now $2^{k-2} < \tilde{p}, \tilde{q} < 2^{k-1}$, $n^2 \approx 2^{4k}$, and $a \overset{u}{\leftarrow} \{1, \ldots, n^2\}$ is uniformly chosen. This implies that the probability of a remainder $c, 0 \leq c < \tilde{p}\tilde{q}$, differs from $1/\tilde{p}\tilde{q}$ by at most $1/n^2 \approx 1/2^{4k}$. This means that the distribution of c is polynomially close to the uniform distribution. This in turn implies that the conditional distribution of x, assuming that g is a generator of QR_n, is polynomially close to the uniform distribution on QR_n. However, $\mathrm{QR}_n \cong \mathbb{Z}_{\tilde{p}\tilde{q}} \cong \mathbb{Z}_{\tilde{p}} \times \mathbb{Z}_{\tilde{q}}$ and therefore $(\tilde{p}-1)(\tilde{q}-1)$ of the $\tilde{p}\tilde{q}$ elements in QR_n are generators. Thus, the probability that g is a generator of QR_n is $\geq 1 - 1/2^{k-3}$, which is exponentially close to 1. Summarizing, we get that the distribution of x is polynomially close to the uniform distribution on QR_n.

We see that A almost perfectly simulates the legitimate signer. The distributions of the keys and the signatures supplied to *Forger* are polynomially close to the distributions when *Forger* interacts with the legitimate signer. By Lemmas B.21 and B.24, we conclude that the probability that *Forger* produces a valid signature, if he interacts with the simulated signer in A, cannot be polynomially distinguished from his probability of success when interacting with the legitimate signer. Thus, *Forger* produces in step 3 of A a valid signature with probability $\geq 1/Q(k)$ for some positive polynomial Q and infinitely many k.

We still have to study the conditional probability that e does not divide f, assuming that *Forger* produces a valid signature. It is sufficient to prove that this probability is non-negligible. We will show that it is $> 1/2$. If we could prove this estimate assuming $h, n, g, x, \tilde{e}$ and the forged signed message (m, σ) fixed, for every $h \in \mathcal{H}_l, n \in I_{SG,k}$, every g, x and $\tilde{e}$ possibly generated by A, and every valid (m, σ) possibly output by *Forger*, then we are done (take the sum over all $h, n, g, x, \tilde{e}, m$ and σ). Therefore, we now assume that h, n, g, x and $\tilde{e}$, and m and σ are fixed. This implies that c and $\tilde{x}$ are also fixed.

Let s be a prime dividing e. Then $s > 2$ and $s \neq \tilde{e}$, because otherwise the verification condition was not satisfied. Moreover, $s \neq e_i$, since the forgery is of type 3. Thus, it suffices to prove that s does not divide $a + h(\tilde{x})$ with probability $\geq 1/2$, assuming $h, n, g, x, \tilde{e}, m$ and σ fixed. Let $a = b\tilde{p}\tilde{q} + c$, as above. $a + h(\tilde{x}) = b\tilde{p}\tilde{q} + c + h(\tilde{x}) = L(b)$, with L a linear function (note that c and $\tilde{x}$ are fixed). The probability that s divides $a + h(\tilde{x})$ is the same as the probability that $L(b) \equiv 0 \bmod s$. Now the conditional distribution of b, assuming c fixed, is also polynomially close to the uniform distribution on $\{0, \ldots, \lfloor n^2 / \tilde{p}\tilde{q} \rfloor\}$. Thus, the distribution of $b \bmod s$ is polynomially close to the uniform distribution. s does not divide $\tilde{p}\tilde{q}$, because $|s| \leq l + 1 < k - 1$ and $|\tilde{p}| = |\tilde{q}| = k - 1$. Thus, $L(b) \equiv 0 \bmod s$ is a non-vanishing linear equation over $\mathbb{Z}_s$. Hence, the probability that $L(b) = 0 \bmod s$ is very close to $1/s$. This means that s and hence e do not divide f, with a probability $\geq 1 - 1/(s-1) \geq 1/2$ (recall that $s > 2$).

Now the proof of case 3 is finished, and the proof of Theorem 10.12 is complete. $\square$

Exercises

1. Consider the construction of collision-resistant hash functions in Section 10.2. Explain how the prefix-free encoding of the messages can be avoided by applying Merkle's meta method (Section 3.4.2).

2. Let $I = (I_k)_{k \in \mathbb{N}}$ be a key set with security parameter k, and let $\mathcal{H} = \left(h_i : \{0,1\}^* \longrightarrow \{0,1\}^{g(k(i))} \right)_{i \in I}$ be a family of collision-resistant hash functions. Let $l_i \geq g(k(i)) + k(i)$ for all $i \in I$, and let $\{0,1\}^{\leq l_i} := \{m \in \{0,1\}^* \mid 1 \leq |m| \leq l_i\}$ be the bit strings of length $\leq l_i$. Show that the family
$$\left(h_i : \{0,1\}^{\leq l_i} \longrightarrow \{0,1\}^{g(k(i))} \right)_{i \in I}$$
is a family of one-way functions (with respect to $\mathcal{H}$'s key generator).

3. The RSA and ElGamal signature schemes are introduced in Chapter 3. There, various attacks against the basic schemes (no hash function is applied to the messages) are discussed. Classify these attacks and their levels of success (according to Section 10.1).

4. The following signature scheme was suggested by Ong, Schnorr and Shamir ([OngSchSha84]). Alice chooses two random large distinct primes p and q, and a random $x \in \mathbb{Z}_n^*$, where $n := pq$. Then she computes $y := -x^{-2} \in \mathbb{Z}_n^*$ and publishes (n, y) as her public key. Her secret is x (she does not need to know the prime factors of n). To sign a message $m \in \mathbb{Z}_n$, she randomly selects an $r \in \mathbb{Z}_n^*$ and calculates (in $\mathbb{Z}_n$)

$$s_1 := 2^{-1} \left(r^{-1} m + r \right) \text{ and } s_2 := 2^{-1} x \left(r^{-1} m - r \right).$$

(s_1, s_2) is the signature of m. To verify a signature (s_1, s_2), Bob checks that $m = s_1^2 + y s_2^2$ (in $\mathbb{Z}_n$):

   a. Prove that retrieving the secret key by a key-only attack is equivalent to the factoring of n.

   b. The scheme is existentially forgeable by a key-only attack.

   c. The probability that a randomly chosen pair (s_1, s_2) is a signature for a given message m is negligibly small.

   d. State the problem, which an adversary has to solve, of forging signatures for messages of his choice by a key-only attack.
   (In fact, Pollard has broken the scheme by an efficient algorithm for this problem, see [PolSch87].)

5. Let $I = (I_k)_{k \in \mathbb{N}}$ be a key set with security parameter k, and let $f_0 = (f_{0,i} : D_i \longrightarrow D_i)_{i \in I}, f_1 = (f_{1,i} : D_i \longrightarrow D_i)_{i \in I}$ be a claw-free pair of trapdoor permutations with key generator K. We consider the following signature scheme. As her public key, Alice randomly chooses an index $i \in I_k$ – by computing $K(1^k)$ – and a reference value $x \overset{u}{\leftarrow} D_i$. Her private key is the trapdoor information of $f_{0,i}$ and $f_{1,i}$. We encode the messages $m \in \{0,1\}^*$ in a prefix-free way (see Section 10.2), and denote the encoded m by $[m]$. Then Alice's signature of a message m is $\sigma(i, x, m) := f_{[m],i}^{-1}(x)$, where $f_{[m],i}$ is defined as in Section 10.2. Bob can verify Alice's signature σ by comparing $f_{[m],i}(\sigma)$ with x. Study the security of the scheme. More precisely, show

   a. A claw of f_0 and f_1 can be computed from $\sigma(i, x, m)$ and $\sigma(i, x, m')$ if the messages m and m' are distinct.

   b. The scheme is secure against existential forgery by a key-only attack.

   c. Assume that the message space has polynomial cardinality. More precisely, let $c \in \mathbb{N}$ and assume that only messages $m \in \{0,1\}^{c \lfloor \log_2(k) \rfloor}$ are signed by the scheme. Then the scheme is secure against adaptively-chosen-message attacks if used as a *one-time signature scheme* (i.e., Alice signs at most one message with her key).

   d. Which problem do you face in the security proof in c if the scheme is used to sign arbitrary messages in $\{0,1\}^*$?

6. We consider the same setting as in Exercise 5. We assume that the generation of the messages $m \in \{0,1\}^*$ to be signed can be uniformly modeled for all users by a probabilistic polynomial algorithm $M(i)$. In particular,

this means that the messages Alice wants to sign do not depend on her reference value x.

Let (i, x) be the public key of Alice, and let m_j be the j-th message to be signed by Alice. The signature $\sigma(i, x, m_j)$ of m_j is defined as $\sigma(i, x, m_j) := (s_j, [m_1] \| \dots \| [m_{j-1}])$, with $s_j := f_{[m_j], i}^{-1}(s_{j-1})$. Here $[m_1] \| \dots \| [m_{j-1}]$ denotes the concatenation of the prefix-free encoded messages, and $s_0 := x$ is Alice's randomly chosen reference value (see [GolBel01]):

a. Show by an example that in order to prevent forging by known signature attacks, the verification procedure has to check whether the bit string $\hat{m}$ in a signature $(s, \hat{m})$ is well formed with respect to the prefix-free encoding.

   Give the complete verification condition.

b. Prove that no one can existentially forge a signature by a known-signature attack.

A. Algebra and Number Theory

Public-key cryptosystems are based on modular arithmetic. In this section, we summarize the concepts and results from algebra and number theory which are necessary for an understanding of the cryptographic methods. Textbooks on number theory and modular arithmetic include [HarWri79], [IreRos82], [Rose94], [Forster96] and [Rosen2000]. This section is also intended to establish notation. We assume that the reader is familiar with the elementary notions of algebra, such as groups, rings and fields.

A.1 The Integers

$\mathbb{Z}$ denotes the ring of integers; $\mathbb{N} = \{z \in \mathbb{Z} \mid z > 0\}$ denotes the subset of natural numbers.

We first introduce the notion of divisors and the fundamental Euclidean algorithm which computes the greatest common divisor of two numbers.

Definition A.1. Let $a, b \in \mathbb{Z}$:

1. a *divides* b if there is some $c \in \mathbb{Z}$, with $b = ac$. We write $a \mid b$ for "a divides b".
2. $d \in \mathbb{N}$ is called the *greatest common divisor* of a and b, if:
   a. $d \mid a$ and $d \mid b$.
   b. If $t \in \mathbb{Z}$ divides both a and b, then t divides d.
   The greatest common divisor is denoted by $\gcd(a, b)$.
3. If $\gcd(a, b) = 1$, then a is called *relatively prime to* b, or *prime to* b for short.

Theorem A.2 (*Division with remainder*). *Let* $z, a \in \mathbb{Z}, a \neq 0$. *Then there are unique numbers* $q, r \in \mathbb{Z}$, *such that* $z = q \cdot a + r$ *and* $0 \leq r < |a|$.

Proof. In the first step, we prove that such q and r exist. If $a > 0$ and $z \geq 0$, we may apply induction on z. For $0 \leq z < a$ we obviously have $z = 0 \cdot a + z$. If $z \geq a$, then, by induction, $z - a = q \cdot a + r$ for some q and $r, 0 \leq r < a$, and hence $z = (q + 1) \cdot a + r$. If $z < 0$ and $a > 0$, then we have just shown the existence of an equation $-z = q \cdot a + r, 0 \leq r < a$. Then $z = -q \cdot a$ if $r = 0$, and $z = -q \cdot a - r = -q \cdot a - a + (a - r) = -(q + 1) \cdot a + (a - r)$ and

$0 < a - r < a$. If $a < 0$, then $-a > 0$. Hence $z = q \cdot (-a) + r = -q \cdot a + r$, with $0 \le r < |a|$.

To prove uniqueness, consider $z = q_1 \cdot a + r_1 = q_2 \cdot a + r_2$. Then $0 = (q_1 - q_2) \cdot a + (r_1 - r_2)$. Hence a divides $(r_1 - r_2)$. Since $|r_1 - r_2| < |a|$, this implies $r_1 = r_2$, and then also $q_1 = q_2$.   □

Remark. r is called the *remainder* of z modulo a. We write z mod a for r. The number q is the *(integer) quotient* of z and a. We write z div a for q.

The Euclidean Algorithm. Let $a, b \in \mathbb{Z}$, $a > b > 0$. The greatest common divisor $\gcd(a, b)$ can be computed by an iterated division with remainder. Let $r_0 := a, r_1 := b$ and

$$
\begin{aligned}
r_0 &= q_1 r_1 + r_2, & 0 &< r_2 < r_1, \\
r_1 &= q_2 r_2 + r_3, & 0 &< r_3 < r_2, \\
&\ \ \vdots \\
r_{k-1} &= q_k r_k + r_{k+1}, & 0 &< r_{k+1} < r_k, \\
&\ \ \vdots \\
r_{n-2} &= q_{n-1} r_{n-1} + r_n, & 0 &< r_n < r_{n-1}, \\
r_{n-1} &= q_n r_n + r_{n+1}, & 0 &= r_{n+1}.
\end{aligned}
$$

By construction, $r_1 > r_2 > \dots$. Therefore, the remainder becomes 0 after a finite number of steps. The last remainder $\neq 0$ is the greatest common divisor, as is shown in the next proposition.

Proposition A.3.

1. $r_n = \gcd(a, b)$.
2. *There are numbers $d, e \in \mathbb{Z}$ with $\gcd(a, b) = da + eb$.*

Proof. 1. From the equations considered in reverse order, we conclude that r_n divides r_k, $k = n-1, n-2 \dots$. In particular, r_n divides $r_1 = b$ and $r_0 = a$. Now let t be a divisor of $a = r_0$ and $b = r_1$. Then $t \mid r_k, k = 2, 3, \dots$, and hence $t \mid r_n$. Thus, r_n is the greatest common divisor.

2. Iteratively substituting r_{k+1} by $r_{k-1} - q_k r_k$, we get

$$
\begin{aligned}
r_n &= r_{n-2} - q_{n-1} \cdot r_{n-1} \\
&= r_{n-2} - q_{n-1} \cdot (r_{n-3} - q_{n-2} \cdot r_{n-2}) \\
&= (1 + q_{n-1} q_{n-2}) \cdot r_{n-2} - q_{n-1} \cdot r_{n-3} \\
&\ \ \vdots \\
&= da + eb,
\end{aligned}
$$

with integers d and e.   □

We have shown that the following algorithm, called *Euclid's algorithm*, outputs the greatest common divisor. abs(a) denotes the absolute value of a.

Algorithm A.4.

   int $gcd(\text{int } a, b)$
1   while $b \neq 0$ do
2      $r \leftarrow a \bmod b$
3      $a \leftarrow b$
4      $b \leftarrow r$
5   return $\text{abs}(a)$

We now extend the algorithm, such that not only $\gcd(a, b)$ but also the coefficients d and e of the linear combination $\gcd(a, b) = da + eb$ are computed. For this purpose, we write the recursion

$$r_{k-1} = q_k r_k + r_{k+1}$$

using matrices

$$\begin{pmatrix} r_k \\ r_{k+1} \end{pmatrix} = Q_k \begin{pmatrix} r_{k-1} \\ r_k \end{pmatrix}, \quad \text{where } Q_k = \begin{pmatrix} 0 & 1 \\ 1 & -q_k \end{pmatrix}, k = 1, \ldots, n.$$

Multiplying the matrices, we get

$$\begin{pmatrix} r_n \\ r_{n+1} \end{pmatrix} = Q_n \cdot Q_{n-1} \cdot \ldots \cdot Q_1 \begin{pmatrix} r_0 \\ r_1 \end{pmatrix}.$$

The first component of this equation yields the desired linear combination for $r_n = \gcd(a, b)$. Therefore, we have to compute $Q_n \cdot Q_{n-1} \cdot \ldots \cdot Q_1$. This is accomplished by iteratively computing the matrices

$$\Lambda_0 = \begin{pmatrix} 1 & 0 \\ 0 & 1 \end{pmatrix}, \quad \Lambda_k = \begin{pmatrix} 0 & 1 \\ 1 & -q_k \end{pmatrix} \Lambda_{k-1}, \quad k = 1, \ldots, n,$$

to finally get $\Lambda_n = Q_n \cdot Q_{n-1} \cdot \ldots \cdot Q_1$. In this way, we have derived the following algorithm, called the *extended Euclidean algorithm*. On inputs a and b it outputs the greatest common divisor and the coefficients d and e of the linear combination $\gcd(a, b) = da + eb$.

Algorithm A.5.

```
int array gcdCoef (int a, b)
 1    λ₁₁ ← 1, λ₂₂ ← 1, λ₁₂ ← 0, λ₂₁ ← 0
 2    while b ≠ 0 do
 3        q ← a div b
 4        r ← a mod b
 5        a ← b
 6        b ← r
 7        t₂₁ ← λ₂₁; t₂₂ ← λ₂₂
 8        λ₂₁ ← λ₁₁ − q · λ₂₁
 9        λ₂₂ ← λ₁₂ − q · λ₂₂
10        λ₁₁ ← t₂₁
11        λ₁₂ ← t₂₂
12    return (abs(a), λ₁₁, λ₁₂)
```

We analyze the running time of the Euclidean algorithm. Here we meet the Fibonacci numbers.

Definition A.6. The *Fibonacci numbers* f_n are recursively defined by

$$f_0 := 0, \quad f_1 := 1,$$
$$f_n := f_{n-1} + f_{n-2}, \text{ for } n \geq 2.$$

Remark. The Fibonacci numbers can be non-recursively computed using the formula

$$f_n = \frac{1}{\sqrt{5}}(g^n - \tilde{g}^n),$$

where g and $\tilde{g}$ are the solutions of the equation $x^2 = x + 1$:

$$g := \frac{1}{2}\left(1 + \sqrt{5}\right) \text{ and } \tilde{g} := 1 - g = -\frac{1}{g} = \frac{1}{2}\left(1 - \sqrt{5}\right).$$

See, for example, [Forster96].

Definition A.7. g is called the *Golden Ratio*.[1]

Lemma A.8. *For $n \geq 2$, $f_n \geq g^{n-2}$. In particular, the Fibonacci numbers grow exponentially fast.*

Proof. The statement is clear for $n = 2$. By induction on n, assuming that the statement holds for $\leq n$, we get

$$f_{n+1} = f_n + f_{n-1} \geq g^{n-2} + g^{n-3} = g^{n-3}(1 + g) = g^{n-3}g^2 = g^{n-1}.$$

□

Proposition A.9. *Let $a, b \in \mathbb{Z}$, $a > b > 0$. Assume that computing $\gcd(a, b)$ by the Euclidean algorithm takes n iterations (i.e., using n divisions with remainder). Then $a \geq f_{n+1}$ and $b \geq f_n$.*

---

[1] It is the proportion of length to width which the Greeks found most beautiful.

Proof. Let $r_0 := a, r_1 := b$ and consider

$$
\begin{aligned}
r_0 &= q_1 r_1 + r_2, \\
r_1 &= q_2 r_2 + r_3, \\
&\ \vdots \\
r_{n-2} &= q_{n-1} r_{n-1} + r_n, \\
r_{n-1} &= q_n r_n,
\end{aligned}
\qquad \text{and} \qquad
\begin{aligned}
f_{n+1} &= f_n + f_{n-1}, \\
f_n &= f_{n-1} + f_{n-2}, \\
&\ \vdots \\
f_3 &= f_2 + f_1, \\
f_2 &= f_1.
\end{aligned}
$$

By induction, starting with $i = n$ and descending, we show that $r_i \geq f_{n+1-i}$. For $i = n$, we have $r_n \geq f_1 = 1$. Now assume the inequality proven for $\geq i$. Then

$$
r_{i-1} = q_i r_i + r_{i+1} \geq r_i + r_{i+1} \geq f_{n+1-i} + f_{n+1-(i+1)} = f_{n+1-(i-1)}.
$$

Hence $a = r_0 \geq f_{n+1}$ and $b = r_1 \geq f_n$. □

Notation. As is common use, we denote by $\lfloor x \rfloor$ the greatest integer less than or equal to x (the *"floor"* of x), and by $\lceil x \rceil$ the smallest integer greater than or equal to x (the *"ceiling"* of x).

Corollary A.10. *Let $a, b \in \mathbb{Z}$. Then the Euclidean algorithm computes $\gcd(a, b)$ in at most $\lfloor \log_g(a) \rfloor + 1$ iterations.*

Proof. Let n be the number of iterations. From $a \geq f_{n+1} \geq g^{n-1}$ (Lemma A.8) we conclude $n - 1 \leq \lfloor \log_g(a) \rfloor$. □

The Binary Encoding of Numbers. Studying algorithms with numbers as inputs and outputs, we need *binary encodings* of numbers (and residues, see below). We always assume that integers $n \geq 0$ are encoded in the standard way as unsigned integers:
The sequence $z_{k-1} z_{k-2} \ldots z_1 z_0$ of bits $z_i \in \{0, 1\}, 0 \leq i \leq k - 1$, is the encoding of

$$
n = z_0 + z_1 \cdot 2^1 + \ldots + z_{k-2} \cdot 2^{k-2} + z_{k-1} \cdot 2^{k-1} = \sum_{i=0}^{k-1} z_i \cdot 2^i.
$$

If the leading digit z_{k-1} is not zero (i.e., $z_{k-1} = 1$), we call n a *k-bit integer*, and k is called the *binary length* of n. The binary length of $n \in \mathbb{N}$ is usually denoted by $|n|$. Of course, we only use this notation if it cannot be confused with the absolute value. The binary length of $n \in \mathbb{N}$ is $\lfloor \log_2(n) \rfloor + 1$. The numbers of binary length k are the numbers $n \in \mathbb{N}$ with $2^{k-1} \leq n \leq 2^k - 1$.

The Big-*O* Notation. To state estimates, the *big-O* notation is useful. Suppose $f(k)$ and $g(k)$ are functions of the positive integers k which take positive (not necessarily integer) values. We say that $f(k) = O(g(k))$ if there is a constant C such that $f(k) \leq C \cdot g(k)$ for all sufficiently large k. For example, $2k^2 + k + 1 = O(k^2)$ because $2k^2 + k + 1 \leq 4k^2$ for all $k \geq 1$. In our examples, the constant C is always "small", and we use the big-O notation for convenience. We do not want to state a precise value of C.

Remark. Applying the classical grade school methods, we see that adding and subtracting two k-bit numbers requires $O(k)$ binary operations. Multiplication and division with remainder can be done with $O(k^2)$ binary operations (see [Knuth98] for a more detailed discussion of time estimates for doing arithmetic). Thus, the greatest common divisor of two k-bit numbers can be computed by the Euclidean algorithm with $O(k^3)$ binary operations.

Next we will show that every natural number can be uniquely decomposed into prime numbers.

Definition A.11. Let $p \in \mathbb{N}, p \geq 2$. p is called a *prime* (or a *prime number*) if 1 and p are the only positive divisors of p. A number $n \in \mathbb{N}$ which is not a prime is called a *composite*.

Remark. If p is a prime and $p \mid ab$, $a, b \in \mathbb{Z}$, then either $p \mid a$ or $p \mid b$.

Proof. Assume that p does not divide a and does not divide b. Then there are $d_1, d_2, e_1, e_2 \in \mathbb{Z}$, with $1 = d_1 p + e_1 a, 1 = d_2 p + e_2 b$ (Proposition A.3). Then $1 = d_1 d_2 p^2 + d_1 e_2 bp + e_1 a d_2 p + e_1 e_2 ab$. If p divided ab, then p would divide 1, which is impossible. Thus, p does not divide ab. $\square$

Theorem A.12 *(Fundamental Theorem of Arithmetic). Let $n \in \mathbb{N}, n \geq 2$. There are pairwise distinct primes $p_1, \ldots, p_r$ and exponents $e_1, \ldots, e_r \in \mathbb{N}, e_i \geq 1, i = 1, \ldots, r$, such that*

$$n = \prod_{i=1}^{r} p_i^{e_i}.$$

The primes $p_1, \ldots, p_r$ and exponents $e_1, \ldots, e_r$ are unique.

Proof. By induction on n we obtain the existence of such a decomposition. $n = 2$ is a prime. Now assume that the existence is proven for numbers $\leq n$. Either $n + 1$ is a prime or $n + 1 = l \cdot m$, with $l, m < n + 1$. By assumption, there are decompositions of l and m and hence also for $n + 1$.
In order to prove uniqueness, we assume that there are two different decompositions of n. Dividing both decompositions by all common primes, we get (not necessarily distinct) primes $p_1, \ldots, p_s$ and $q_1, \ldots, q_t$, with $\{p_1, \ldots, p_s\} \cap \{q_1, \ldots, q_t\} = \emptyset$ and $p_1 \cdot \ldots \cdot p_s = q_1 \cdot \ldots \cdot q_t$. Since $p_1 \mid q_1 \cdot \ldots \cdot q_t$, we conclude from the preceding remark that there is an $i, 1 \leq i \leq t$, with $p_1 \mid q_i$. This is a contradiction. $\square$

A.2 Residues

In public-key cryptography, we usually have to compute with remainders modulo n. This means that the computations take place in the residue class ring $\mathbb{Z}_n$.

Definition A.13. Let $n \in \mathbb{N}, n \geq 2$:

1. $a, b \in \mathbb{Z}$ are *congruent modulo* n, written as

$$a \equiv b \bmod n,$$

   if n divides $a - b$. This means that a and b have the same remainder when divided by n: $a \bmod n = b \bmod n$.
2. Let $a \in \mathbb{Z}$. $[a] := \{x \in \mathbb{Z} \mid x \equiv a \bmod n\}$ is called the *residue class* of a modulo m.
3. $\mathbb{Z}_n := \{[a] \mid a \in \mathbb{Z}\}$ is the set of residue classes modulo n.

Remark. As is easily seen, "congruent modulo n" is a symmetric, reflexive and transitive relation, i.e., it is an equivalence relation. The residue classes are the equivalence classes. A residue class $[a]$ is completely determined by one of its members. If $a' \in [a]$, then $[a] = [a']$. An element $x \in [a]$ is called a *representative* of $[a]$. Division with remainder by n yields the remainders $0, \ldots, n - 1$. Therefore, there are n residue classes in $\mathbb{Z}_n$:

$$\mathbb{Z}_n = \{[0], \ldots, [n - 1]\}.$$

The integers $0, \ldots, n - 1$ are called the *natural representatives*. The natural representative of $[x] \in \mathbb{Z}_n$ is just the remainder $(x \bmod n)$ of x modulo n (see division with remainder, Theorem A.2). If, in the given context, no confusion is possible, we sometimes identify the residue classes with their natural representatives.

Since we will study algorithms whose inputs and outputs are residue classes, we need *binary encodings* of the residue classes. The binary encoding of $[x] \in \mathbb{Z}_n$ is the binary encoding of the natural representative $x \bmod n$ as an unsigned integer (see our remark on the binary encoding of non-negative integers in Section A.1).

Definition A.14. By defining addition and multiplication as

$$[a] + [b] = [a + b] \text{ and } [a] \cdot [b] = [a \cdot b],$$

$\mathbb{Z}_n$ becomes a commutative ring, with unit element $[1]$. It is called the *residue class ring* modulo n.

Remark. The sum $[a] + [b]$ and the product $[a] \cdot [b]$ do not depend on the choice of the representatives by which they are computed, as straightforward computations show. For example, let $a' \in [a]$ and $b' \in [b]$. Then $n \mid a' - a$ and $n \mid b' - b$. Hence $n \mid a' + b' - (a + b)$, and therefore $[a + b] = [a' + b']$.

Doing multiplications in a ring, we are interested in those elements which have a multiplicative inverse. They are called the units.

Definition A.15. Let R be a commutative ring with unit element e. An element $x \in R$ is called a *unit* if there is an element $y \in R$ with $x \cdot y = e$. We call y a *multiplicative inverse* of x. The subset of units is denoted by R^*.

Remark. The multiplicative inverse of a unit x is uniquely determined, and we denote it by x^{-1}. The set of units R^* is a subgroup of R with respect to multiplication.

Example. In $\mathbb{Z}$, elements a and b satisfy $a \cdot b = 1$ if and only if both a and b are equal to 1, or both are equal to -1. Thus, 1 and -1 are the only units in $\mathbb{Z}$. The residue class rings $\mathbb{Z}_n$ contain many more units, as the subsequent considerations show. For example, if p is a prime then every residue class in $\mathbb{Z}_p$ different from $[0]$ is a unit. An element $[x] \in \mathbb{Z}_n$ in a residue class ring is a unit if there is a residue class $[y] \in \mathbb{Z}_n$ with $[x] \cdot [y] = [1]$, i.e., n divides $x \cdot y - 1$.

Proposition A.16. *An element* $[x] \in \mathbb{Z}_n$ *is a unit if and only if* $\gcd(x, n) = 1$. *The multiplicative inverse* $[x]^{-1}$ *of a unit* $[x]$ *can be computed using the extended Euclidean algorithm.*

Proof. If $\gcd(x, n) = 1$, then there is an equation $xb + nc = 1$ in $\mathbb{Z}$, and the coefficients $b, c \in \mathbb{Z}$ can be computed using the extended Euclidean algorithm A.5. The residue class $[b]$ is an inverse of $[x]$. Conversely, if $[x]$ is a unit, then there are $y, k \in \mathbb{Z}$ with $x \cdot y = 1 + k \cdot n$. This implies $\gcd(x, n) = 1$. □

Corollary A.17. *Let* p *be a prime. Then every* $[x] \neq [0]$ *in* $\mathbb{Z}_p$ *is a unit. Thus,* $\mathbb{Z}_p$ *is a field.*

Definition A.18. The subgroup

$$\mathbb{Z}_n^* := \{x \in \mathbb{Z}_n \mid x \text{ is a unit in } \mathbb{Z}_n\}$$

of units in $\mathbb{Z}_n$ is called the *prime residue class group modulo n*.

Definition A.19. Let M be a finite set. The number of elements in M is called the *cardinality* or *order* of M. It is denoted by $|M|$.

We introduce the Euler phi function, which gives the number of units modulo n.

Definition A.20.
$$\varphi : \mathbb{N} \longrightarrow \mathbb{N}, \; n \longmapsto |\mathbb{Z}_n^*|$$
is called the *Euler phi function* or the *Euler totient function*.

Proposition A.21 *(Euler).*

$$\sum_{d \mid n} \varphi(d) = n.$$

Proof. If d is a divisor of n, let $Z_d := \{x \mid 1 \leq x \leq n, \gcd(x, n) = d\}$. Each $k \in \{1, \dots, n\}$ belongs to exactly one Z_d. Thus $n = \sum_{d \mid n} |Z_d|$. Since $x \mapsto x/d$ is a bijective map from Z_d to $\mathbb{Z}^*_{n/d}$, we have $|Z_d| = \varphi(n/d)$, and hence $n = \sum_{d \mid n} \varphi(n/d) = \sum_{d \mid n} \varphi(d)$. □

Corollary A.22. *Let p be a prime and $k \in \mathbb{N}$. Then $\varphi(p^k) = p^{k-1}(p-1)$.*

Proof. By Euler's result, $\varphi(1) + \varphi(p) + \dots + \varphi(p^k) = p^k$ and $\varphi(1) + \varphi(p) + \dots + \varphi(p^{k-1}) = p^{k-1}$. Subtracting both equations yields $\varphi(p^k) = p^k - p^{k-1} = p^{k-1}(p-1)$. □

Remarks:

1. By using the Chinese Remainder Theorem below (Section A.3), we will also get a formula for $\varphi(n)$ if n is not a power of a prime (Corollary A.30).
2. At some points in the book we need a lower bound for the fraction $\varphi(n)/n$ of units in $\mathbb{Z}_n$. In [RosSch62] it is proven that

$$\varphi(n) > \frac{n}{e^\gamma \log(\log(n)) + \frac{2.6}{\log(\log(n))}}, \text{ with Euler's constant } \gamma = 0.5772\dots.$$

This inequality implies, for example, that

$$\varphi(n) > \frac{n}{6 \log(\log(n))} \text{ for } n \geq 1.3 \cdot 10^6,$$

as a straightforward computation shows.

The RSA cryptosystem is based on old results by Fermat and Euler.[2] These results are special cases of the following proposition.

Proposition A.23. *Let G be a finite group and e be the unit element of G. Then $x^{|G|} = e$ for all $x \in G$.*

Proof. Since we apply this result only to Abelian groups, we assume in our proof that the group G is Abelian. A proof for the general case may be found in most introductory textbooks on algebra.

The map $\mu_x : G \longrightarrow G$, $g \longmapsto xg$, multiplying group elements by x, is a bijective map (multiplying by x^{-1} is the inverse map). Hence,

$$\prod_{g \in G} g = \prod_{g \in G} xg = x^{|G|} \prod_{g \in G} g,$$

and this implies $x^{|G|} = e$. □

As a first corollary of Proposition A.23, we get Fermat's Little Theorem.

Proposition A.24 *(Fermat). Let p be a prime and $a \in \mathbb{Z}$ be a number that is prime to p (i.e., p does not divide a). Then*

$$a^{p-1} \equiv 1 \bmod p.$$

---

[2] Pierre de Fermat (1601–1665) and Leonhard Euler (1707–1783).

Proof. The residue class $[a]$ of a modulo p is a unit, because a is prime to p (Proposition A.16). Since $|\mathbb{Z}_p^*| = p - 1$ (Corollary A.17), we have $[a]^{p-1} = 1$ by Proposition A.23. $\qquad\square$

Remark. Fermat stated a famous conjecture known as Fermat's Last Theorem. It says that the equation $x^n + y^n = z^n$ has no solutions with non-zero integers x, y and z, for $n \geq 3$. For more than 300 years, Fermat's conjecture was one of the outstanding challenges of mathematics. It was finally proven in 1995 by Andrew Wiles.

Euler generalized Fermat's Little Theorem.

Proposition A.25 *(Euler). Let $n \in \mathbb{N}$ and let $a \in \mathbb{Z}$ be a number that is prime to n. Then*
$$a^{\varphi(n)} \equiv 1 \bmod n.$$

Proof. It follows from Proposition A.23, in the same way as Proposition A.24. The residue class $[a]$ of a modulo n is a unit and $|\mathbb{Z}_n^*| = \varphi(n)$. $\qquad\square$

Fast Modular Exponentiation. In cryptography, we often have to compute a power x^e or a modular power $x^e \bmod n$. This can be done efficiently by the fast exponentiation algorithm. The idea is that if the exponent e is a power of 2, say $e = 2^k$, then we can exponentiate by successively squaring:

$$x^e = x^{2^k} = ((((\ldots (x^2)^2)^2 \ldots)^2)^2)^2.$$

In this way we compute x^e by k squarings. For example, $x^{16} = (((x^2)^2)^2)^2$.

If the exponent is not a power of 2, then we use its binary representation. Assume that e is a k-bit number, $2^{k-1} \leq e < 2^k$. Then

$$
\begin{aligned}
e &= 2^{k-1}e_{k-1} + 2^{k-2}e_{k-2} + \ldots + 2^1 e_1 + 2^0 e_0, \quad \text{(with } e_{k-1} = 1) \\
&= (2^{k-2}e_{k-1} + 2^{k-3}e_{k-2} + \ldots + e_1) \cdot 2 + e_0 \\
&= (\ldots((2e_{k-1} + e_{k-2}) \cdot 2 + e_{k-3}) \cdot 2 + \ldots + e_1) \cdot 2 + e_0.
\end{aligned}
$$

Hence,

$$
\begin{aligned}
x^e &= x^{(\ldots((2e_{k-1}+e_{k-2})\cdot 2 + e_{k-3})\cdot 2 + \ldots + e_1)\cdot 2 + e_0} = \\
&= \left(x^{(\ldots((2e_{k-1}+e_{k-2}))\cdot 2 + e_{k-3})\cdot 2 + \ldots + e_1}\right)^2 \cdot x^{e_0} = \\
&= (\ldots(((x^2 \cdot x^{e_{k-2}})^2 \cdot x^{e_{k-3}})^2 \cdot \ldots)^2 \cdot x^{e_1})^2 \cdot x^{e_0}.
\end{aligned}
$$

We see that x^e can be computed in $k - 1$ steps, with each step consisting of squaring the intermediate result and, if the corresponding binary digit e_i of e is 1, an additional multiplication by x. If we want to compute the modular power $x^e \bmod n$, then we take the remainder modulo n after each squaring and multiplication:

$x^e \bmod n =$

$(\ldots(((x^2 \cdot x^{e_{k-2}} \bmod n)^2 \cdot x^{e_{k-3}} \bmod n)^2 \cdot \ldots)^2 \cdot x^{e_1} \bmod n)^2 \cdot x^{e_0} \bmod n.$

We obtain the following algorithm for fast modular exponentiation.

Algorithm A.26.
  int $ModPower(\text{int } x, e, n)$
   1 $y \leftarrow x;$
   2 for $i \leftarrow BitLength(e) - 2$ downto 0 do
   3     $y \leftarrow y^2 \cdot x^{Bit(e,i)} \bmod n$
   4 return y

In particular, we get

Proposition A.27. *Let $l = \lfloor \log_2 e \rfloor$. The computation of $x^e \bmod n$ can be done by l squarings, l multiplications and l divisions.*

Proof. The binary length k of e is $\lfloor \log_2(e) \rfloor + 1$. $\square$

A.3 The Chinese Remainder Theorem

The Chinese Remainder Theorem provides a method of solving systems of congruences. The solutions can be found using an easy and efficient algorithm.

Theorem A.28. *Let $n_1, \ldots, n_r \in \mathbb{N}$ be pairwise relatively prime numbers, i.e., $\gcd(n_i, n_j) = 1$ for $i \neq j$. Let $b_1, b_2, \ldots, b_r$ be arbitrary integers. Then there is an integer b such that*

$$b \equiv b_i \bmod n_i, \quad i = 1, \ldots, r.$$

Furthermore, the remainder $b \bmod n$ is unique, where $n = n_1 \cdot \ldots \cdot n_r$.

The statement means that there is a one-to-one correspondence between the residue classes modulo n and tuples of residue classes modulo $n_1, \ldots, n_r$. This one-to-one correspondence preserves the additive and multiplicative structure. Therefore, we have the following ring-theoretic formulation of Theorem A.28.

Theorem A.29 *(Chinese Remainder Theorem). Let $n_1, \ldots, n_r \in \mathbb{N}$ be pairwise relatively prime numbers, i.e., $\gcd(n_i, n_j) = 1$, for $i \neq j$. Let $n = n_1 \cdot \ldots \cdot n_r$. Then the map*

$$\psi : \mathbb{Z}_n \longrightarrow \mathbb{Z}_{n_1} \times \ldots \times \mathbb{Z}_{n_r}, \ [x] \longmapsto ([x \bmod n_1], \ldots, [x \bmod n_r])$$

is an isomorphism of rings.

Remark. Before we give a proof, we review the notion of an *"isomorphism"*. It means that ψ is a homomorphism and bijective. "Homomorphism" means that ψ preserves the additive and multiplicative structure. More precisely, a map $f : R \longrightarrow R'$ between rings with unit elements e and e' is called a *(ring) homomorphism* if

$$f(e) = e' \text{ and } f(a + b) = f(a) + f(b), f(a \cdot b) = f(a) \cdot f(b) \text{ for all } a, b \in R.$$

If f is a bijective homomorphism, then, automatically, the inverse map $g = f^{-1}$ is also a homomorphism. Namely, let $a', b' \in R'$. Then $a' = f(a)$ and $b' = f(b)$, and $g(a' \cdot b') = g(f(a) \cdot f(b)) = g(f(a \cdot b)) = a \cdot b = g(a') \cdot g(b')$ (analogously for $+$ instead of $\cdot$).

Being an isomorphism, as ψ is, is an extremely nice feature. It means, in particular, that a is a unit in R if and only if $f(a)$ is a unit R' (to see this, compute $e' = f(e) = f(a \cdot a^{-1}) = f(a) \cdot f(a^{-1})$, hence $f(a^{-1})$ is an inverse of $f(a)$). And the "same" equations hold in domain and range. For example, we have $a^2 = b$ in R if and only if $f(a)^2 = f(b)$ (note that $f(a)^2 = f(a^2)$). Thus, b is a square if and only if $f(b)$ is a square (we will use this example in Section A.7).

Isomorphism means that the domain and range may be considered to be the same for all questions concerning addition and multiplication.

Proof (of Theorem A.29). Since each n_i divides n, the map is well defined, and it obviously is a ring homomorphism. The domain and range of the map have the same cardinality (i.e., they contain the same number of elements). Thus, it suffices to prove that ψ is surjective.

Let $t_i := n/n_i = \prod_{k \neq i} n_k$. Then $t_i \equiv 0 \bmod n_k$ for all $k \neq i$, and $\gcd(t_i, n_i) = 1$. Hence, there is a $d_i \in \mathbb{Z}$ with $d_i \cdot t_i \equiv 1 \bmod n_i$ (Proposition A.16). Setting $u_i := d_i \cdot t_i$, we have

$$u_i \equiv 0 \bmod n_k, \text{ for all } k \neq i, \text{ and } u_i \equiv 1 \bmod n_i.$$

This means that the element $(0, \ldots, 0, 1, 0, \ldots, 0)$ (the i-th component is 1, all other components are 0) is in the image of ψ. If $([x_1], \ldots, [x_r]) \in \mathbb{Z}_{n_1} \times \ldots \times \mathbb{Z}_{n_r}$ is an arbitrary element, then $\psi(\sum_{i=1}^{r} x_i \cdot u_i) = ([x_1], \ldots, [x_r])$. □

Remarks:

1. Actually, the proof describes an efficient algorithm for computing a number b, with $b \equiv b_i \bmod n_i$, $i = 1, \ldots, r$ (recall our first formulation of the Chinese Remainder Theorem in Theorem A.28). In a preprocessing step, the inverse elements $[d_i] = [t_i]^{-1}$ are computed modulo n_i using the extended Euclidean algorithm (Proposition A.16). Then b can be computed as $b = \sum_{i=1}^{r} b_i \cdot d_i \cdot t_i$, for any given integers b_i, $1 \leq i \leq r$.
   We mainly apply the Chinese Remainder Theorem with $r = 2$ (for example, in the RSA cryptosystem). Here we simply compute coefficients d and e with $1 = d \cdot n_1 + e \cdot n_2$ (using the extended Euclidean algorithm A.5), and then $b = d \cdot n_1 \cdot b_2 + e \cdot n_2 \cdot b_1$.
2. The Chinese Remainder Theorem can be used to make arithmetic computations modulo n easier and (much) more efficient. We map the operands to $\mathbb{Z}_{n_1} \times \ldots \times \mathbb{Z}_{n_r}$ by ψ and do our computation there. $\mathbb{Z}_{n_1} \times \ldots \times \mathbb{Z}_{n_r}$ is a direct product of rings. Addition and multiplication are done componentwise, i.e., we perform the computation modulo n_i, for $i = 1, \ldots, r$.

Here we work with (much) smaller numbers.[3] Finally, we map the result back to $\mathbb{Z}_n$ by ψ^{-1} (which is easily done, as we have seen in the preceding remark).

As a corollary of the Chinese Remainder Theorem, we get a formula for Euler's phi function for composite inputs.

Corollary A.30. *Let $n \in \mathbb{N}$ and $n = p_1^{e_1} \cdot \ldots \cdot p_r^{e_r}$ be the decomposition of n into primes (as stated in Theorem A.12). Then:*

1. $\mathbb{Z}_n$ *is isomorphic to* $\mathbb{Z}_{p_1^{e_1}} \times \ldots \times \mathbb{Z}_{p_r^{e_r}}$.
2. $\mathbb{Z}_n^*$ *is isomorphic to* $\mathbb{Z}_{p_1^{e_1}}^* \times \ldots \times \mathbb{Z}_{p_r^{e_r}}^*$.

In particular, we have for Euler's phi function that

$$\varphi(n) = n \prod_{i=1}^{r} \left(1 - \frac{1}{p_i}\right).$$

Proof. The ring isomorphism of Theorem A.29 induces, in particular, an isomorphism on the units. Hence,

$$\varphi(n) = \varphi(p_1^{e_1}) \cdot \ldots \cdot \varphi(p_r^{e_r}),$$

and the formula follows from Corollary A.22.   □

A.4 Primitive Roots and the Discrete Logarithm

Definition A.31. Let G be a finite group and let e be the unit element of G. Let $x \in G$. The smallest $n \in \mathbb{N}$ with $x^n = e$ is called the *order of x*. We write this as $\mathrm{ord}(x)$.

Remark. There are exponents $n \in \mathbb{N}$, with $x^n = e$. Namely, since G is finite, there are exponents m and $m', m < m'$, with $x^m = x^{m'}$. Then $m' - m > 0$ and $x^{m'-m} = e$.

Lemma A.32. *Let G be a finite group and $x \in G$. Let $n \in \mathbb{N}$ with $x^n = e$. Then $\mathrm{ord}(x)$ divides n.*

Proof. Let $n = q \cdot \mathrm{ord}(x) + r, 0 \le r < \mathrm{ord}(x)$ (division with remainder). Then $x^r = e$. Since $0 \le r < \mathrm{ord}(x)$, this implies $r = 0$.   □

Corollary A.33. *Let G be a finite group and $x \in G$. Then $\mathrm{ord}(x)$ divides the order $|G|$ of G.*

Proof. By Proposition A.23, $x^{|G|} = e$.   □

Lemma A.34. *Let G be a finite group and $x \in G$. Let $l \in \mathbb{Z}$ and $d = \gcd(l, \mathrm{ord}(x))$. Then $\mathrm{ord}(x^l) = \mathrm{ord}(x)/d$.*

---

[3] For example, if $n = pq$ (as in an RSA scheme) with 512-bit numbers p and q, then we compute with 512-bit numbers instead of with 1024-bit numbers.

Proof. Let $r = \mathrm{ord}(x^l)$. From $(x^l)^{\mathrm{ord}(x)/d} = (x^{\mathrm{ord}(x)})^{l/d} = e$ we conclude $r \leq \mathrm{ord}(x)/d$. Choose numbers a and b with $d = a \cdot l + b \cdot \mathrm{ord}(x)$ (Proposition A.3). From $x^{r \cdot d} = x^{r \cdot a \cdot l + r \cdot b \cdot \mathrm{ord}(x)} = x^{l \cdot r \cdot a} = e$, we derive $\mathrm{ord}(x) \leq r \cdot d$.   □

Definition A.35. Let G be a finite group. G is called *cyclic* if there is an $x \in G$ which generates G, i.e., $G = \{x, x^2, x^3, \ldots, x^{\mathrm{ord}(x)-1}, x^{\mathrm{ord}(x)} = e\}$. Such an element x is called a *generator* of G.

Theorem A.36. *Let p be a prime. Then $\mathbb{Z}_p^*$ is cyclic, and the number of generators is $\varphi(p-1)$.*

Proof. For $1 \leq d \leq p-1$, let $S_d = \{x \in \mathbb{Z}_p^* \mid \mathrm{ord}(x) = d\}$ be the units of order d. If $S_d \neq \emptyset$, let $a \in S_d$. The equation $X^d - 1$ has at most d solutions in $\mathbb{Z}_p$, since $\mathbb{Z}_p$ is a field (Corollary A.17). Hence, the solutions of $X^d - 1$ are just the elements of $A := \{a, a^2, \ldots, a^d\}$. Each $x \in S_d$ is a solution of $X^d - 1$, and therefore $S_d \subset A$. Using Lemma A.34 we derive that $S_d = \{a^c \mid 1 \leq c < d, \gcd(c,d) = 1\}$. In particular, we conclude that $|S_d| = \varphi(d)$ if $S_d \neq \emptyset$ (and an $a \in S_d$ exists).

By Fermat's Little Theorem (Proposition A.24), $\mathbb{Z}_p^*$ is the disjoint union of the sets S_d, $d \mid p-1$. Hence $|\mathbb{Z}_p^*| = p - 1 = \sum_{d \mid p-1} |S_d|$. On the other hand, $p - 1 = \sum_{d \mid p-1} \varphi(d)$ (Proposition A.21), and we see that $|S_d| = \varphi(d)$ must hold for all divisors d of $p-1$. In particular, $|S_{p-1}| = \varphi(p-1)$. This means that there are $\varphi(p-1)$ generators of $\mathbb{Z}_p^*$.   □

Definition A.37. Let p be a prime. A generator g of the cyclic group $\mathbb{Z}_p^*$ is called a *primitive root* of $\mathbb{Z}_p^*$ or a *primitive root modulo* p.

Remark. It can be proven that $\mathbb{Z}_n^*$ is cyclic if and only if n is one of the following numbers: $1, 2, 4, p^k$ or $2p^k$; p a prime, $p \geq 3$, $k \geq 1$.

Proposition A.38. *Let p be a prime. Then $x \in \mathbb{Z}_p^*$ is a primitive root if and only if $x^{(p-1)/q} \neq [1]$ for every prime q which divides $p-1$.*

Proof. An element x is a primitive root if and only if x has order $p-1$. Since $\mathrm{ord}(x)$ divides $p-1$ (Corollary A.33), either $x^{(p-1)/q} = [1]$ for some prime divisor q of $p-1$ or $\mathrm{ord}(x) = p-1$.   □

We may use Proposition A.38 to generate a primitive root for those primes p for which we know (or can efficiently compute) the prime factors of $p-1$.

Algorithm A.39.

```
int PrimitiveRoot(prime p)
1    Randomly choose an integer g, with 0 < g < p − 1
2    if g^(p−1) div q ≢ 1 mod p, for all primes q dividing p − 1
3        then return g
4        else  go to 1
```

Since $\varphi(p-1) > (p-1)/6\log(\log(p-1))$ (see Section A.2), we expect to find a primitive element after $O(\log(\log(p)))$ iterations (see Lemma B.12).

No efficient algorithm is known for the computation of primitive roots for arbitrary primes. The problem is to compute the prime factors of $p-1$, which we need in Algorithm A.39. Often there are primitive roots which are small.

Algorithm A.39 is used, for example, in the key-generation procedure of the ElGamal cryptosystem (see Section 3.5.1). There the primes p are chosen in such a way that the prime factors of $p-1$ can be derived efficiently.

Lemma A.40. *Let p be a prime and let q be a prime that divides $p-1$. Then the set*

$$G_q = \{x \in \mathbb{Z}_p^* \mid \operatorname{ord}(x) = q \text{ or } x = [1]\},$$

which consists of the unit element $[1]$ and the elements of order q, is a subgroup of $\mathbb{Z}_p^*$. G_q is a cyclic group, and every element $x \in \mathbb{Z}_p^*$ of order q, i.e., every element $x \in G_q$, $x \neq [1]$, is a generator. G_q is generated, for example, by $g^{(p-1)/q}$, where g is a primitive root modulo p. G_q is the only subgroup of G of order q.

Proof. Let $x, y \in G_q$. Then $(xy)^q = x^q y^q = [1]$, and therefore $\operatorname{ord}(xy)$ divides q. Since q is a prime, we conclude that $\operatorname{ord}(xy)$ is 1 or q. Thus $xy \in G_q$, and G_q is a subgroup of $\mathbb{Z}_p^*$. Let $h \in \mathbb{Z}_p^*$ be an element of order q, for example, $h := g^{p-1/q}$, where g is a primitive root modulo p. Then $\{h^0, h^1, h^2, \ldots h^{q-1}\} \subseteq G_q$. The elements of G_q are solutions of the equation $X^q - 1$ in $\mathbb{Z}_p$. This equation has at most q solutions in $\mathbb{Z}_p$, since $\mathbb{Z}_p$ is a field (Corollary A.17). Therefore $\{h^0, h^1, h^2, \ldots h^{q-1}\} = G_q$, and h is a generator of G_q. If H is any subgroup of order q and $z \in H$, $z \neq [1]$, then $\operatorname{ord}(z)$ divides q, and hence $\operatorname{ord}(z) = q$, because q is a prime. Thus $z \in G_q$, and we conclude that $H = G_q$. $\square$

Computing Modulo a Prime. The security of many cryptographic schemes is based on the discrete logarithm assumption, which says that $x \mapsto g^x \bmod p$ is a one-way function. Here p is a large prime and the base element g is

1. either a primitive root modulo p, i.e., a generator of $\mathbb{Z}_p^*$, or
2. it is an element of order q in $\mathbb{Z}_p^*$, i.e., a generator of the subgroup G_q of order q, and q is a (large) prime that divides $p-1$.

Examples of such schemes which we discuss in this book are ElGamal's encryption and digital signatures, the digital signature standard DSS (see Section 3.5), commitment schemes (see Section 4.3.2), electronic elections (see Section 4.4) and digital cash (see Section 4.5).

When setting up such schemes, generators g of $\mathbb{Z}_p^*$ or G_q have to be selected. This can be difficult or even infeasible in the first case, because we must know the prime factors of $p-1$ in order to test whether a given element g is a primitive root (see Algorithm A.39 above). On the other hand, it is easy to find a generator g of G_q. We simply take a random element $h \in \mathbb{Z}_p^*$ and set $g := h^{(p-1)/q}$. The order of g divides q, because $g^q = h^{p-1} = [1]$.

Since q is a prime, we conclude that $\text{ord}(g) = 1$ or $\text{ord}(g) = q$. Therefore, if $g \neq [1]$, then $\text{ord}(g) = q$ and g is a generator of G_q.

To implement cryptographic operations, we have to compute in $\mathbb{Z}_p^*$ or in the subgroup G_q. The following rules simplify these computations.

1. Let $x \in \mathbb{Z}_p^*$. Then $x^k = x^{k'}$, if $k \equiv k' \bmod (p-1)$.

   In particular, $x^k = x^{k \bmod (p-1)}$, i.e., exponents can be reduced by modulo $(p-1)$, and $x^{-k} = x^{p-1-k}$.
2. Let $x \in \mathbb{Z}_p^*$ be an element of order q, i.e., $x \in G_q$. Then $x^k = x^{k'}$, if $k \equiv k' \bmod q$.

   In particular, $x^k = x^{k \bmod q}$, i.e., exponents can be reduced by modulo q, and $x^{-k} = x^{q-k}$.

The rules state that the exponents are added and multiplied modulo $(p-1)$ or modulo q. The rules hold, because $x^{p-1} = [1]$ for $x \in \mathbb{Z}_p^*$ (Proposition A.24) and $x^q = [1]$ for $x \in G_q$, which implies that

$$x^{k+l\cdot(p-1)} = x^k x^{l\cdot(p-1)} = x^k \left(x^{p-1}\right)^l = x^k[1]^l = x^k \text{ for } x \in \mathbb{Z}_p^*$$

$$\text{and } x^{k+l\cdot q} = x^k x^{l\cdot q} = x^k \left(x^q\right)^l = x^k[1]^l = x^k \text{ for } x \in G_q.$$

These rules can be very useful in computations. For example, let $x \in \mathbb{Z}_p^*$ and $k \in \{0, 1, \ldots, p-2\}$. Then you can compute the inverse x^{-k} of x^k by raising x to the $(p-1-k)$-th power, $x^{-k} = x^{p-1-k}$, without explicitly computing an inverse by using, for example, the Euclidean algorithm. Note that $(p-1-k)$ is a positive exponent. Powers of x are efficiently computed by the fast exponentiation algorithm (Algorithm A.26).

In many cases it is also possible to compute the k-th root of elements in $\mathbb{Z}_p^*$.

1. Let $x \in \mathbb{Z}_p^*$ and $k \in \mathbb{N}$ with $\gcd(k, p-1) = 1$, i.e., k is a unit modulo $p-1$. Let k^{-1} be the inverse of k modulo $p-1$, i.e., $k \cdot k^{-1} \equiv 1 \bmod (p-1)$.

   Then $\left(x^{k^{-1}}\right)^k = x$, i.e., $x^{k^{-1}}$ is a k-th root of x in $\mathbb{Z}_p^*$.
2. Let $x \in \mathbb{Z}_p^*$ be an element of order q, i.e., $x \in G_q$, and $k \in \mathbb{N}$ with $1 \leq k < q$. Let k^{-1} be the inverse of k modulo q, i.e., $k \cdot k^{-1} \equiv 1 \bmod q$.

   Then $\left(x^{k^{-1}}\right)^k = x$, i.e., $x^{k^{-1}}$ is a k-th root of x in $\mathbb{Z}_p^*$.

It is common practice to denote the k-th root $x^{k^{-1}}$ by $x^{1/k}$.

You can apply analogous rules of computation to elements g^k in any finite group G. Proposition A.23, which says that $g^{|G|}$ is the unit element, implies that exponents k are added and multiplied modulo the order $|G|$ of G.

A.5 Polynomials and Finite Fields

A *finite field* is a field with a finite number of elements. In Section A.2, we met examples of finite fields: The residue class ring $\mathbb{Z}_n$ is a field, if and only if n is a prime. The fields $\mathbb{Z}_p$, p a prime number, are called the finite *prime fields*, and they are also denoted by $\mathbb{F}_p$. Finite fields are extensions of these prime fields. Field extensions are constructed by using polynomials. So we first study the ring of polynomials with coefficients in a field k.

A.5.1 The Ring of Polynomials

Let $k[X]$ be the ring of polynomials in one variable X over a (not necessarily finite) field k. The elements of $k[X]$ are the polynomials

$$F(X) = a_0 + a_1 X + a_2 X^2 + \ldots a_d X^d = \sum_{i=0}^{d} a_i X^i,$$

with coefficients $a_i \in k$, $0 \le i \le d$.

If we assume that $a_d \ne 0$, then the *leading term* $a_d X^d$ really appears in the polynomial, and we call d the *degree* of F, $\deg(F)$ for short. The polynomials of degree 0 are just the elements of k.

The polynomials in $k[X]$ are added and multiplied as usual:

1. We add two polynomials $F = \sum_{i=0}^{d} a_i X^i$ and $G = \sum_{i=0}^{e} b_i X^i$, assume $d \le e$, by adding the coefficients (set $a_i = 0$ for $d < i \le e$):

$$F + G = \sum_{i=0}^{e} (a_i + b_i) X^i.$$

2. The product of two polynomials $F = \sum_{i=0}^{d} a_i X^i$ and $G = \sum_{i=0}^{e} b_i X^i$ is

$$F \cdot G = \sum_{i=0}^{de} \left(\sum_{k=0}^{i} a_k b_{i-k} \right) X^i.$$

With this addition and multiplication, $k[X]$ becomes a commutative ring with unit element. The unit element of $k[X]$ is the unit element 1 of k. The ring $k[X]$ has no zero divisors, i.e., if F and G are non-zero polynomials, then the product $F \cdot G$ is also non-zero.

The algebraic properties of the ring $k[X]$ of polynomials are analogous to the algebraic properties of the ring of integers.

Analogously to Definition A.1, we define for polynomials F and G what it means that F divides G and the *greatest common divisor* of F and G. The greatest common divisor is unique up to a factor $c \in k, c \ne 0$, i.e., if A is a greatest common divisor of F and G, then $c \cdot A$ is also a greatest common divisor, for $c \in k^* = k \setminus \{0\}$.

A polynomial F is *(relatively) prime* to G if the only common divisors of F and G are the units k^* of k.

Division with remainder works as with the integers. The difference is that the "size" of a polynomial is measured by using the degree, whereas the absolute value was used for an integer.

Theorem A.41 *(Division with remainder).* *Let* $F, G \in k[X], G \neq 0$. *Then there are unique polynomials* $Q, R \in k[X]$, *such that* $F = Q \cdot G + R$ *and* $0 \leq deg(R) < deg(G)$.

Proof. The proof runs exactly in the same way as the proof of Theorem A.2: Replace the absolute value with the degree. $\square$

R is called the *remainder* of F modulo G. We write F mod G for R. The polynomial Q is the *quotient* of F and G. We write F div G for Q.

You can compute a greatest common divisor of polynomials F and G by using the *Euclidean algorithm*, and the *extended Euclidean algorithm* yields the coefficients $C, D \in k[X]$ of a linear combination

$$A = C \cdot F + D \cdot G,$$

with A a greatest common divisor of F and G.

If you have obtained such a linear combination for one greatest common divisor, then you immediately get a linear combination for any other greatest common divisor by multiplying with a unit from k^*.

In particular, if F is prime to G, then the extended Euclidean algorithm computes a linear combination

$$1 = C \cdot F + D \cdot G.$$

We also have the analogue of prime numbers.

Definition A.42. Let $P \in k[X], P \notin k$. P is called *irreducible* (or a *prime*) if the only divisors of P are the elements $c \in k^*$ and $c \cdot P, c \in k^*$, or, equivalently, if whenever one can write $P = F \cdot G$ with $F, G \in k[X]$, then $F \in k^*$ or $G \in k^*$. A polynomial $Q \in k[X]$ which is not irreducible is called *reducible* or a *composite*.

As the ring $\mathbb{Z}$ of integers, the ring $k[X]$ of polynomials is factorial, i.e., every element has a unique decomposition into irreducible elements.

Theorem A.43. *Let* $F \in k[X], F \neq 0$, *be a non-zero polynomial. There are pairwise distinct irreducible polynomials* $P_1, \ldots, P_r, r \geq 0$, *exponents* $e_1, \ldots, e_r \in \mathbb{N}, e_i \geq 1, i = 1, \ldots, r$, *and a unit* $u \in k^*$, *such that*

$$F = u \prod_{i=1}^{r} P_i^{e_i}.$$

This factorization is unique in the following sense: If

$$F = v \prod_{i=1}^{s} Q_i^{f_i}$$

is another factorization of F, *then we have* $r = s$, *and after a permutation of the indices* i *we have* $Q_i = u_i P_i$, *with* $u_i \in k^*$, *and* $e_i = f_i$ *for* $1 \leq i \leq r$.

Proof. The proof runs in the same way as the proof of the Fundamental Theorem of Arithmetic (Theorem A.12).   □

A.5.2 Residue Class Rings

As in the ring of integers, we can consider residue classes in $k[X]$ and residue class rings.

Definition A.44. Let $P \in k[X]$ be a polynomial of degree ≥ 1:

1. $F, G \in k[X]$ are *congruent modulo* P, written as

$$F \equiv G \bmod P,$$

   if P divides $F - G$. This means that F and G have the same remainder when divided by P, i.e., $F \bmod P = G \bmod P$.
2. Let $F \in k[X]$. $[F] := \{G \in k[X] \mid G \equiv F \bmod P\}$ is called the *residue class* of F modulo P.

As before, "congruent modulo" is an equivalence relation, the equivalence classes are the residue classes, and the set of residue classes

$$k[X]/Pk[X] := \{[F] \mid F \in k[X]\}$$

is a ring. Residue classes are added and multiplied by adding and multiplying a representative:

$$[F] + [G] := [F + G], \quad [F] \cdot [G] := [F \cdot G].$$

We also have a natural representative of $[F]$, the remainder $F \bmod P$ of F modulo P: $[F] = [F \bmod P]$. As remainders modulo P, we get all the polynomials which have a degree $< \deg(P)$. Therefore, we have a one-to-one correspondence between $k[X]/Pk[X]$ and the set of residues $\{F \in k[X] \mid \deg(F) < \deg(P)\}$. We often identify both sets:

$$k[X]/Pk[X] = \{F \in k[X] \mid \deg(F) < \deg(P)\}.$$

Two residues F and G are added or multiplied by first adding or multiplying them as polynomials and then taking the residue modulo P. Since the sum of two residues F and G has a degree $< \deg(P)$, it is a residue, and we do not have to reduce. After a multiplication, we have, in general, to take the remainder.

Addition : $(F, G) \longmapsto F + G$,   Multiplication : $(F, G) \longmapsto F \cdot G \bmod P.$

Let $n := \deg(P)$ be the degree of P. The residue class ring $k[X]/Pk[X]$ is an n-dimensional vector space over k. A basis of this vector space is given by the elements $[1], [X], [X^2], \ldots, [X^{n-1}]$. If k is a finite field with q elements, then $k[X]/Pk[X]$ consists of q^n elements.

Example. Let $k = \mathbb{F}_2 = \mathbb{Z}_2 = \{0, 1\}$ be the field with two elements 0 and 1 consisting of the residues modulo 2, and $P := X^8 + X^4 + X^3 + X + 1 \in k[X]$. The elements of $k[X]/Pk[X]$ may be identified with the binary polynomials $b_7 X^7 + b_6 X^6 + \ldots + b_1 X + b_0, b_i \in \{0, 1\}, 0 \leq i \leq 7$, of degree ≤ 7. The ring $k[X]/Pk[X]$ contains $2^8 = 256$ elements. We have, for example,

$$
\begin{aligned}
(X^6 &+ X^3 + X^2 + 1) \cdot (X^5 + X^2 + 1) \\
&= X^{11} + X^7 + X^6 + X^4 + X^3 + 1 \\
&= X^3 \cdot (X^8 + X^4 + X^3 + X + 1) + 1 \\
&\equiv 1 \bmod (X^8 + X^4 + X^3 + X + 1).
\end{aligned}
$$

Thus, $X^6 + X^3 + X^2 + 1$ is a unit in $k[X]/Pk[X]$, and its inverse is $X^5 + X^2 + 1$.

We may characterize units as in the integer case.

Proposition A.45. *An element $[F] \in k[X]/Pk[X]$ is a unit if and only if F is prime to P. The multiplicative inverse $[F]^{-1}$ of a unit $[F]$ can be computed using the extended Euclidean algorithm.*

Proof. The proof is the same as the proof in the integer case (see Proposition A.16). Recall that the inverse may be calculated as follows: If F is prime to P, then the extended Euclidean algorithm produces a linear combination

$$
C \cdot F + D \cdot P = 1, \text{ with polynomials } C, D \in k[X].
$$

We see that $C \cdot F \equiv 1 \bmod P$. Hence, $[C]$ is the inverse $[F]^{-1}$.   □

If the polynomial P is irreducible, then all residues modulo P, i.e., all polynomials with a degree $< \deg(P)$, are prime to P. So we get the same corollary as in the integer case.

Corollary A.46. *Let P be irreducible. Then every $[F] \neq [0]$ in $k[X]/Pk[X]$ is a unit. Thus, $k[X]/Pk[X]$ is a field.*

Remarks:

1. Let P be an irreducible polynomial of degree n. The field k is a subset of the larger field $k[X]/Pk[X]$. We therefore call $k[X]/Pk[X]$ an *extension field* of k of degree n.
2. If P is reducible, then $P = F \cdot G$, with polynomials F, G of degree $< \deg(P)$. Then $[F] \neq [0]$ and $[G] \neq [0]$, but $[F] \cdot [G] = [P] = [0]$. $[F]$ and $[G]$ are "zero divisors". They have no inverse, and we see that $k[X]/Pk[X]$ is not a field.

A.5.3 Finite Fields

Now, let $k = \mathbb{Z}_p = \mathbb{F}_p$ be the prime field of residues modulo p, $p \in \mathbb{Z}$ a prime number, and let $P \in \mathbb{F}_p[X]$ be an irreducible polynomial of degree

n. Then $k[X]/Pk[X] = \mathbb{F}_p[X]/P\mathbb{F}_p[X]$ is an extension field of $\mathbb{F}_p$. It is an n-dimensional vector space over $\mathbb{F}_p$, and it contains p^n elements.

In general, there is more than one irreducible polynomial of degree n over $\mathbb{F}_p$. Therefore there are more finite fields with p^n elements. For example, if $Q \in \mathbb{F}_p[X]$ is another irreducible polynomial of degree n, $Q \neq cP$ for all $c \in k$, then $\mathbb{F}_p[X]/Q\mathbb{F}_p[X]$ is a field with p^n elements, different from $k[X]/Pk[X]$. But one can show that all the finite fields with p^n elements are isomorphic to each other in a very natural way. As the mathematicians state it, up to canonical isomorphism, there is only one finite field with p^n elements. It is denoted by $\mathbb{F}_{p^n}$ or by $\mathrm{GF}(p^n)$.[4]

If you need a concrete representation of $\mathbb{F}_{p^n}$, then you choose an irreducible polynomial $P \in \mathbb{F}_p[X]$ of degree n, and you have $\mathbb{F}_{p^n} = \mathbb{F}_p[X]/P\mathbb{F}_p[X]$. But there are different representations, reflecting your degrees of freedom when choosing the irreducible polynomial.

One can also prove that in every finite field k, the number $|k|$ of elements in k must be a power p^n of a prime number p. Therefore, the fields $\mathbb{F}_{p^n}$ are all the finite fields that exist.

In cryptography, finite fields play an important role in many places. For example, the classical ElGamal cryptosystems are based on the discrete logarithm problem in a finite prime field (see Section 3.5), the elliptic curves used in cryptography are defined over finite fields, and the basic encryption operations of the Advanced Encryption Standard AES are algebraic operations in the field $\mathbb{F}_{2^8}$ with 2^8 elements. The AES is discussed in this book (see Section 2.2.2). This motivates the following closer look at the fields $\mathbb{F}_{2^n}$.

We identify $\mathbb{F}_2 = \mathbb{Z}_2 = \{0, 1\}$. Let $P = X^n + a_{n-1}X^{n-1} + \ldots + a_1 X + a_0$, $a_i \in \{0, 1\}$, $0 \leq i \leq n-1$ be a binary irreducible polynomial of degree n. Then $\mathbb{F}_{2^n} = \mathbb{F}_p[X]/P\mathbb{F}_p[X]$, and we may consider the binary polynomials $A = b_{n-1}X^{n-1} + b_{n-2}X^{n-2} + \ldots + b_1 X + b_0$ of degree $\leq n-1$ ($b_i \in \{0, 1\}$, $0 \leq i \leq n-1$) as the elements of $\mathbb{F}_{2^n}$. Adding two of these polynomials in $\mathbb{F}_{2^n}$ means to add them as polynomials, and multiplying them means to first multiply them as polynomials and then take the remainder modulo P.

Now we can represent the polynomial A by the n-dimensional vector $b_{n-1}b_{n-2}\ldots b_1 b_0$ of its coefficients. In this way, we get a binary representation of the elements of $\mathbb{F}_{2^n}$; the elements of $\mathbb{F}_{2^n}$ are just the bit strings of length n. To add two of these elements means to add them as binary vectors, i.e., you add them bitwise modulo 2, which is the same as bitwise XORing:

$$b_{n-1}b_{n-2}\ldots b_1 b_0 \quad + \quad c_{n-1}c_{n-2}\ldots c_1 c_0$$

$$= (b_{n-1} \oplus c_{n-1})(b_{n-2} \oplus c_{n-2})\ldots(b_1 \oplus c_1)(b_0 \oplus c_0).$$

To multiply two elements is more complicated: You have to convert the bit strings to polynomials, multiply them as polynomials, reduce modulo P and

---

[4] Finite fields are also called *Galois fields*, in honor of the French mathematician Évariste Galois (1811–1832).

take the coefficients of the remainder. The 0-element of $\mathbb{F}_{2^n}$ is $00\ldots00$ and the 1-element is $00\ldots001$.

In the Advanced Encryption Standard AES, encryption depends on algebraic operations in the finite field $\mathbb{F}_{2^8}$. The irreducible binary polynomial $P := X^8 + X^4 + X^3 + X + 1$ is taken to represent $\mathbb{F}_{2^8}$ as $\mathbb{F}_2[X]/P\mathbb{F}_2[X]$ (we already used this polynomial in an example above). Then the elements of $\mathbb{F}_{2^8}$ are just strings of 8 bits. In this way, a byte is an element of $\mathbb{F}_{2^8}$ and vice versa. One of the core operations of AES is the so-called S-Box. The AES S-Box maps a byte x to its inverse x^{-1} in $\mathbb{F}_{2^8}$ and then modifies the result by an $\mathbb{F}_2$-affine transformation (see Section 2.2.2). We conclude this section with examples for adding, multiplying and inverting bytes in $\mathbb{F}_{2^8}$.

$$01001101 + 00100101 = 01101000,$$
$$10111101 \cdot 01101001 = 11111100,$$
$$01001101 \cdot 00100101 = 00000001,$$
$$01001101^{-1} = 00100101.$$

As is common practice, we sometimes represent a byte and hence an element of $\mathbb{F}_{2^8}$ by two hexadecimal digits. Then the examples read as follows:

$$4D + 25 = 68, \ BD \cdot 69 = FC, \ 4D \cdot 25 = 01, \ 4D^{-1} = 25.$$

A.6 Quadratic Residues

We will study the question as to which of the residues modulo n are squares.

Definition A.47. Let $n \in \mathbb{N}$ and $x \in \mathbb{Z}$. We call that x is a *quadratic residue modulo* n if there is an element $y \in \mathbb{Z}$ with $x \equiv y^2 \bmod n$. Otherwise, x is called a *quadratic non-residue modulo* n.

Examples:

1. The numbers $0, 1, 4, 5, 6$ and 9 are the quadratic residues modulo 10.
2. The numbers $0, 1, 3, 4, 5$ and 9 are the quadratic residues modulo 11.

Remark. The property of being a quadratic residue depends only on the residue class $[x] \in \mathbb{Z}_n$ of x modulo n. An integer x is a quadratic residue modulo n if and only if its residue class $[x]$ is a square in the residue class ring $\mathbb{Z}_n$ (i.e., if and only if there is some $[y] \in \mathbb{Z}_n$ with $[x] = [y]^2$). The residue class $[x]$ is often also called a quadratic residue.

In most cases we are only interested in the quadratic residues x which are units modulo n (i.e., x and n are relatively prime, see Proposition A.16).

Definition A.48. The subgroup of $\mathbb{Z}_n^*$ that consists of the residue classes represented by a quadratic residue is denoted by QR_n:

$$\mathrm{QR}_n = \{[x] \in \mathbb{Z}_n^* \mid \text{There is a } [y] \in \mathbb{Z}_n^* \text{ with } [x] = [y]^2\}.$$

It is called the *subgroup of quadratic residues* or the *subgroup of squares*. The complement of QR_n is denoted by $\mathrm{QNR}_n := \mathbb{Z}_n^* \setminus \mathrm{QR}_n$. It is called the *subset of quadratic non-residues*.

We give a criterion for determining the quadratic residues modulo a prime.

Lemma A.49. *Let p be a prime > 2 and $g \in \mathbb{Z}_p^*$ be a primitive root of $\mathbb{Z}_p^*$. Let $x \in \mathbb{Z}_p^*$. Then $[x] \in \mathrm{QR}_p$ if and only if $x \equiv g^t \bmod p$ for some even number $t, 0 \leq t \leq p - 2$.*

Proof. Recall that $\mathbb{Z}_p^*$ is a cyclic group generated by g (Theorem A.36). If $[x] \in \mathrm{QR}_p$, then $x \equiv y^2 \bmod p$, and $y \equiv g^s \bmod p$ for some s. Then $x = g^{2s} \bmod p \equiv g^t \bmod p$, with $t := 2s \bmod (p - 1)$ (the order of g is $p - 1$) and $0 \leq t \leq p - 2$. Since $p - 1$ is even, t is also even.
Conversely, if $x \equiv g^t \bmod p$, and t is even, then $x \equiv (g^{t/2})^2 \bmod p$, which means that $x \in \mathrm{QR}_p$. $\qquad\square$

Proposition A.50. *Let p be a prime > 2. Exactly half of the elements of $\mathbb{Z}_p^*$ are squares, i.e., $|\mathrm{QR}_p| = (p-1)/2$.*

Proof. Since half of the integers x with $0 \leq x \leq p-2$ are even, the proposition follows from the preceding lemma. $\qquad\square$

Definition A.51. Let p be a prime > 2, and let $x \in \mathbb{Z}$ be prime to p.

$$\left(\frac{x}{p}\right) := \begin{cases} +1 & \text{if } [x] \in \mathrm{QR}_p, \\ -1 & \text{if } [x] \notin \mathrm{QR}_p, \end{cases}$$

is called the *Legendre symbol* of x mod p. For $x \in \mathbb{Z}$ with $p \mid x$, we set $\left(\frac{x}{p}\right) := 0$.

Proposition A.52 *(Euler's criterion). Let p be a prime > 2, and let $x \in \mathbb{Z}$. Then*

$$\left(\frac{x}{p}\right) \equiv x^{(p-1)/2} \bmod p.$$

Proof. If p divides x, then both sides are congruent 0 modulo p. Suppose p does not divide x. Let $[g] \in \mathbb{Z}_p^*$ be a primitive element.
We first observe that $g^{(p-1)/2} \equiv -1 \bmod p$. Namely, $[g]^{(p-1)/2}$ is a solution of the equation $X^2 - 1$ over the field $\mathbb{Z}_p^*$. Hence, $g^{(p-1)/2} \equiv \pm 1 \bmod p$. However, $g^{(p-1)/2} \bmod p \neq 1$, because the order of $[g]$ is $p - 1$.
Let $[x] = [g]^t, 0 \leq t \leq p - 2$. By Lemma A.49, $[x] \in \mathrm{QR}_p$ if and only if t is even. On the other hand, $x^{(p-1)/2} \equiv g^{t(p-1)/2} \equiv \pm 1 \bmod p$, and it is $\equiv 1 \bmod p$ if and only if t is even. This completes the proof. $\qquad\square$

Remarks:

1. The Legendre symbol is multiplicative in x:

$$\left(\frac{xy}{p}\right) = \left(\frac{x}{p}\right) \cdot \left(\frac{y}{p}\right).$$

This immediately follows, for example, from Euler's criterion. It means that $[xy] \in QR_p$ if and only if either both $[x], [y] \in QR_p$ or both $[x], [y] \notin QR_p$.

2. The Legendre symbol $\left(\frac{x}{p}\right)$ depends only on $x \bmod p$, and the map

$$\mathbb{Z}_p^* \longrightarrow \{1, -1\}, \ x \longmapsto \left(\frac{x}{p}\right)$$

is a homomorphism of groups.

We do not give proofs of the following two important results. Proofs may be found, for example, in [HarWri79], [Rosen2000], [Koblitz94] and [Forster96].

Theorem A.53. *Let p be a prime > 2. Then:*

1. $\left(\frac{-1}{p}\right) = (-1)^{(p-1)/2} = \begin{cases} +1 & \text{if } p \equiv 1 \bmod 4, \\ -1 & \text{if } p \equiv 3 \bmod 4. \end{cases}$

2. $\left(\frac{2}{p}\right) = (-1)^{(p^2-1)/8} = \begin{cases} +1 & \text{if } p \equiv \pm 1 \bmod 8, \\ -1 & \text{if } p \equiv \pm 3 \bmod 8. \end{cases}$

Theorem A.54 *(Law of Quadratic Reciprocity). Let p and q be primes > 2, $p \neq q$. Then*

$$\left(\frac{p}{q}\right)\left(\frac{q}{p}\right) = (-1)^{(p-1)(q-1)/4}.$$

We generalize the Legendre symbol for composite numbers.

Definition A.55. Let $n \in \mathbb{Z}$ be a positive odd number and $n = \prod_{i=1}^{r} p_i^{e_i}$ be the decomposition of n into primes. Let $x \in \mathbb{Z}$.

$$\left(\frac{x}{n}\right) := \prod_{i=1}^{r} \left(\frac{x}{p_i}\right)^{e_i}$$

is called the *Jacobi symbol* of $x \bmod n$.

Remarks:

1. The value of $\left(\frac{x}{n}\right)$ only depends on the residue class $[x] \in \mathbb{Z}_n$.

2. If $[x] \in \mathrm{QR}_n$, then $[x] \in \mathrm{QR}_p$ for all primes p that divide n. Hence, $\left(\frac{x}{n}\right) = 1$. The converse is not true, in general. For example, let $n = pq$ be the product of two primes. Then $\left(\frac{x}{n}\right) = \left(\frac{x}{p}\right) \cdot \left(\frac{x}{q}\right)$ can be 1, whereas both $\left(\frac{x}{p}\right)$ and $\left(\frac{x}{q}\right)$ are -1. This means that $x \bmod p$ (and $x \bmod q$), and hence $x \bmod n$ are not squares.

3. The Jacobi symbol is multiplicative in both arguments:
$$\left(\frac{xy}{n}\right) = \left(\frac{x}{n}\right) \cdot \left(\frac{y}{n}\right) \text{ and } \left(\frac{x}{mn}\right) = \left(\frac{x}{m}\right) \cdot \left(\frac{x}{n}\right).$$

4. The map $\mathbb{Z}_n^* \longrightarrow \{1, -1\}$, $[x] \longmapsto \left(\frac{x}{n}\right)$ is a homomorphism of groups.

5. $\mathrm{J}_n^{+1} := \{[x] \in \mathbb{Z}_n^* | \left(\frac{x}{n}\right) = 1\}$ is a subgroup of $\mathbb{Z}_n^*$.

Lemma A.56. *Let $n \geq 3$ be an odd integer. If n is a square (in $\mathbb{Z}$), then $\left(\frac{x}{n}\right) = 1$ for all x. Otherwise, half of the elements of $\mathbb{Z}_n^*$ have a Jacobi symbol of 1, i.e., $|\mathrm{J}_n^{+1}| = \varphi(n)/2$.*

Proof. If n is a square, then the exponents e_i in the prime factorization of n are all even (notation as above), and the Jacobi symbol is always 1. If n is not a square, then there is an odd e_i, say e_1. By the Chinese Remainder Theorem (Theorem A.29), we find a unit x which is a quadratic non-residue modulo p_1 and a quadratic residue modulo p_i for $i = 2, \ldots, r$. Then $\left(\frac{x}{n}\right) = -1$, and mapping $[y]$ to $[y \cdot x]$ yields a one-to-one map from J_n^{+1} to $\mathbb{Z}_n^* \setminus \mathrm{J}_n^{+1}$. $\square$

Theorem A.57. *Let $n \geq 3$ be an odd integer. Then:*

1. $\left(\frac{-1}{n}\right) = (-1)^{(n-1)/2} = \begin{cases} +1 & \text{if } n \equiv 1 \bmod 4, \\ -1 & \text{if } n \equiv 3 \bmod 4. \end{cases}$

2. $\left(\frac{2}{n}\right) = (-1)^{(n^2-1)/8} = \begin{cases} +1 & \text{if } n \equiv \pm 1 \bmod 8, \\ -1 & \text{if } n \equiv \pm 3 \bmod 8. \end{cases}$

Proof. Let $f(n) = (-1)^{(n-1)/2}$ for statement 1 and $f(n) = (-1)^{(n^2-1)/8}$ for statement 2. You can easily check that $f(n_1 n_2) = f(n_1)f(n_2)$ for odd numbers n_1 and n_2 (for statement 2, consider the different cases of $n_1, n_2 \bmod 8$). Thus, both sides of the equations $\left(\frac{-1}{n}\right) = (-1)^{(n-1)/2}$ and $\left(\frac{2}{n}\right) = (-1)^{(n^2-1)/8}$ are multiplicative in n, and the proposition follows from Theorem A.53. $\square$

Theorem A.58 *(Law of Quadratic Reciprocity). Let $n, m \geq 3$ be odd integers. Then*
$$\left(\frac{m}{n}\right) = (-1)^{(n-1)(m-1)/4} \left(\frac{n}{m}\right).$$

Proof. If m and n have a common factor, then both sides are zero by the definition of the symbols. So we can suppose that m is prime to n. We write $m = p_1 p_2 \ldots p_r$ and $n = q_1 q_2 \ldots q_s$ as a product of primes. Converting from $\left(\frac{m}{n}\right) = \prod_{i,j} \left(\frac{p_i}{q_j}\right)$ to $\left(\frac{n}{m}\right) = \prod_{i,j} \left(\frac{q_j}{p_i}\right)$, we apply the reciprocity law for the Legendre symbol (Theorem A.54) for each of the factors. We get rs multipliers $\varepsilon_{ij} = (-1)^{(p_i-1)(q_j-1)/4}$. As in the previous proof, we use that $f(n) = (-1)^{(n-1)/2}$ is multiplicative in n and get

$$\prod_{i,j} (-1)^{(p_i-1)(q_j-1)/4} = \prod_j \left(\prod_i (-1)^{(p_i-1)/2}\right)^{(q_j-1)/2}$$

$$= \prod_j \left((-1)^{(m-1)/2}\right)^{(q_j-1)/2} = \left(\prod_j (-1)^{(q_j-1)/2}\right)^{(m-1)/2}$$

$$= \left((-1)^{(n-1)/2}\right)^{(m-1)/2} = (-1)^{(n-1)(m-1)/4},$$

as desired.   □

Remark. Computing a Jacobi symbol $\left(\frac{m}{n}\right)$ simply by using the definition requires knowing the prime factors of n. No algorithm is known that can compute the prime factors in polynomial time. However, using the Law of Quadratic Reciprocity (Theorem A.58) and Theorem A.57, we can efficiently compute $\left(\frac{m}{n}\right)$ using the following algorithm, without knowing the factorization of n.

Algorithm A.59.
    Jac(int m, n)
      1 Replace m by $m \bmod n$.
      2 If $m = 0$, then $\left(\frac{m}{n}\right) = 0$, and if $m = 1$, then $\left(\frac{m}{n}\right) = 1$.
      3 Else, set $m = 2^t r$, with r odd.
        Compute $\left(\frac{2^t}{n}\right) = \left(\frac{2}{n}\right)^t$ by Theorem A.57.
        If $r = 1$, we are finished.
      4 If $r \geq 3$, we still need to compute $\left(\frac{r}{n}\right)$.
        Apply the Law of Quadratic Reciprocity and compute $\left(\frac{r}{n}\right)$ by $\left(\frac{n}{r}\right)$.
      5 Now set $m = n$ and $n = r$ and go to 1.

We have $r \leq m \bmod n$, and (n, r) becomes the pair (m, n) in the next iteration. An analysis similar to that of the Euclidean algorithm (Algorithm A.4) shows that the algorithm terminates after at most $O(\log_2(n))$ iterations (see Corollary A.10).

Example. We want to determine whether the prime 7331 is a quadratic residue modulo the prime 9859. For this purpose, we have to compute the Legendre symbol $\left(\frac{7331}{9859}\right)$ and could do that using Euler's criterion (Proposition A.52). However, applying Algorithm A.59 is much more efficient:

$$\left(\frac{7331}{9859}\right) = -\left(\frac{9859}{7331}\right) = -\left(\frac{2528}{7331}\right) = -\left(\frac{2}{7331}\right)^5 \cdot \left(\frac{79}{7331}\right)$$

$$= -(-1)^5 \cdot \left(\frac{79}{7331}\right) = \left(\frac{79}{7331}\right) = -\left(\frac{7331}{79}\right) = -\left(\frac{63}{79}\right)$$

$$= \left(\frac{79}{63}\right) = \left(\frac{16}{63}\right) = \left(\frac{2}{63}\right)^4 = 1.$$

Thus, 7331 is a quadratic residue modulo 9859.

A.7 Modular Square Roots

We now discuss how to get the square root of a quadratic residue. Computing square roots modulo n can be a difficult or even an infeasible task if n is a composite number (and, e.g., the Rabin cryptosystem is based on this; see Section 3.6). However, if n is a prime, we can determine square roots using an efficient algorithm.

Proposition A.60. *There is a (probabilistic) polynomial algorithm Sqrt which, given as inputs a prime p and an $a \in QR_p$, computes a square root $x \in \mathbb{Z}_p^*$ of a: $Sqrt(p, a) = x$ and $x^2 = a$ (in $\mathbb{Z}_p$).*

Remarks:

1. The square roots of $a \in QR_p$ are the solutions of the equation $X^2 - a = 0$ over $\mathbb{Z}_p$. Hence a has two square roots (for $p > 2$). If x is a square root, then $-x$ is the other root.
2. "Probabilistic" means that random choices[5] are included in the algorithm. Polynomial means that the running time (the number of binary operations) of the algorithm is bounded by a polynomial in the binary length of the inputs. *Sqrt* is a so-called Las Vegas algorithm, i.e., we expect *Sqrt* to return a correct result in polynomial time (for a detailed discussion of the notion of probabilistic polynomial algorithms, see Chapter 5).

Proof. Let $a \in QR_p$. By Euler's criterion (Proposition A.52), $a^{(p-1)/2} = 1$. Hence $a^{(p+1)/2} = a$. We first consider the (easy) case of $p \equiv 3 \bmod 4$. Since 4 divides $p + 1$, $(p + 1)/4$ is an integer, and $x := a^{(p+1)/4}$ is a square root of a.

Now assume $p \equiv 1 \bmod 4$. The straightforward computation of the square root as in the first case does not work, since $(p + 1)/2$ is not divisible by 2. We choose a quadratic non-residue $b \in QNR_p$ (here the random choices come into play, see below). By Proposition A.52, $b^{(p-1)/2} = -1$. We have $a^{(p-1)/2} = 1$, and $(p - 1)/2$ is even. Let $(p - 1)/2 = 2^l r$, with r odd and $l \geq 1$. We will

---

[5] In this chapter, all random choices are with respect to the uniform distribution.

compute an exponent s, such that $a^r b^{2s} = 1$. Then we are finished. Namely, $a^{r+1} b^{2s} = a$ and $a^{(r+1)/2} b^s$ is a square root of a.

We obtain s in l steps. The intermediate result after step i is a representation $a^{2^{l-i}r} \cdot b^{2s_i} = 1$. We start with $a^{(p-1)/2} \cdot b^0 = a^{(p-1)/2} = 1$ and $s_0 = 0$. Let $y_i = a^{2^{l-i}r} \cdot b^{2s_i}$. In the i-th step we take the square root $y'_i := a^{2^{l-i}r} \cdot b^{s_{i-1}}$ of $y_{i-1} = a^{2^{l-i+1}r} \cdot b^{2s_{i-1}}$. The value of y'_i is either 1 or -1. If $y'_i = 1$, then we take $y_i := y'_i$. If $y'_i = -1$, then we set $y_i := y'_i \cdot b^{(p-1)/2}$. The first time that b appears with an exponent > 0 in the representation is after the first step (if ever), and then b's exponent is $(p - 1)/2 = 2^l r$. This implies that s_{i-1} is indeed an even number for $i = 1, \ldots, l$.

Thus, we may compute a square root using the following algorithm.

Algorithm A.61.

```
int Sqrt(int a, prime p)
 1   if p ≡ 3 mod 4
 2      then return a^(p+1) div 4 mod p
 3   else
 4      randomly choose b ∈ QNR_p
 5      i ← (p − 1) div 2; j ← 0
 6      repeat
 7         i ← i div 2; j ← j div 2
 8         if a^i b^j ≡ −1 mod p
 9            then j ← j + (p − 1) div 2
10      until i ≡ 1 mod 2
11      return a^((i+1) div 2) b^(j div 2) mod p
```

In the algorithm we get a quadratic non-residue by a random choice. For this purpose, we randomly choose an element b of $\mathbb{Z}_p^*$ and test (by Euler's criterion) whether b is a non-residue. Since half of the elements in $\mathbb{Z}_p^*$ are non-residues, we expect (on average) to get a non-residue after 2 random choices (see Lemma B.12). □

Now let n be a composite number. If n is a product of distinct primes and if we know these primes, we can apply the Chinese Remainder Theorem (Theorem A.29) and reduce the computation of square roots in $\mathbb{Z}_n^*$ to the computation of square roots modulo a prime. There we can apply Algorithm A.61. We discuss this procedure in detail for the RSA and Rabin settings, where $n = pq$, with p and q being distinct primes. The extended Euclidean algorithm yields numbers $d, e \in \mathbb{Z}$ with $1 = dp + eq$. By the Chinese Remainder Theorem, the map

$$\psi_{p,q} : \mathbb{Z}_n \longrightarrow \mathbb{Z}_p \times \mathbb{Z}_q, \quad [x] \longmapsto ([x \bmod p], [x \bmod q])$$

is an isomorphism. The inverse map is given by

$$\chi_{p,q} : \mathbb{Z}_p \times \mathbb{Z}_q \longrightarrow \mathbb{Z}_n, \quad ([x_1], [x_2]) \longmapsto [dpx_2 + eqx_1].$$

Addition and multiplication in $\mathbb{Z}_p \times \mathbb{Z}_q$ are done component-wise:

$$([x_1], [x_2]) + ([x_1'], [x_2']) = ([x_1 + x_1'], [x_2 + x_2']),$$

$$([x_1], [x_2]) \cdot ([x_1'], [x_2']) = ([x_1 \cdot x_1'], [x_2 \cdot x_2']).$$

For $[x] \in \mathbb{Z}_n$ let $([x_1], [x_2]) := \psi_{p,q}([x])$. We have

$$[x]^2 = [a] \text{ if and only if } [x_1]^2 = [a_1] \text{ and } [x_2]^2 = [a_2]$$

(see the remark after Theorem A.29). Thus, in order to compute the roots of $[a]$, we can compute the square roots of $[a_1]$ and $[a_2]$ and apply the inverse Chinese remainder map $\chi_{p,q}$. In $\mathbb{Z}_p$ and $\mathbb{Z}_q$, we can efficiently compute square roots by using Algorithm A.61.

Recall that $\mathbb{Z}_p$ and $\mathbb{Z}_q$ are fields, and over a field a quadratic equation $X^2 - [a_1] = 0$ has at most 2 solutions. Hence, $[a_1] \in \mathbb{Z}_p$ has at most two square roots. The zero-element $[0]$ has the only square root $[0]$. For $p = 2$, the only non-zero element $[1]$ of $\mathbb{Z}_2$ has the only square root $[1]$. For $p > 2$, a non-zero element $[a_1]$ has either no or two distinct square roots. If $[x_1]$ is a square root of $[a_1]$, then $-[x_1]$ is also a root and for $p > 2$, $[x_1] \neq -[x_1]$.

Combining the square roots of $[a_1]$ in $\mathbb{Z}_p$ with the square roots of $[a_2]$ in $\mathbb{Z}_q$, we get the square roots of $[a]$ in $\mathbb{Z}_n$. We summarize the results in the following proposition.

Proposition A.62. *Let p and q be distinct primes > 2 and $n := pq$. Assume that the prime factors p and q of n are known. Then the square roots of a quadratic residue $[a] \in \mathbb{Z}_n$ can be efficiently computed. Moreover, if $[a] = ([a_1], [a_2])^6$, then:*

1. *$[x] = ([x_1], [x_2])$ is a square root of $[a]$ in $\mathbb{Z}_n$ if and only if $[x_1]$ is a square root of $[a_1]$ in $\mathbb{Z}_p$ and $[x_2]$ is a square root of $[a_2]$ in $\mathbb{Z}_q$.*
2. *If $[x_1]$ and $-[x_1]$ are the square roots of $[a_1]$, and $[x_2]$ and $-[x_2]$ are the square roots of $[a_2]$, then $[u] = ([x_1], [x_2]), [v] = ([x_1], -[x_2]), -[v] = (-[x_1], [x_2])$ and $-[u] = (-[x_1], -[x_2])$ are all the square roots of $[a]$.*
3. *$[0]$ is the only square root of $[0] = ([0], [0])$.*
4. *If $[a_1] \neq [0]$ and $[a_2] \neq [0]$, which means that $[a]$ is a unit in $\mathbb{Z}_n$, then the square roots of $[a]$ given in 2. are pairwise distinct, i.e., $[a]$ has four square roots.*
5. *If $[a_1] = [0]$ (i.e., p divides a) and $[a_2] \neq [0]$ (i.e., q does not divide a), then $[a]$ has only two distinct square roots, $[u]$ and $[v]$.*

Remark. If one of the primes is 2, say $p = 2$, then statements 1–3 of Proposition A.62 are also true. Statements 4 and 5 have to be modified. $[a]$ has only one or two roots, because $[x_1] = -[x_1]$. If $[a_2] = 0$, then there is only one root. If $[a_2] \neq 0$, then the roots $[u]$ and $[v]$ are distinct.

---

[6] We identify $\mathbb{Z}_n$ with $\mathbb{Z}_p \times \mathbb{Z}_q$ via the Chinese remainder isomorphism $\psi_{p,q}$.

Conversely, the ability to compute square roots modulo n implies the ability to factorize n.

Lemma A.63. *Let $n := pq$, with distinct primes $p, q > 2$. Let $[u]$ and $[v]$ be square roots of $[a] \in \mathrm{QR}_n$ with $[u] \neq \pm[v]$. Then the prime factors of n can be computed from $[u]$ and $[v]$ using the Euclidean algorithm.*

Proof. We have $n \mid u^2 - v^2 = (u + v)(u - v)$, but n does not divide $u + v$ and n does not divide $u - v$. Hence, the computation of $\gcd(u + v, n)$ yields one of the prime factors of n. $\qquad\square$

Proposition A.64. *Let $I := \{n \in \mathbb{N} \mid n = pq, \ p, q \text{ distinct primes}\}$. Then the following statements are equivalent:*

1. *There is a probabilistic polynomial algorithm A_1 that on inputs $n \in I$ and $a \in \mathrm{QR}_n$ returns a square root of a in $\mathbb{Z}_n^*$.*
2. *There is a probabilistic polynomial algorithm A_2 that on input $n \in I$ yields the prime factors of n.*

Proof. Let A_1 be a probabilistic polynomial algorithm that on inputs n and a returns a square root of a modulo n. Then we can find the factors of n in the following way. We randomly select an $x \in \mathbb{Z}_n^*$ and compute $y = A_1(n, x^2)$. Since a has four distinct roots by Proposition A.62, the probability that $x \neq \pm y$ is $1/2$. If $x \neq \pm y$, we easily compute the factors by Lemma A.63. Otherwise we choose a new random x. We expect to be successful after two iterations.

Conversely, if we can compute the prime factors of n using a polynomial algorithm A_2, we can also compute (all the) square roots of arbitrary quadratic residues in polynomial time, as we have seen in Proposition A.62. $\qquad\square$

The Chinese Remainder isomorphism can also be used to determine the number of quadratic residues modulo n.

Proposition A.65. *Let p and q be distinct primes, and $n := pq$. Then $|\mathrm{QR}_n| = (p-1)(q-1)/4$.*

Proof. $[a] = ([a_1], [a_2]) \in \mathrm{QR}_n$ if and only if $[a_1] \in \mathrm{QR}_p$ and $[a_2] \in \mathrm{QR}_q$. By Proposition A.50, $|\mathrm{QR}_p| = (p-1)/2$ and $|\mathrm{QR}_q| = (q-1)/2$. $\qquad\square$

Proposition A.66. *Let p and q be distinct primes with $p, q \equiv 3 \bmod 4$, and $n := pq$. Let $[a] \in \mathrm{QR}_n$, and $[u] = ([x_1], [x_2]), [v] = ([x_1], -[x_2]), -[v] = (-[x_1], [x_2])$ and $-[u] = (-[x_1], -[x_2])$ be the four square roots of $[a]$ (see Proposition A.62). Then:*

1. $\left(\frac{u}{n}\right) = -\left(\frac{v}{n}\right)$.
2. *One and only one of the four square roots is in QR_n.*

Proof. 1. We have $u \equiv v \bmod p$ and $u \equiv -v \bmod q$, hence we conclude by Theorem A.53 that

$$\left(\frac{u}{n}\right) = \left(\frac{u}{p}\right)\left(\frac{u}{q}\right) = \left(\frac{v}{p}\right)\left(\frac{v}{q}\right)\left(\frac{-1}{q}\right) = -\left(\frac{v}{p}\right)\left(\frac{v}{q}\right) = -\left(\frac{v}{n}\right).$$

2. By Theorem A.53, $\left(\frac{-1}{p}\right) = \left(\frac{-1}{q}\right) = -1$. Thus, exactly one of the roots $[x_1]$ or $-[x_1]$ is in QR_p, say $[x_1]$, and exactly one of the roots $[x_2]$ or $-[x_2]$ is in QR_q, say $[x_2]$. Then $[u] = ([x_1], [x_2])$ is the only square root of $[a]$ that is in QR_n. $\qquad\square$

A.8 Primes and Primality Tests

Theorem A.67 *(Euclid's Theorem). There are infinitely many primes.*

Proof. Assume that there are only a finite number of primes $p_1, \ldots, p_r$. Let $n = 1 + p_1 \cdot \ldots \cdot p_r$. Then p_i does not divide n, $1 \leq i \leq r$. Thus, either n is a prime or it contains a new prime factor different from p_i, $1 \leq i \leq r$. This is a contradiction. $\qquad\square$

There is the following famous result on the distribution of primes. It is called the *Prime Number Theorem* and was proven by Hadamard and de la Vallée Poussin.

Theorem A.68.
Let $\pi(x) = |\{p \text{ prime} \mid p \leq x\}|$. Then for large x,

$$\pi(x) \approx \frac{x}{\ln(x)}.$$

A proof can be found, for example, in [HarWri79] or [Newman80].

Corollary A.69. *The frequency of primes among the numbers in the magnitude of x is approximately $1/\ln(x)$.*

Sometimes we are interested in primes which have a given remainder c modulo b.

Theorem A.70 *(Dirichlet's Theorem). Let $b, c \in \mathbb{N}$ and $\gcd(b, c) = 1$.*
Let $\pi_{b,c}(x) = |\{p \text{ prime} \mid p \leq x, p = kb + c, k \in \mathbb{N}\}|$. *Then for large x,*

$$\pi_{b,c}(x) \approx \frac{1}{\varphi(b)} \frac{x}{\ln(x)}.$$

Corollary A.71. *Let $b, c \in \mathbb{N}$ and $\gcd(b, c) = 1$. The frequency of primes among the numbers a with $a \bmod b = c$ in the magnitude of x is approximately $b/\varphi(b)\ln(x)$.*

Our goal in this section is to give criteria for prime numbers which can be efficiently checked by an algorithm. We will use the fact that a proper subgroup H of $\mathbb{Z}_n^*$ contains at most $|\mathbb{Z}_n^*|/2 = \varphi(n)/2$ elements. More generally, we prove the following basic result on groups.

Proposition A.72. *Let G be a finite group and $H \subset G$ be a subgroup. Then $|H|$ is a divisor of $|G|$.*

Proof. We consider the following equivalence relation ($\sim$) on G: $g_1 \sim g_2$ if and only if $g_1 \cdot g_2^{-1} \in H$. The equivalence class of an element $g \in G$ is $gH = \{gh \mid h \in H\}$. Thus, all equivalence classes contain the same number, namely $|H|$, of elements. Since G is the disjoint union of the equivalence classes, we have $|G| = |H| \cdot r$, where r is the number of equivalence classes, and we see that $|H|$ divides $|G|$.   □

Fermat's Little Theorem (Theorem A.24) yields a necessary condition for primes. Let $n \in \mathbb{N}$ be an odd number. If n is a prime and $a \in \mathbb{N}$ with $\gcd(a, n) = 1$, then $a^{n-1} \equiv 1 \bmod n$. If there is an $a \in \mathbb{N}$ with $\gcd(a, n) = 1$ and $a^{n-1} \not\equiv 1 \bmod n$, then n is not a prime.
Unfortunately, the converse is not true: there are composite (i.e., non-prime) numbers n, such that $a^{n-1} \equiv 1 \bmod n$, for all $a \in \mathbb{N}$ with $\gcd(a, n) = 1$. The smallest n with this property is $561 = 3 \cdot 11 \cdot 17$.

Definition A.73. *Let $n \in \mathbb{N}, n \geq 3$, be a composite number. We call n a Carmichael number if*
$$a^{n-1} \equiv 1 \bmod n,$$
for all $a \in \mathbb{N}$ with $\gcd(a, n) = 1$.

Proposition A.74. *Let n be a Carmichael number, and let p be a prime that divides n. Then p^2 does not divide n. In other words, the factorization of n does not contain squares.*

Proof. Assume $n = p^k m$, with $k \geq 2$ and p does not divide m. Let $b := 1 + pm$. From $b^p = (1 + pm)^p = 1 + p^2 \cdot \alpha$ we derive that $b^p \equiv 1 \bmod p^2$. Since p does not divide m, we have $b \not\equiv 1 \bmod p^2$ and conclude that b has order p modulo p^2. Now, b is prime to n, because it is prime to p and m, and n is a Carmichael number. Hence $b^{n-1} \equiv 1 \bmod n$, and then, in particular, $b^{n-1} \equiv 1 \bmod p^2$. Thus $p \mid n - 1$ (by Lemma A.32), a contradiction to $p \mid n$.   □

Proposition A.75. *Let n be an odd, composite number that does not contain squares. Then n is a Carmichael number if and only if $(p - 1) \mid (n - 1)$ for all prime divisors p of n.*

Proof. Let $n = p_1 \cdot \ldots \cdot p_r$, with p_i being a prime, $i = 1, \ldots, r$, and $p_i \neq p_j$ for $i \neq j$. n is a Carmichael number if and only if $a^{n-1} \equiv 1 \bmod n$ for all a that are prime to n and, by the Chinese Remainder Theorem this in turn is equivalent to $a^{n-1} \equiv 1 \bmod p_i$, for all a which are not divided by p_i,

$i = 1, \ldots, r$. This is the case if and only if $(p_i - 1) \mid (n - 1)$, for $i = 1, \ldots, r$. The last equivalence follows from Proposition A.23 and Corollary A.33, since $\mathbb{Z}^*_{p_i}$ is a cyclic group of order $p_i - 1$ (Theorem A.36). $\qquad \square$

Corollary A.76. *Every Carmichael number n contains at least three distinct primes.*

Proof. Assume $n = p_1 \cdot p_2$, with $p_1 < p_2$. Then $n - 1 = p_1(p_2 - 1) + (p_1 - 1) \equiv (p_1 - 1) \bmod (p_2 - 1)$. However, $(p_1 - 1) \not\equiv 0 \bmod (p_2 - 1)$, since $0 < p_1 - 1 < p_2 - 1$. Hence, $p_2 - 1$ does not divide $n - 1$. This is a contradiction. $\qquad \square$

Though Carmichael numbers are extremely rare (there are only 2163 Carmichael numbers below $25 \cdot 10^9$), the Fermat condition $a^{n-1} \equiv 1 \bmod p$ is not reliable for a primality test.[7] We are looking for other criteria.

Let $n \in \mathbb{N}$ be an odd number. If n is a prime, by Euler's criterion (Proposition A.52), $\left(\frac{a}{n}\right) \equiv a^{(n-1)/2} \bmod n$ for every $a \in \mathbb{N}$ with $\gcd(a, n) = 1$. Here the converse is also true. More precisely:

Proposition A.77. *Let n be an odd and composite number. Let*

$$\overline{E_n} = \left\{ [a] \in \mathbb{Z}^*_n \,\Big|\, \left(\frac{a}{n}\right) \not\equiv a^{(n-1)/2} \bmod n \right\}.$$

Then $|\overline{E_n}| \geq \varphi(n)/2$, i.e., for more than half of the $[a]$, we have $\left(\frac{a}{n}\right) \not\equiv a^{(n-1)/2} \bmod n$.

Proof. Let $E_n := \mathbb{Z}^*_n \setminus \overline{E_n}$ be the complement of $\overline{E_n}$. We have $E_n = \{[a] \in \mathbb{Z}^*_n \mid \left(\frac{a}{n}\right) \equiv a^{(n-1)/2} \bmod n\} = \{[a] \in \mathbb{Z}^*_n \mid a^{(n-1)/2} \cdot \left(\frac{a}{n}\right)^{-1} \equiv 1 \bmod n\}$. Since E_n is a subgroup of $\mathbb{Z}^*_n$, we could infer $|E_n| \leq \varphi(n)/2$ if E_n were a proper subset of $\mathbb{Z}^*_n$ (Proposition A.72). Then $|\overline{E_n}| = |\mathbb{Z}^*_n| - |E_n| \geq \varphi(n) - \varphi(n)/2 = \varphi(n)/2$.

Thus it suffices to prove: if $E_n = \mathbb{Z}^*_n$, then n is a prime.

Assume $E_n = \mathbb{Z}^*_n$ and n is not a prime. From $\left(\frac{a}{n}\right) \equiv a^{(n-1)/2} \bmod n$, it follows that $a^{n-1} \equiv 1 \bmod n$. Thus n is a Carmichael number, and hence does not contain squares (Proposition A.74). Let $n = p_1 \cdot \ldots \cdot p_k, k \geq 3$, be the decomposition of n into distinct primes. Let $[v] \in \mathbb{Z}^*_{p_1}$ be a quadratic non-residue, i.e., $\left(\frac{v}{p_1}\right) = -1$. Using the Chinese Remainder Theorem, choose an $[x] \in \mathbb{Z}^*_n$ with $x \equiv v \bmod p_1$ and $x \equiv 1 \bmod n/p_1$. Then $\left(\frac{x}{n}\right) = \left(\frac{x}{p_1}\right) \cdot \ldots \cdot \left(\frac{x}{p_n}\right) = -1$. Since $E_n = \mathbb{Z}^*_n$, $\left(\frac{x}{n}\right) \equiv x^{(n-1)/2} \bmod n$, and hence $x^{(n-1)/2} \equiv -1 \bmod n$, in particular $x^{(n-1)/2} \equiv -1 \bmod p_2$. This is a contradiction. $\qquad \square$

The following considerations lead to a necessary condition for primes that does not require the computation of Jacobi symbols. Let $n \in \mathbb{N}$ be an odd number, and let $n - 1 = 2^t m$, with m odd. Suppose that n is a prime. Then $\mathbb{Z}_n$ is a field (Corollary A.17), and hence ± 1 are the only square roots of 1,

---

[7] Numbers n satisfying $a^{n-1} \equiv 1 \bmod n$ are called pseudoprimes for the base a.

i.e., the only solutions of $X^2 - 1$, modulo n. Moreover, $a^{n-1} \equiv 1 \bmod n$ for every $a \in \mathbb{N}$ that is prime to n (Theorem A.24). Thus
$a^{n-1} \equiv 1 \bmod n$ and $a^{(n-1)/2} \equiv \pm 1 \bmod n$, and
if $a^{(n-1)/2} \equiv 1 \bmod n$, then $a^{(n-1)/4} \equiv \pm 1 \bmod n$, and
if $a^{(n-1)/4} \equiv 1 \bmod n$, then ....
We see: if n is a prime, then for every $a \in \mathbb{N}$ with $\gcd(a, n) = 1$, either $a^m \equiv 1 \bmod n$, or there is a $j \in \{0, \ldots, t-1\}$ with $a^{2^j m} \equiv -1 \bmod n$. The converse is also true, i.e., if n is composite, then there exists an $a \in \mathbb{N}$ with $\gcd(a, n) = 1$, such that $a^m \not\equiv 1 \bmod n$ and $a^{2^j m} \not\equiv -1 \bmod n$ for $0 \leq j \leq t - 1$. More precisely:

Proposition A.78. *Let $n \in \mathbb{N}$ be a composite odd number. Let $n - 1 = 2^t m$, with m odd. Let*

$$\overline{W_n} = \{[a] \in \mathbb{Z}_n^* \mid a^m \not\equiv 1 \bmod n \text{ and } a^{2^j m} \not\equiv -1 \bmod n \text{ for } 0 \leq j \leq t-1\}.$$

Then $|\overline{W_n}| \geq \varphi(n)/2$.

Proof. Let $W_n := \mathbb{Z}_n^* \setminus \overline{W_n}$ be the complement of $\overline{W_n}$. We will show that W_n is contained in a proper subgroup U of $\mathbb{Z}_n^*$. Then the desired estimate follows by Proposition A.72, as in the proof of Proposition A.77. We distinguish two cases.
Case 1: There is an $[a] \in \mathbb{Z}_n^*$ with $a^{n-1} \not\equiv 1 \bmod n$.
Then $U = \{[a] \in \mathbb{Z}_n^* \mid a^{n-1} \equiv 1 \bmod n\}$ is a proper subgroup of $\mathbb{Z}_n^*$, which contains W_n, and the proof is finished.
Case 2: We have $a^{n-1} \equiv 1 \bmod n$ for all $[a] \in \mathbb{Z}_n^*$.
Then n is a Carmichael number. Hence n does not contain any squares (Proposition A.74). Let $n = p_1 \cdot \ldots \cdot p_k, k \geq 3$, be the decomposition into distinct primes. We set

$$W_n^i = \{[a] \in \mathbb{Z}_n^* \mid a^{2^i m} \equiv -1 \bmod n\}.$$

W_n^0 is not empty, since $[-1] \in W_n^0$. Let $r = \max\{i \mid W_n^i \neq \emptyset\}$ and

$$U := \{[a] \in \mathbb{Z}_n^* \mid a^{2^r m} \equiv \pm 1 \bmod n\}.$$

U is a subgroup of $\mathbb{Z}_n^*$ and $W_n \subset U$. Let $[a] \in W_n^r$. Using the Chinese Remainder Theorem, we get a $[w] \in \mathbb{Z}_n^*$ with $w \equiv a \bmod p_1$ and $w \equiv 1 \bmod n/p_1$. Then $w^{2^r m} \equiv -1 \bmod p_1$ and $w^{2^r m} \equiv +1 \bmod p_2$, hence $w^{2^r m} \not\equiv \pm 1 \bmod n$. Thus $w \notin U$, and we see that U is indeed a proper subgroup of $\mathbb{Z}_n^*$. $\square$

Remark. The set $\overline{E_n}$ from Proposition A.77 is a subset of $\overline{W_n}$, and it can even be proven that $|\overline{W_n}| \geq \frac{3}{4}\varphi(n)$ (see, e.g., [Koblitz94]).

Probabilistic Primality Tests. The preceding propositions are the basis of probabilistic algorithms which test whether a given odd number is prime. Proposition A.77 yields the *Solovay-Strassen primality test*, and Proposition A.78 yields the *Miller-Rabin primality test*. The basic procedure is the same in both tests: we define a set W of *witnesses* for the fact that n is composite. Set $W := \overline{E_n}$ (Solovay-Strassen) or $W := \overline{W_n}$ (Miller-Rabin). If we can find a $w \in W$, then $W \neq \emptyset$, and n is a composite number.

To find a witness $w \in W$, we randomly choose (with respect to the uniform distribution) an element $a \in \mathbb{Z}_n^*$ and check whether $a \in W$. Since $|W| \geq \varphi(n)/2$ (Propositions A.77 and A.78), the probability that we get a witness by the random choice, if n is composite, is $\geq 1/2$. By repeating the random choice k times, we can increase the probability of finding a witness if n is composite. The probability is then $\geq 1 - 1/2^k$. If we do not find a witness, n is considered to be a prime.

The tests are probabilistic and the result is not necessarily correct in all cases. However, the error probability is $\leq 1/2^k$ and hence very small, even for moderate values of k.

Remark. The primality test of Miller-Rabin is the better choice:

1. The test condition is easier to compute.
2. A witness for the Solovay-Strassen test is also a witness for the Miller-Rabin test.
3. In the Miller-Rabin test, the probability of obtaining a witness by one random choice is $\geq 3/4$ (we only proved the weaker bound of $1/2$).

Here follows the algorithm for the Miller-Rabin test (with error probability $\leq 1/4^k$).

Algorithm A.79.

```
boolean MillerRabinTest(int n, k)
 1   if n is even
 2      then return false
 3   m ← (n − 1) div 2; t ← 1
 4   while m is even do
 5       m ← m div 2; t ← t + 1
 6   for i ← 1 to k do
 7       a ← Random() mod n
 8       u ← a^m mod n
 9       if u ≠ 1
10          then j ← 1
11               while u ≠ −1 and j < t do
12                   u ← u² mod n; j ← j + 1
13               if u ≠ −1
14                  then return false
15   return true ;
```

B. Probabilities and Information Theory

We review some basic notions and results using in this book concerning probability, probability spaces and random variables. This chapter is also intended to establish notation. There are many textbooks on probability theory, including [Bauer96], [Feller68], [GanYlv67], [Gordon97] and [Rényi70].

B.1 Finite Probability Spaces and Random Variables

First we summarize basic concepts and notations. At the end of this section we derive a couple of elementary results. They are useful in our computations with probabilities. We consider only finite probability spaces.

Definition B.1.

1. A *probability distribution* (or simply *distribution*) $p = (p_1, \ldots, p_n)$ is a tuple of elements $p_i \in \mathbb{R}, 0 \leq p_i \leq 1$, called *probabilities*, such that $\sum_{i=1}^{n} p_i = 1$.
2. A *probability space* (X, p_X) is a finite set $X = \{x_1, \ldots, x_n\}$ equipped with a probability distribution $p_X = (p_1, \ldots, p_n)$. p_i is called the probability of $x_i, 1 \leq i \leq n$. We also write $p_X(x_i) := p_i$ and consider p_X as a map $X \to [0,1]$, called the *probability measure* on X, associating with $x \in X$ its probability.
3. An *event* $\mathcal{E}$ in a probability space (X, p_X) is a subset $\mathcal{E}$ of X. The probability measure is extended to events:

$$p_X(\mathcal{E}) = \sum_{y \in \mathcal{E}} p_X(y).$$

Example. Let X be a finite set. The *uniform distribution* $p_{X,u}$ is defined by $p_{X,u}(x) := 1/|X|$, for all $x \in X$. All elements of X have the same probability.

Notation. If the probability measure is determined by the context, we often do not specify it explicitly and simply write X instead of (X, p_X) and $\text{prob}(x)$ or $\text{prob}(\mathcal{E})$ instead of $p_X(x)$ or $p_X(\mathcal{E})$. If $\mathcal{E}$ and $\mathcal{F}$ are events, we write $\text{prob}(\mathcal{E}, \mathcal{F})$ instead of $\text{prob}(\mathcal{E} \cap \mathcal{F})$ for the probability that both events $\mathcal{E}$ and $\mathcal{F}$ occur. Separating events by commas means combining them with

AND. The events $\{x\}$ containing a single element are simply denoted by x. For example, $\mathrm{prob}(x, \mathcal{E})$ means the probability of $\{x\} \cap \mathcal{E}$ (which is 0 if $x \notin \mathcal{E}$ and $\mathrm{prob}(x)$ otherwise).

Remark. Let X be a probability space. The following properties of the probability measure are immediate consequences of Definition B.1:

1. $\mathrm{prob}(X) = 1$, $\mathrm{prob}(\emptyset) = 0$.
2. $\mathrm{prob}(\mathcal{A} \cup \mathcal{B}) = \mathrm{prob}(\mathcal{A}) + \mathrm{prob}(\mathcal{B})$, if $\mathcal{A}$ and $\mathcal{B}$ are disjoint events in X.
3. $\mathrm{prob}(X \setminus \mathcal{A}) = 1 - \mathrm{prob}(\mathcal{A})$.

Definition B.2. Let $S : X \longrightarrow Y$ be a map from a probability space (X, p_X) to a set Y. Then S and p_X induce a distribution $S(p_X)$ on Y:

$$S(p_X)(y) := p_X(S^{-1}(y)) := p_X(\{x \in X \mid S(x) = y\}).$$

The distribution $S(p_X)$ is called the *image (distribution)* of p_X under S.

Definition B.3. Let (X, p_X) be a probability space. A map $S : X \longrightarrow Y$ is called a Y-valued *random variable* on X. The *distribution* p_S *of a random variable* S is the image of p_X under S:

$$p_S(y) := S(p_X)(y) = p_X(\{x \in X \mid S(x) = y\}) \text{ for } y \in Y.$$

We call S a *real-valued random variable* if $Y \subset \mathbb{R}$, and a *binary random variable* if $Y = \{0, 1\}$. Binary random variables are also called *Boolean predicates*.

Remark. Considering the distribution of a random variable $S : X \longrightarrow Y$ means considering the distribution of the probability space induced as image on Y by S.

Vice versa, it is sometimes convenient (and common practice in probability theory) to look at a probability space as if it were a random variable. Namely, consider (X, p_X) as a random variable S_X that is defined on some not-further-specified probability space (Ω, p_Ω) and samples over X according to the given distribution:

$$S_X : \Omega \longrightarrow X \text{ and } p_X = S_X(p_\Omega).$$

Definition B.4. Let S be a real-valued random variable on a finite probability space X. The probabilistic average

$$E(S) := \sum_{x \in X} \mathrm{prob}(x) \cdot S(x)$$

is called the *expected value* of S.

Remark. The expected value $E(S)$ is a weighted average. The weight of each value is its probability.

Definition B.5. Let (X, p_X) be a probability space and $\mathcal{A}, \mathcal{B} \subseteq X$ be events, with $p_X(\mathcal{B}) > 0$. The *conditional probability* of $\mathcal{A}$ assuming $\mathcal{B}$ is

$$p_X(\mathcal{A} \mid \mathcal{B}) := \frac{p_X(\mathcal{A}, \mathcal{B})}{p_X(\mathcal{B})}.$$

In particular, we have

$$p_X(x \mid \mathcal{B}) = \begin{cases} p_X(x)/p_X(\mathcal{B}) & \text{if } x \in \mathcal{B}, \\ 0 & \text{if } x \notin \mathcal{B}. \end{cases}$$

The conditional probabilities $p_X(x \mid \mathcal{B}), x \in X$, define a probability distribution on X. They describe the probability of x assuming that event $\mathcal{B}$ occurs.

If $\mathcal{C}$ is a further event, then $p_X(\mathcal{A} \mid \mathcal{B}, \mathcal{C})$ is the conditional probability of $\mathcal{A}$ assuming $\mathcal{B} \cap \mathcal{C}$. Separating events by commas in a condition means combining them with AND.

Definition B.6. Let $\mathcal{A}, \mathcal{B} \subseteq X$ be events in a probability space (X, p_X). $\mathcal{A}$ and $\mathcal{B}$ are called *independent* if and only if $\mathrm{prob}(\mathcal{A}, \mathcal{B}) = \mathrm{prob}(\mathcal{A}) \cdot \mathrm{prob}(\mathcal{B})$. If $\mathrm{prob}(\mathcal{B}) > 0$, then this condition is equivalent to $\mathrm{prob}(\mathcal{A} \mid \mathcal{B}) = \mathrm{prob}(\mathcal{A})$.

Definition B.7. A probability space (X, p_X) is called a *joint probability space* with factors $(X_1, p_1), \ldots, (X_r, p_r)$, denoted for short by $X_1 X_2 \ldots X_r$, if:

1. The set X is the Cartesian product of the sets $X_1, \ldots, X_r$:

$$X = X_1 \times X_2 \times \ldots \times X_r.$$

2. The distribution $p_i, 1 \leq i \leq r$, is the image of p_X under the projection

$$\pi_i : X \longrightarrow X_i, \ (x_1, \ldots, x_r) \longmapsto x_i,$$

which means

$$p_i(x) = p_X(\pi_i^{-1}(x)), \text{ for } 1 \leq i \leq r \text{ and } x \in X_i.$$

The probability spaces $X_1, \ldots, X_r$ are called *independent* if and only if

$$p_X(x_1, \ldots, x_r) = \prod_{i=1}^{r} p_i(x_i), \text{ for all } (x_1, \ldots, x_r) \in X$$

(or, equivalently, if the fibers $\pi_i^{-1}(x_i), 1 \leq i \leq r,$[1] are independent events in X, in the sense of Definition B.6). In this case X is called the *direct product* of the X_i, denoted for short by $X = X_1 \times X_2 \times \ldots \times X_r$.

---

[1] The set $\pi^{-1}(x)$ of pre-images of a single element x under a projection map π is called the *fiber* over x.

Analogously, random variables $S_1, \ldots, S_r$ are called *jointly distributed* if there is a joint probability distribution p_S of $S_1, \ldots, S_r$:

$$\text{prob}(S_1 = x_1, S_2 = x_2, \ldots, S_r = x_r) = p_S(x_1, \ldots, x_r).$$

They are called *independent* if and only if

$$\text{prob}(S_1 = x_1, S_2 = x_2, \ldots, S_r = x_r) = \prod_{i=1}^{r} \text{prob}(S_i = x_i),$$

for all $x_1, \ldots, x_r$.

Remark. Recall that the distribution of a random variable may be considered as the distribution of a probability space, and vice versa (see above). In this way, joint probability spaces correspond to jointly distributed random variables.

Notation. Let (XY, p_{XY}) be a joint probability space with factors (X, p_X) and (Y, p_Y). Let $x \in X$ and $y \in Y$. Then we denote by $\text{prob}(y \,|\, x)$ the conditional probability $p_{XY}((x, y) \,|\, \{x\} \times Y)$ of (x, y), assuming that the first component is x. Thus, $\text{prob}(y \,|\, x)$ is the probability that y occurs as the second component if the first component is x.

Remark. In this book we often meet joint probability spaces in the following way: a set X and a family $W = (W_x)_{x \in X}$ of sets are given. Then we may join the set X with the family W, to get

$$X \bowtie W := \{(x, w) \mid x \in X, w \in W_x\} = \bigcup_{x \in X} \{x\} \times W_x.$$

The set W_x becomes the fiber over x.

Assume that probability distributions p_X on X and p_{W_x} on $W_x, x \in X$, are given. Then we get a joint probability distribution p_{XW} on $X \bowtie W$ using

$$p_{XW}(x, w) := p_X(x) \cdot p_{W_x}(w).$$

Conversely, given a distribution p_{XW} on $X \bowtie W$, we can project $X \bowtie W$ to X. We get the image distribution p_X on X using

$$p_X(x) := \sum_{w \in W_x} p_{XW}(x, w)$$

and the probability distributions p_{W_x} on $W_x, x \in X$, using

$$p_{W_x}(w) = p_{XW}((x, w) \,|\, \{x\} \times W_x) = {p_{XW}(x, w)}/{p_X(x)}.$$

The probabilities p_{W_x} are conditional probabilities.
$\text{prob}(w \,|\, x) := p_{W_x}(w)$ is the conditional probability that w occurs as the second component, assuming that the first component x is given.

We write XW for short for the probability space $(X \bowtie W, p_{XW})$, as before, and we call it a joint probability space. At first glance it does not meet Definition B.7, because the underlying set is not a Cartesian product (except in the case where all sets W_x are equal). However, all sets are assumed to be finite. Therefore, we can easily embed all the sets W_x into one larger set $\tilde{W}$. Then $X \bowtie W \subseteq X \times \tilde{W}$, and p_{XW} may also be considered as a probability distribution on $X \times \tilde{W}$ (extend it by zero, so that all elements outside $X \bowtie W$ have a probability of 0). In this way we get a joint probability space $X\tilde{W}$ in the strict sense of Definition B.7. XW and $X\tilde{W}$ are practically the "same" as probability spaces (the difference has a measure of 0).

As an example, think of the domain of the modular squaring one-way function (Section 3.2) $(n, x) \mapsto x^2 \bmod n$, where $n \in I_k = \{pq \mid p, q$ distinct primes, $|p| = |q| = k\}$ and $x \in \mathbb{Z}_n$, and assume that the moduli n (the keys, see Rabin encryption, Section 3.6.1) and the elements x (the messages) are selected according to some probability distributions (e.g. the uniform ones). Here $X = I_k$ and $W = (\mathbb{Z}_n)_{n \in I_k}$.

The joining may be iterated. Let $X_1 = (X_1, p_1)$ be a probability space and let $X_j = (X_{j,x}, p_{j,x})_{x \in X_1 \bowtie \ldots \bowtie X_{j-1}}, 2 \leq j \leq r$, be families of probability spaces. Then by iteratively joining the fibers we get a joint probability space $(X_1 X_2 \ldots X_r, p_{X_1 X_2 \ldots X_r})$.

Notation. We introduce some notation which turns out to be very useful in many situations.

Let (X, p_X) be a probability space and $B : X \longrightarrow \{0, 1\}$ be a Boolean predicate. Then

$$\text{prob}(B(x) = 1 : x \overset{p_X}{\leftarrow} X) := p_X(\{x \in X \mid B(x) = 1\}).$$

The notation suggests that $p_X(\{x \in X \mid B(x) = 1\})$ is the probability for $B(x) = 1$ if x is randomly selected from X according to p_X. If the distribution p_X is clear from the context, we simply write

$$\text{prob}(B(x) = 1 : x \leftarrow X)$$

instead of $\text{prob}(B(x) : x \overset{p_X}{\leftarrow} X)$. If p_X is the uniform distribution, we write

$$\text{prob}(B(x) = 1 : x \overset{u}{\leftarrow} X).$$

Sometimes, we denote the probability distribution on X by $x \leftarrow X$ and the uniform distribution by $x \overset{u}{\leftarrow} X$. We emphasize in this way that the members of X are chosen randomly.

If $Y \subset X$ and p_Y is a distribution on Y, then the notation $x \leftarrow Y$ is not only used for the distribution p_Y, but also for the image of p_Y on X (under the inclusion map). This intuitively suggests that elements x are randomly chosen from the subset Y, whereas elements outside of Y have zero probability and are not chosen.

As before, let $S(p_X)$ denote the image distribution under a map $S : X \longrightarrow Y$. Then

$$\text{prob}(S(x) = y : x \leftarrow X) = S(p_X)(y) = p_X(\{x \in X \mid S(x) = y\}), \text{ for } y \in Y.$$

Sometimes we denote the image distribution by $\{S(x) : x \leftarrow X\}$. This notation indicates that the probability measure is concentrated on the image of X – only the elements $S(x)$ in Y can have a probability > 0 – and that the probability of $y \in Y$ is given by the probability for y appearing as $S(x)$, if x is randomly selected from X.

Now let (XW, p_{XW}) be a joint probability space, $W = (W_x)_{x \in X}$, as introduced above, and let p_X and $p_{W_x}, x \in X$, be the probability distributions induced on X and the fibers W_x. We write

$$\text{prob}(B(x, w) = 1 : x \overset{p_X}{\leftarrow} X, w \overset{p_{W_x}}{\leftarrow} W_x), \text{ or simply}$$

$$\text{prob}(B(x, w) = 1 : x \leftarrow X, w \leftarrow W_x)$$

instead of $p_{XW}(\{(x, w) \mid B(x, w) = 1\})$. Here, B is a Boolean predicate on $X \bowtie W$. The notation suggests that we mean the probability for $B(x, w) = 1$, if first x is randomly selected and then w is randomly selected from W_x.

More generally, if $(X_1 X_2 \ldots X_r, p_{X_1 X_2 \ldots X_r})$ is formed by iteratively joining fibers (see above; we also use the notation from above), then we write

$$\text{prob}(B(x_1, \ldots, x_r) = 1 : x_1 \leftarrow X_1, x_2 \leftarrow X_{2, x_1},$$
$$x_3 \leftarrow X_{3, x_1 x_2}, \quad \ldots \quad , x_r \leftarrow X_{r, x_1 \ldots x_{r-1}})$$

instead of $p_{X_1 X_2 \ldots X_r}(\{(x_1, \ldots, x_r) \mid B(x_1, \ldots, x_r) = 1\})$. Again we write more precisely $\overset{u}{\leftarrow}$ (or $\overset{p}{\leftarrow}$) instead of $\leftarrow$ if the distribution is the uniform one (or not clear from the context).

The distribution $x_j \leftarrow X_{j, x_1 \ldots x_{j-1}}$ is the conditional distribution of $x_j \in X_{j, x_1 \ldots x_{j-1}}$, assuming $x_1, \ldots, x_{j-1}$; i.e., it gives the conditional probabilities $\text{prob}(x_j \mid x_1, \ldots, x_{j-1})$. We have (for $r = 3$) that

$$\text{prob}(B(x_1, x_2, x_3) = 1 : x_1 \leftarrow X_1, x_2 \leftarrow X_{2, x_1}, x_3 \leftarrow X_{3, x_1 x_2})$$
$$= \sum_{(x_1, x_2, x_3) : B(x_1, x_2, x_3) = 1} \text{prob}(x_1) \cdot \text{prob}(x_2 \mid x_1) \cdot \text{prob}(x_3 \mid x_2, x_1).$$

We now derive a couple of elementary results. They are used in our computations with probabilities.

Proposition B.8. *Let X be a finite probability space, and let X be the disjoint union of events $\mathcal{E}_1, \ldots, \mathcal{E}_r \subseteq X$, with $\text{prob}(\mathcal{E}_i) > 0$ for $i = 1 \ldots r$. Then*

$$\text{prob}(\mathcal{A}) = \sum_{i=1}^{r} \text{prob}(\mathcal{E}_i) \cdot \text{prob}(\mathcal{A} \mid \mathcal{E}_i)$$

for every event $\mathcal{A} \subseteq X$.

Proof.

$$\text{prob}(\mathcal{A}) = \sum_{i=1}^{r} \text{prob}(\mathcal{A} \cap \mathcal{E}_i) = \sum_{i=1}^{r} \text{prob}(\mathcal{E}_i) \cdot \text{prob}(\mathcal{A} \,|\, \mathcal{E}_i).$$

□

Lemma B.9. *Let $\mathcal{A}, \mathcal{B}, \mathcal{E} \subseteq X$ be events in a probability space X, with $\text{prob}(\mathcal{E}) > 0$. Suppose that $\mathcal{A}$ and $\mathcal{B}$ have the same conditional probability assuming $\mathcal{E}$, i.e., $\text{prob}(\mathcal{A} \,|\, \mathcal{E}) = \text{prob}(\mathcal{B} \,|\, \mathcal{E})$. Then*

$$|\text{prob}(\mathcal{A}) - \text{prob}(\mathcal{B})| \leq \text{prob}(X \setminus \mathcal{E}).$$

Proof. By Proposition B.8, we have

$$\text{prob}(\mathcal{A}) = \text{prob}(\mathcal{E}) \cdot \text{prob}(\mathcal{A} \,|\, \mathcal{E}) + \text{prob}(X \setminus \mathcal{E}) \cdot \text{prob}(\mathcal{A} \,|\, X \setminus \mathcal{E})$$

and an analogous equality for $\text{prob}(\mathcal{B})$. Subtracting both equalities, we get the desired inequality. □

Lemma B.10. *Let $\mathcal{A}, \mathcal{E} \subseteq X$ be events in a probability space X, with $\text{prob}(\mathcal{E}) > 0$. Then*

$$|\,\text{prob}(\mathcal{A}) - \text{prob}(\mathcal{A} \,|\, \mathcal{E})\,| \leq \text{prob}(X \setminus \mathcal{E}).$$

Proof. By Proposition B.8, we have

$$\text{prob}(\mathcal{A}) = \text{prob}(\mathcal{E}) \cdot \text{prob}(\mathcal{A} \,|\, \mathcal{E}) + \text{prob}(X \setminus \mathcal{E}) \cdot \text{prob}(\mathcal{A} \,|\, X \setminus \mathcal{E}).$$

Hence,

$$\begin{aligned}
|\,\text{prob}(\mathcal{A}) &- \text{prob}(\mathcal{A} \,|\, \mathcal{E})\,| \\
&= \text{prob}(X \setminus \mathcal{E}) \cdot |\,\text{prob}(\mathcal{A} \,|\, X \setminus \mathcal{E}) - \text{prob}(\mathcal{A} \,|\, \mathcal{E})\,| \\
&\leq \text{prob}(X \setminus \mathcal{E}),
\end{aligned}$$

as desired. □

We continue with two results on expected values of random variables.

Proposition B.11. *Let R and S be real-valued random variables:*

1. $E(R + S) = E(R) + E(S)$.
2. If R and S are independent, then $E(R \cdot S) = E(R) \cdot E(S)$.

Proof. Let $W_R, W_S \subset \mathbb{R}$ be the images of R and S. Since we consider only finite probability spaces, W_R and W_S are finite sets. We have

$$E(R + S) = \sum_{x \in W_R, y \in W_S} \text{prob}(R = x, S = y) \cdot (x + y)$$

$$= \sum_{x \in W_R, y \in W_S} \text{prob}(R = x, S = y) \cdot x$$

$$+ \sum_{x \in W_R, y \in W_S} \text{prob}(R = x, S = y) \cdot y$$

$$= \sum_{x \in W_R} \text{prob}(R = x) \cdot x + \sum_{y \in W_S} \text{prob}(S = y) \cdot y$$

$$= E(R) + E(S),$$

and if R and S are independent,

$$E(R \cdot S) = \sum_{x \in W_R, y \in W_S} \text{prob}(R = x, S = y) \cdot x \cdot y$$

$$= \sum_{x \in W_R, y \in W_S} \text{prob}(R = x) \cdot \text{prob}(S = y) \cdot x \cdot y$$

$$= \left(\sum_{x \in W_R} \text{prob}(R = x) \cdot x \right) \cdot \left(\sum_{y \in W_S} \text{prob}(S = y) \cdot y \right)$$

$$= E(R) \cdot E(S),$$

as desired.   □

A probability space X is the model of a random experiment. n independent repetitions of the random experiment are modeled by the direct product $X^n = X \times \ldots \times X$. The following lemma answers the question as to how often we have to repeat the experiment until a given event is expected to occur.

Lemma B.12. *Let $\mathcal{E}$ be an event in a probability space X, with $\text{prob}(\mathcal{E}) = p > 0$. Repeatedly, we execute the random experiment X independently. Let G be the number of executions of X, until $\mathcal{E}$ occurs the first time. Then the expected value $E(G)$ of the random variable G is $1/p$.*

Proof. We have $\text{prob}(G = t) = p \cdot (1 - p)^{t-1}$. Hence,

$$E(G) = \sum_{t=1}^{\infty} t \cdot p \cdot (1 - p)^{t-1} = -p \cdot \frac{d}{dp} \sum_{t=1}^{\infty} (1 - p)^t = -p \cdot \frac{d}{dp} \frac{1}{p} = \frac{1}{p}.$$

□

Remark. G is called the *geometric random variable* with respect to p.

Lemma B.13. *Let R, S and B be jointly distributed random variables with values in $\{0, 1\}$. Assume that B and S are independent and that B is uniformly distributed: $\text{prob}(B = 0) = \text{prob}(B = 1) = 1/2$. Then*

$$\text{prob}(R = S) = \frac{1}{2} + \text{prob}(R = B \,|\, S = B) - \text{prob}(R = B).$$

Proof. We denote by $\overline{B}$ the complementary value $1 - B$ of B. First we observe that $\mathrm{prob}(S = B) = \mathrm{prob}(S = \overline{B}) = 1/2$, since B is uniformly distributed and independent of S:

$$\mathrm{prob}(S = B)$$
$$= \mathrm{prob}(S = 0) \cdot \mathrm{prob}(B = 0 \,|\, S = 0) + \mathrm{prob}(S = 1) \cdot \mathrm{prob}(B = 1 \,|\, S = 1)$$
$$= \mathrm{prob}(S = 0) \cdot \mathrm{prob}(B = 0) + \mathrm{prob}(S = 1) \cdot \mathrm{prob}(B = 1)$$
$$= (\mathrm{prob}(S = 0) + \mathrm{prob}(S = 1)) \cdot \frac{1}{2} = \frac{1}{2}.$$

Now, we compute

$$\mathrm{prob}(R = S)$$
$$= \frac{1}{2} \cdot \mathrm{prob}(R = B \,|\, S = B) + \frac{1}{2} \cdot \mathrm{prob}(R = \overline{B} \,|\, S = \overline{B})$$
$$= \frac{1}{2} \cdot (\mathrm{prob}(R = B \,|\, S = B) + 1 - \mathrm{prob}(R = B \,|\, S = \overline{B}))$$
$$= \frac{1}{2} + \frac{1}{2} \cdot (\mathrm{prob}(R = B \,|\, S = B) - \mathrm{prob}(R = B \,|\, S = \overline{B}))$$
$$= \frac{1}{2} + \frac{1}{2} \cdot \left(\mathrm{prob}(R = B \,|\, S = B) \right.$$
$$\left. - \frac{\mathrm{prob}(R = B) - \mathrm{prob}(S = B) \cdot \mathrm{prob}(R = B \,|\, S = B)}{\mathrm{prob}(S = \overline{B})} \right)$$
$$= \frac{1}{2} + \frac{1}{2} \cdot (\mathrm{prob}(R = B \,|\, S = B)$$
$$- (2 \cdot \mathrm{prob}(R = B) - \mathrm{prob}(R = B \,|\, S = B)))$$
$$= \frac{1}{2} + \mathrm{prob}(R = B \,|\, S = B) - \mathrm{prob}(R = B),$$

and the lemma is proven. □

B.2 The Weak Law of Large Numbers

The variance of a random variable S measures how much the values of S on average differ from the expected value.

Definition B.14. Let S be a real-valued random variable. The expected value $E((S - E(S))^2)$ of $(S - E(S))^2$ is called the *variance* of S or, for short, $Var(S)$.

If the variance is small, the probability that the value of S is far from $E(S)$ is low. This fundamental fact is stated precisely in Chebyshev's inequality:

Theorem B.15 *(Chebyshev's inequality). Let S be a real-valued random variable with expected value α and variance σ^2. Then for every $\varepsilon > 0$,*

$$\mathrm{prob}(|S - \alpha| \geq \varepsilon) \leq \frac{\sigma^2}{\varepsilon^2}.$$

Proof. Let $W_S \subset \mathbb{R}$ be the image of S. Since we consider only finite probability spaces, W_S is a finite set. Computing

$$\sigma^2 = E((S - \alpha)^2) = \sum_{x \in W_S} \mathrm{prob}(S = x) \cdot (x - \alpha)^2$$

$$\geq \sum_{x \in W_S, |x - \alpha| \geq \varepsilon} \mathrm{prob}(S = x) \cdot \varepsilon^2 = \varepsilon^2 \cdot \mathrm{prob}(|S - \alpha| \geq \varepsilon),$$

we obtain Chebyshev's inequality. $\qquad\square$

Proposition B.16 *(weak law of large numbers). Let $S_1, \ldots, S_t$ be pairwise-independent real-valued random variables, with a common expected value and a common variance: $E(S_i) = \alpha$ and $Var(S_i) = \sigma^2, i = 1, \ldots, t$. Then for every $\varepsilon > 0$,*

$$\mathrm{prob}\left(\left| \frac{1}{t} \sum_{i=1}^{t} S_i - \alpha \right| < \varepsilon \right) \geq 1 - \frac{\sigma^2}{t\varepsilon^2}.$$

Proof. Let $Z = \frac{1}{t} \sum_{i=1}^{t} S_i$. Then we have $E(Z) = \alpha$ and

$$Var(Z) = E((Z - \alpha)^2) = E\left(\left(\frac{1}{t} \sum_{i=1}^{t} (S_i - \alpha) \right)^2 \right)$$

$$= \frac{1}{t^2} E\left(\sum_{i=1}^{t} (S_i - \alpha)^2 + \sum_{i \neq j} (S_i - \alpha) \cdot (S_j - \alpha) \right)$$

$$= \frac{1}{t^2} \sum_{i=1}^{t} E((S_i - \alpha)^2) + \sum_{i \neq j} E((S_i - \alpha) \cdot (S_j - \alpha))$$

$$= \frac{1}{t^2} \sigma^2 t = \frac{\sigma^2}{t}.$$

Here observe that for $i \neq j$, $E((S_i - \alpha) \cdot (S_j - \alpha)) = E(S_i - \alpha) \cdot E(S_j - \alpha)$, since S_i and S_j are independent, and $E(S_i - \alpha) = E(S_i) - \alpha = 0$ (Proposition B.11).

From Chebyshev's inequality (Theorem B.15), we conclude that

$$\mathrm{prob}(|Z - \alpha| \geq \varepsilon) \leq \frac{\sigma^2}{t\varepsilon^2},$$

and the weak law of large numbers is proven. $\qquad\square$

Remark. In particular, the weak law of large numbers may be applied to t executions of the random experiment underlying a random variable S. It says that the mean of the observed values of S comes close to the expected value of S if the executions are pairwise-independent (or even independent) and its number t is sufficiently large.

If S is a binary random variable observing whether an event occurs or not, there is an upper estimate for the variance, and we get the following corollary.

Corollary B.17. *Let $S_1, \ldots, S_t$ be pairwise-independent binary random variables, with a common expected value and a common variance: $E(S_i) = \alpha$ and $Var(S_i) = \sigma^2, i = 1, \ldots, t$. Then for every $\varepsilon > 0$,*

$$\mathrm{prob}\left(\left| \frac{1}{t} \sum_{i=1}^{t} S_i - \alpha \right| < \varepsilon \right) \geq 1 - \frac{1}{4t\varepsilon^2}.$$

Proof. Since S_i is binary, we have $S_i = S_i^2$ and get

$$Var(S_i) = E((S_i - \alpha)^2) = E(S_i^2 - 2\alpha S_i + \alpha^2)$$
$$= E(S_i^2) - 2\alpha E(S_i) + \alpha^2 = \alpha - 2\alpha^2 + \alpha^2 = \alpha(1 - \alpha) \leq \frac{1}{4}.$$

Now apply Proposition B.16. □

Corollary B.18. *Let $S_1, \ldots, S_t$ be pairwise-independent binary random variables, with a common expected value and a common variance: $E(S_i) = \alpha$ and $Var(S_i) = \sigma^2, i = 1, \ldots, t$. Assume $E(S_i) = \alpha = 1/2 + \varepsilon, \varepsilon > 0$. Then*

$$\mathrm{prob}\left(\sum_{i=1}^{t} S_i > \frac{t}{2} \right) \geq 1 - \frac{1}{4t\varepsilon^2}.$$

Proof.

$$\text{If } \left| \frac{1}{t} \sum_{i=1}^{t} S_i - \alpha \right| < \varepsilon, \text{ then } \frac{1}{t} \sum_{i=1}^{t} S_i > \frac{1}{2}.$$

We conclude from Corollary B.17 that

$$\mathrm{prob}\left(\sum_{i=1}^{t} S_i > \frac{t}{2} \right) \geq \mathrm{prob}\left(\left| \frac{1}{t} \sum_{i=1}^{t} S_i - \alpha \right| < \varepsilon \right) \geq 1 - \frac{1}{4t\varepsilon^2},$$

and the corollary is proven. □

B.3 Distance Measures

We define the distance between probability distributions. Further, we prove some statements on the behavior of negligible probabilities when the probability distribution is varied a little.

Definition B.19. Let p and $\tilde{p}$ be probability distributions on a finite set X. The *statistical distance* between p and $\tilde{p}$ is

$$\text{dist}(p, \tilde{p}) := \frac{1}{2} \sum_{x \in X} |p(x) - \tilde{p}(x)|.$$

Remark. The statistical distance defines a metric on the set of distributions on X.

Lemma B.20. *The statistical distance between probability distributions p and $\tilde{p}$ on a finite set X is the maximal distance between the probabilities of events in X, i.e.,*

$$\text{dist}(p, \tilde{p}) = \max_{\mathcal{E} \subseteq X} |p(\mathcal{E}) - \tilde{p}(\mathcal{E})|.$$

Proof. Recall that the events in X are the subsets of X. Let

$$\mathcal{E}_1 := \{x \in X \mid p(x) > \tilde{p}(x)\}, \quad \mathcal{E}_2 := \{x \in X \mid p(x) < \tilde{p}(x)\},$$

$$\mathcal{E}_3 := \{x \in X \mid p(x) = \tilde{p}(x)\}.$$

Then

$$0 = p(X) - \tilde{p}(X) = \sum_{i=1}^{3} p(\mathcal{E}_i) - \tilde{p}(\mathcal{E}_i),$$

and $p(\mathcal{E}_3) - \tilde{p}(\mathcal{E}_3) = 0$. Hence, $p(\mathcal{E}_2) - \tilde{p}(\mathcal{E}_2) = -(p(\mathcal{E}_1) - \tilde{p}(\mathcal{E}_1))$, and we obviously get

$$\max_{\mathcal{E} \subseteq X} |p(\mathcal{E}) - \tilde{p}(\mathcal{E})| = p(\mathcal{E}_1) - \tilde{p}(\mathcal{E}_1) = -(p(\mathcal{E}_2) - \tilde{p}(\mathcal{E}_2)).$$

We compute

$$
\begin{aligned}
\text{dist}(p, \tilde{p}) &= \frac{1}{2} \sum_{x \in X} |p(x) - \tilde{p}(x)| \\
&= \frac{1}{2} \left(\sum_{x \in \mathcal{E}_1} (p(x) - \tilde{p}(x)) - \sum_{x \in \mathcal{E}_2} (p(x) - \tilde{p}(x)) \right) \\
&= \frac{1}{2} (p(\mathcal{E}_1) - \tilde{p}(\mathcal{E}_1) - (p(\mathcal{E}_2) - \tilde{p}(\mathcal{E}_2))) \\
&= \max_{\mathcal{E} \subseteq X} |p(\mathcal{E}) - \tilde{p}(\mathcal{E})|.
\end{aligned}
$$

The lemma follows.   □

Lemma B.21. *Let (XW, p_{XW}) be a joint probability space (see Section B.1, p. 327 and p. 328). Let p_X be the induced distribution on X, and $\operatorname{prob}(w\,|\,x)$ be the conditional probability of w, assuming x ($x \in X$, $w \in W_x$). Let $\tilde{p}_X$ be another distribution on X. Setting $\operatorname{prob}(x, w) := \tilde{p}_X(x) \cdot \operatorname{prob}(w\,|\,x)$, we get another probability distribution $\tilde{p}_{XW}$ on XW (see Section B.1). Then*

$$\operatorname{dist}(p_{XW}, \tilde{p}_{XW}) \leq \operatorname{dist}(p_X, \tilde{p}_X).$$

Proof. We have

$$|p_{XW}(x, w) - \tilde{p}_{XW}(x, w)| = |(p_X(x) - \tilde{p}_X(x)) \cdot \operatorname{prob}(w\,|\,x)| \leq |p_X(x) - \tilde{p}_X(x)|,$$

and the lemma follows immediately from Definition B.19.   □

Throughout the book we consider families of sets, probability distributions on these sets (often the uniform one) and events concerning maps and probabilistic algorithms between these sets. The index sets J are partitioned index sets with security parameter k, $J = \bigcup_{k \in \mathbb{N}} J_k$, usually written as $J = (J_k)_{k \in \mathbb{N}}$ (see Definition 6.2). The indexes $j \in J$ are assumed to be binarily encoded, and k is a measure for the binary length $|j|$ of an index j. Recall that, by definition, there is an $m \in \mathbb{N}$, with $k^{1/m} \leq |j| \leq k^m$ for $j \in J_k$. As an example, think of the family of RSA functions (see Chapters 3 and 6):

$$(\operatorname{RSA}_{n,e} : \mathbb{Z}_n^* \longrightarrow \mathbb{Z}_n^*, \ x \longmapsto x^e)_{(n,e) \in I},$$

where $I_k = \{(n, e) \mid n = pq, p \neq q \text{ primes}, |p| = |q| = k, e \text{ prime to } \varphi(n)\}$. We are often interested in asymptotic statements, i.e., statements holding for sufficiently large k.

Definition B.22. *Let $J = (J_k)_{k \in \mathbb{N}}$ be an index set with security parameter k, and let $(X_j)_{j \in J}$ be a family of sets. Let $p = (p_j)_{j \in J}$ and $\tilde{p} = (\tilde{p}_j)_{j \in J}$ be families of probability distributions on $(X_j)_{j \in J}$.*
p and $\tilde{p}$ are called polynomially close,[2] if for every positive polynomial P there is a $k_0 \in \mathbb{N}$, such that for all $k \geq k_0$ and $j \in J_k$

$$\operatorname{dist}(p_j, \tilde{p}_j) \leq \frac{1}{P(k)}.$$

Remarks:

1. "Polynomially close" defines an equivalence relation between distributions.
2. Polynomially close distributions cannot be distinguished by a statistical test implemented as a probabilistic polynomial algorithm (see [Luby96], Lecture 7). We do not consider probabilistic polynomial algorithms in this appendix. Probabilistic polynomial statistical tests for pseudorandom sequences are studied in Chapter 8 (see, e.g., Definition 8.2).

---

[2] Also called (ε)-*statistically indistinguishable* (see, e.g., [Luby96]) or *statistically close* (see, e.g., [Goldreich01]).

The following lemma gives an example of polynomially close distributions.

Lemma B.23. *Let* $J_k := \{n \mid n = rs, r, s \text{ primes}, |r| = |s| = k, r \neq s\}$ [3] *and* $J := \bigcup_{k \in \mathbb{N}} J_k$. *The distributions* $x \xleftarrow{u} \mathbb{Z}_n$ *and* $x \xleftarrow{u} \mathbb{Z}_n^*$ *are polynomially close. In other words, uniformly choosing any* x *from* $\mathbb{Z}_n$ *is polynomially close to choosing only units.*

Proof. Let p_n be the uniform distribution on $\mathbb{Z}_n$ and let $\tilde{p}_n$ be the distribution $x \xleftarrow{u} \mathbb{Z}_n^*$. Then $p_n(x) = 1/n$ for all $x \in \mathbb{Z}_n$, $\tilde{p}_n(x) = 1/\varphi(n)$ if $x \in \mathbb{Z}_n^*$, and $\tilde{p}_n(x) = 0$ if $x \in \mathbb{Z}_n \setminus \mathbb{Z}_n^*$. We have $|\mathbb{Z}_n^*| = \varphi(n) = n \prod_{p|n} \frac{p-1}{p}$, where the product is taken over the distinct primes p dividing n (Corollary A.30). Hence,

$$\text{dist}(p_n, \tilde{p}_n) = \frac{1}{2} \sum_{x \in \mathbb{Z}_n} |p_n(x) - \tilde{p}_n(x)|$$

$$= \frac{1}{2} \left(\sum_{x \in \mathbb{Z}_n^*} \left(\frac{1}{\varphi(n)} - \frac{1}{n} \right) + \sum_{x \in \mathbb{Z}_n \setminus \mathbb{Z}_n^*} \frac{1}{n} \right) = 1 - \frac{\varphi(n)}{n} = 1 - \prod_{p|n} \frac{p-1}{p}.$$

If $n = rs \in J_k$, then

$$\text{dist}(p_n, \tilde{p}_n) = 1 - \frac{(r-1)(s-1)}{rs} = \frac{1}{r} + \frac{1}{s} - \frac{1}{rs} \leq 2 \cdot \frac{1}{2^{k-1}} = \frac{1}{2^{k-2}},$$

and the lemma follows.   □

Remark. The lemma does not hold for arbitrary n, i.e., with $\tilde{J}_k := \{n \mid |n| = k\}$ instead of J_k. Namely, if n has a small prime factor q, then $1 - \prod_{p|n} \frac{p-1}{p} \geq 1 - \frac{q-1}{q}$, which is not close to 0.

Example. The distributions $x \xleftarrow{u} \mathbb{Z}_n$ and $x \xleftarrow{u} \mathbb{Z}_n \cap \text{Primes}$ are not polynomially close.[4] Their statistical distance is almost 1 (for large n).

Namely, let $k = |n|$, and let p_1 and p_2 be the uniform distributions on $\mathbb{Z}_n$ and $\mathbb{Z}_n \cap \text{Primes}$. As usual, we extend p_2 by 0 to a distribution on $\mathbb{Z}_n$. The number $\pi(x)$ of primes $\leq x$ is approximately $x/\ln(x)$ (Theorem A.68). Thus, we get (with $c = \ln(2)$)

$$\text{dist}(p_1, p_2) = \frac{1}{2} \sum_{x \in \mathbb{Z}_n} |p_1(x) - p_2(x)|$$

$$= \frac{1}{2} \left(\pi(n) \left(\frac{1}{\pi(n)} - \frac{1}{n} \right) + (n - \pi(n)) \frac{1}{n} \right)$$

$$= \left(1 - \frac{\pi(n)}{n} \right) \approx 1 - \frac{1}{\ln(n)}$$

$$= 1 - \frac{1}{c \log_2(n)} \geq 1 - \frac{1}{c(k-1)},$$

---

[3] If $r \in \mathbb{N}$, we denote, as usual, the binary length of r by $|r|$.
[4] As always, we consider $\mathbb{Z}_n$ as the set $\{0, \ldots, n-1\}$.

and see that the statistical distance is close to 1 for large k.

Lemma B.24. *Let $J = (J_k)_{k \in \mathbb{N}}$ be an index set with security parameter k, and let $(X_j)_{j \in J}$ be a family of sets. Let $p = (p_j)_{j \in J}$ and $\tilde{p} = (\tilde{p}_j)_{j \in J}$ be families of probability distributions on $(X_j)_{j \in J}$ which are polynomially close. Let $(\mathcal{E}_j)_{j \in J}$ be a family of events $\mathcal{E}_j \subseteq X_j$.*
Then for every positive polynomial P, there is a $k_0 \in \mathbb{N}$ such that for all $k \geq k_0$

$$|p_j(\mathcal{E}_j) - \tilde{p}_j(\mathcal{E}_j)| \leq \frac{1}{P(k)},$$

for all $j \in J_k$.

Proof. This is an immediate consequence of Lemma B.20.   □

Definition B.25. Let $J = (J_k)_{k \in \mathbb{N}}$ be an index set with security parameter k, and let $(X_j)_{j \in J}$ be a family of sets. Let $p = (p_j)_{j \in J}$ and $\tilde{p} = (\tilde{p}_j)_{j \in J}$ be families of probability distributions on $(X_j)_{j \in J}$.
$\tilde{p}$ is *polynomially bounded* by p if there is a positive polynomial Q such that $p_j(x)Q(k) \geq \tilde{p}_j(x)$ for all $k \in \mathbb{N}, j \in J_k$ and $x \in X_j$.

Examples. In both examples, let $J = (J_k)_{k \in \mathbb{N}}$ be an index set with security parameter k:

1. Let $(X_j)_{j \in J}$ and $(Y_j)_{j \in J}$ be families of sets with $Y_j \subseteq X_j, j \in J$. Assume there is a polynomial Q, such that $|Y_j|Q(k) \geq |X_j|$ for all $k, j \in J_k$. Then the image of the uniform distributions on $(Y_j)_{j \in J}$ under the inclusions $Y_j \subseteq X_j$ is polynomially bounded by the uniform distributions on $(X_j)_{j \in J}$. This is obvious, since $1/|Y_j| \leq Q(k)/|X_j|$ by assumption.
    For example, $x \xleftarrow{u} \mathbb{Z}_n \cap \text{Primes}$ is polynomially bounded by $x \xleftarrow{u} \mathbb{Z}_n$, because the number of primes $\leq n$ is of the order n/k, with $k = |n|$ being the binary length of n (by the Prime Number Theorem, Theorem A.68).
2. Let $(X_j)_{j \in J}$ and $(Y_j)_{j \in J}$ be families of sets. Let $f = (f_j : Y_j \longrightarrow X_j)_{j \in J}$ be a family of surjective maps, and assume that for $j \in J_k$, each $x \in X_j$ has at most $Q(k)$ pre-images, Q a polynomial. Then the image of the uniform distributions on $(Y_j)_{j \in J}$ under f is polynomially bounded by the uniform distributions on $(X_j)_{j \in J}$.
    Namely, let p_u be the uniform probability distribution and p_f be the image distribution. We have for $j \in J_k$ and $x \in X_j$ that

$$p_f(x) \leq \frac{Q(k)}{|Y_j|} \leq \frac{Q(k)}{|X_j|} = Q(k)p_u(x).$$

Proposition B.26. *Let $J = (J_k)_{k \in \mathbb{N}}$ be an index set with security parameter k. Let $(X_j)_{j \in J}$ be a family of sets. Let $p = (p_j)_{j \in J}$ and $\tilde{p} = (\tilde{p}_j)_{j \in J}$ be families of probability distributions on $(X_j)_{j \in J}$. Assume that $\tilde{p}$ is polynomially bounded by p. Let $(\mathcal{E}_j)_{j \in J}$ be a family of events $\mathcal{E}_j \subseteq X_j$, whose probability is negligible with respect to p; i.e., for every positive polynomial P there is a*

$k_0 \in \mathbb{N}$, such that $p_j(\mathcal{E}_j) \leq {}^1\!/_{P(k)}$ for $k \geq k_0$ and $j \in J_k$.
Then the events $(\mathcal{E}_j)_{j \in J}$ have negligible probability also with respect to $\tilde{p}$.

Proof. There is a polynomial Q, such that $\tilde{p}_j \leq Q(k) \cdot p_j$ for $j \in J_k$. Now let R be a positive polynomial. Then there is some k_0, such that for $k \geq k_0$ and $j \in J_k$

$$p_j(\mathcal{E}_j) \leq \frac{1}{R(k)Q(k)} \text{ and hence } \tilde{p}_j(\mathcal{E}_j) \leq Q(k) \cdot p_j(\mathcal{E}_j) \leq \frac{1}{R(k)},$$

and the proposition follows. □

B.4 Basic Concepts of Information Theory

Information theory and the classical notion of provable security for encryption algorithms go back to Shannon and his famous papers [Shannon48] and [Shannon49].

We give a short introduction to some basic concepts and facts from information theory that are needed in this book. Textbooks on the subject include [Ash65], [Hamming86], [CovTho92] and [GolPeiSch94].

Changing the point of view from probability spaces to random variables (see Section B.1), all the following definitions and results can be formulated and are valid for (jointly distributed) random variables as well.

Definition B.27. Let X be a finite probability space. The *entropy* or *uncertainty* of X is defined by

$$H(X) := \sum_{x \in X, prob(x) \neq 0} prob(x) \cdot \log_2\left(\frac{1}{prob(x)}\right)$$

$$= -\sum_{x \in X, prob(x) \neq 0} prob(x) \cdot \log_2(prob(x)).$$

Remark. The probability space X models a random experiment. The possible outcomes of the experiments are the elements $x \in X$. If we execute the experiment and observe the event x, we gain information. The amount of information we obtain with the occurrence of x (or, equivalently, our uncertainty whether x will occur) – measured in bits – is given by

$$\log_2\left(\frac{1}{prob(x)}\right) = -\log_2(prob(x)).$$

The lower the probability of x, the higher the uncertainty. For example, tossing a fair coin we have $prob(heads) = prob(tails) = {}^1\!/_2$. Thus, the amount of information obtained with the outcome *heads* (or *tails*) is 1 bit. If you throw a fair die, each outcome has probability $1/6$. Therefore, the amount of information associated with each outcome is $\log_2(6) \approx 2.6$ bits.

The entropy $H(X)$ measures the average (i.e., the expected) amount of information arising from executing the experiment X. For example, toss a coin which is unfair, say prob(*heads*) $= 3/4$ and prob(*tails*) $= 1/4$. Then we obtain $\log_2(4/3) \approx 0.4$ bits of information with outcome *heads*, and 2 bits of information with outcome *tails*; so the average amount of information resulting from tossing this coin – the entropy – is $3/4 \cdot \log_2(4/3) + 1/4 \cdot 2 \approx 0.8$ bits.

Proposition B.28. *Let X be a finite probability space which contains n elements, $X = \{x_1, \ldots, x_n\}$:*

1. *$0 \leq H(X) \leq \log_2(n)$.*
2. *$H(X) = 0$ if and only if there is some $x \in X$ with prob$(x) = 1$ (and hence all other elements in X have a probability of 0).*
3. *$H(X) = \log_2(n)$ if and only if the distribution on X is uniform.*

For the proof of Proposition B.28, we need the following technical lemma.

Lemma B.29. *Let $(p_1, \ldots, p_n)$ and $(q_1, \ldots, q_n)$ be probability distributions (i.e., all p_i and q_j are ≥ 0 and $\sum_{i=1}^{n} p_i = \sum_{j=1}^{n} q_j = 1$, see Definition B.1). Assume $p_k \neq 0$ and $q_k \neq 0$ for $k = 1, \ldots, n$. Then:*

1.

$$\sum_{k=1}^{n} p_k \log_2\left(\frac{1}{p_k}\right) \leq \sum_{k=1}^{n} p_k \log_2\left(\frac{1}{q_k}\right). \tag{B.1}$$

2. *Equality holds in (B.1) if and only if $(p_1, \ldots, p_n) = (q_1, \ldots, q_n)$.*

Proof. Since $\log_2(x) = \log_2(e) \cdot \ln(x)$ and $\log_2(e) > 0$, it suffices to prove the statements for $\ln$ instead of $\log_2$. We have $\ln(x) \leq x - 1$ for all x, and $\ln(x) = x - 1$ if and only if $x = 1$. Therefore,

$$\ln\left(\frac{q_k}{p_k}\right) \leq \frac{q_k}{p_k} - 1, \text{ hence } p_k \ln\left(\frac{q_k}{p_k}\right) \leq q_k - p_k, \text{ hence}$$

$$\sum_{k=1}^{n} p_k \ln\left(\frac{q_k}{p_k}\right) \leq \sum_{k=1}^{n} (q_k - p_k), \text{ and this implies}$$

$$\sum_{k=1}^{n} p_k (\ln(q_k) - \ln(p_k)) \leq \sum_{k=1}^{n} q_k - \sum_{k=1}^{n} p_k = 0.$$

Obviously, we have equality if and only if $q_k/p_k = 1$ for $k = 1, \ldots, n$, which means $q_k = p_k$ for $k = 1, \ldots, n$.   □

Proof (of Proposition B.28). Since $-\text{prob}(x) \cdot \log_2(\text{prob}(x)) \geq 0$ for every $x \in X$, the first inequality in statement 1, $H(X) \geq 0$, and statement 2 are immediate consequences of the definition of $H(X)$.

To prove statements 1 and 3, set

$$p_k := \text{prob}(x_k), q_k := \frac{1}{n}, 1 \leq k \leq n.$$

Applying Lemma B.29 we get

$$\sum_{k=1}^{n} p_k \log_2\left(\frac{1}{p_k}\right) \leq \sum_{k=1}^{n} p_k \log_2\left(\frac{1}{q_k}\right) = \sum_{k=1}^{n} p_k \log_2(n) = \log_2(n).$$

Equality holds instead of $\leq$ if and only if

$$p_k = q_k = \frac{1}{n}, k = 1, \ldots, n,$$

by statement 2 of Lemma B.29.   □

In the following we assume without loss of generality that all elements of probability spaces have a probability > 0.

We consider joint probability spaces XY and will see how to specify the amount of information gathered about X when learning Y. We will often use the intuitive notation $\text{prob}(y|x)$, for $x \in X$ and $y \in Y$. Recall that $\text{prob}(y|x)$ is the probability that y occurs as the second component, if the first component is x (see Section B.1, p. 328).

Definition B.30. Let X and Y be finite probability spaces with joint distribution XY.

The *joint entropy* $H(XY)$ of X and Y is the entropy of the joint distribution XY of X and Y:

$$H(XY) := - \sum_{x \in X, y \in Y} \text{prob}(x, y) \cdot \log_2\left(\text{prob}(x, y)\right).$$

Conditioning X over $y \in Y$, we define

$$H(X|y) := - \sum_{x \in X} \text{prob}(x|y) \cdot \log_2(\text{prob}(x|y)).$$

The *conditional entropy* (or *conditional uncertainty*) of X assuming Y is

$$H(X|Y) := \sum_{y \in Y} \text{prob}(y) \cdot H(X|y).$$

The *mutual information* of X and Y is the reduction of the uncertainty of X when Y is learned:

$$I(X;Y) = H(X) - H(X|Y).$$

$H(XY)$ measures the average amount of information gathered by observing both X and Y. $H(X|Y)$ measures the average amount of information arising from (an execution of) the experiment X, knowing the result of experiment Y. $I(X;Y)$ measures the amount of information about X obtained by learning Y.

Proposition B.31. *Let X and Y be finite probability spaces with joint distribution XY. Then:*

1. $H(X|Y) \geq 0$.
2. $H(XY) = H(X) + H(Y|X)$.
3. $H(XY) \leq H(X) + H(Y)$.
4. $H(Y) \geq H(Y|X)$.
5. $I(X;Y) = I(Y;X) = H(X) + H(Y) - H(XY)$.
6. $I(X;Y) \geq 0$.

Proof. Statement 1 is true since $H(X|y) \geq 0$ by Proposition B.28. The other statements are a special case of the more general Proposition B.36. $\square$

Proposition B.32. *Let X and Y be finite probability spaces with joint distribution XY. The following statements are equivalent:*

1. X *and* Y *are independent.*
2. $\mathrm{prob}(y|x) = \mathrm{prob}(y)$, *for* $x \in X$ *and* $y \in Y$.
3. $\mathrm{prob}(x|y) = \mathrm{prob}(x)$, *for* $x \in X$ *and* $y \in Y$.
4. $\mathrm{prob}(x|y) = \mathrm{prob}(x|y')$, *for* $x \in X$ *and* $y, y' \in Y$.
5. $H(XY) = H(X) + H(Y)$.
6. $H(Y) = H(Y|X)$.
7. $I(X;Y) = 0$.

Proof. The equivalence of statements 1, 2 and 3 is an immediate consequence of the definitions of independence and conditional probabilities (see Definition B.6). Statement 3 obviously implies statement 4. Conversely, statement 3 follows from statement 4 using

$$\mathrm{prob}(x) = \sum_{y \in Y} \mathrm{prob}(y) \cdot \mathrm{prob}(x|y).$$

The equivalence of the latter statements follows as a special case from Proposition B.37. $\square$

Next we study the mutual information of two probability spaces conditioned on a third one.

Definition B.33. *Let X, Y and Z be finite probability spaces with joint distribution XYZ. The* conditional mutual information $I(X;Y|Z)$ *is given by*

$$I(X;Y|Z) := H(X|Z) - H(X|YZ).$$

The conditional mutual information $I(X;Y|Z)$ is the average amount of information about X obtained by learning Y, assuming that Z is known.

When studying entropies and mutual informations for jointly distributed finite probability spaces X, Y and Z, it is sometimes useful to consider the conditional situation where $z \in Z$ is fixed. Therefore we need the following definition.

Definition B.34. Let X, Y and Z be finite probability spaces with joint distribution XYZ and $z \in Z$.

$$H(X|Y,z) := \sum_{y \in Y} \text{prob}(y|z) \cdot H(X|y,z),$$

$$I(X;Y|z) := H(X|z) - H(X|Y,z).$$

(See Definition B.30 for the definition of $H(X|y,z)$.)

Proposition B.35. Let X, Y and Z be finite probability spaces with joint distribution XYZ and $z \in Z$:

1. $H(X|YZ) = \sum_{z \in Z} \text{prob}(z) \cdot H(X|Y,z)$.
2. $I(X;Y|Z) = \sum_{z \in Z} \text{prob}(z) \cdot I(X;Y|z)$.

Proof.

$$H(X|YZ) = \sum_{y,z} \text{prob}(y,z) \cdot H(X|y,z) \ (\text{by Definition B.30})$$

$$= \sum_{z} \text{prob}(z) \sum_{y} \text{prob}(y|z) \cdot H(X|y,z) = \sum_{z} \text{prob}(z) \cdot H(X|Y,z).$$

This proves statement 1.

$$I(X;Y|Z) = H(X|Z) - H(X|YZ)$$

$$= \sum_{z} \text{prob}(z) \cdot H(X|z) - \sum_{z} \text{prob}(z) \cdot H(X|Y,z) \ (\text{by Def. B.30 and 1.})$$

$$= \sum_{z} \text{prob}(z) \cdot I(X;Y|z).$$

This proves statement 2.   □

Proposition B.36. Let X, Y and Z be finite probability spaces with joint distribution XYZ.

1. $H(XY|Z) = H(X|Z) + H(Y|XZ)$.
2. $H(XY|Z) \leq H(X|Z) + H(Y|Z)$.
3. $H(Y|Z) \geq H(Y|XZ)$.
4. $I(X;Y|Z) = I(Y;X|Z) = H(X|Z) + H(Y|Z) - H(XY|Z)$.
5. $I(X;Y|Z) \geq 0$.

6. $I(X;YZ) = I(X;Z) + I(X;Y|Z)$.
7. $I(X;YZ) \geq I(X;Z)$.

Remark. We get Proposition B.31 from Proposition B.36 by taking $Z := \{z_0\}$ with $\mathrm{prob}(z_0) := 1$ and $XYZ := XY \times Z$.

Proof.

1. We compute

$$H(X|Z) + H(Y|XZ)$$
$$= -\sum_{z \in Z} \mathrm{prob}(z) \sum_{x \in X} \mathrm{prob}(x|z) \cdot \log_2(\mathrm{prob}(x|z))$$
$$\quad - \sum_{x \in X, z \in Z} \mathrm{prob}(x,z) \sum_{y \in Y} \mathrm{prob}(y|x,z) \cdot \log_2(\mathrm{prob}(y|x,z))$$
$$= -\sum_{x,y,z} \mathrm{prob}(z)\mathrm{prob}(x,y|z) \cdot \log_2(\mathrm{prob}(x|z))$$
$$\quad - \sum_{x,y,z} \mathrm{prob}(x,z)\mathrm{prob}(y|x,z) \cdot \log_2(\mathrm{prob}(y|x,z))$$
$$= -\sum_{x,y,z} \mathrm{prob}(x,y,z) \cdot \left(\log_2(\mathrm{prob}(x|z)) + \log_2\left(\frac{\mathrm{prob}(x,y|z)}{\mathrm{prob}(x|z)} \right) \right)$$
$$= -\sum_{x,y,z} \mathrm{prob}(x,y,z) \cdot \log_2(\mathrm{prob}(x,y|z))$$
$$= -\sum_{z} \mathrm{prob}(z) \sum_{x,y} \mathrm{prob}(x,y|z) \cdot \log_2(\mathrm{prob}(x,y|z))$$
$$= H(XY|Z).$$

2. We have

$$H(X|Z) = -\sum_{z \in Z} \mathrm{prob}(z) \sum_{x \in X} \mathrm{prob}(x|z) \cdot \log_2(\mathrm{prob}(x|z))$$
$$= -\sum_{z \in Z} \mathrm{prob}(z) \sum_{y \in Y} \sum_{x \in X} \mathrm{prob}(x,y|z) \cdot \log_2(\mathrm{prob}(x|z)),$$
$$H(Y|Z) = -\sum_{z \in Z} \mathrm{prob}(z) \sum_{y \in Y} \mathrm{prob}(y|z) \cdot \log_2(\mathrm{prob}(y|z))$$
$$= -\sum_{z \in Z} \mathrm{prob}(z) \sum_{x \in X} \sum_{y \in Y} \mathrm{prob}(x,y|z) \cdot \log_2(\mathrm{prob}(y|z)).$$

Hence,

$$H(X|Z) + H(Y|Z)$$
$$= -\sum_{z \in Z} \mathrm{prob}(z) \sum_{x,y} \mathrm{prob}(x,y|z) \cdot \log_2(\mathrm{prob}(x|z)\mathrm{prob}(y|z)).$$

By definition,

$$H(XY\,|\,Z) = -\sum_{z\in Z} \text{prob}(z) \sum_{x,y} \text{prob}(x,y\,|\,z) \cdot \log_2(\text{prob}(x,y\,|\,z)).$$

Since $(\text{prob}(x,y\,|\,z))_{(x,y)\in XY}$ and $(\text{prob}(x\,|\,z) \cdot \text{prob}(y\,|\,z))_{(x,y)\in XY}$ are probability distributions, the inequality follows from Lemma B.29.

3. Follows from statements 1 and 2.
4. Follows from the definition of the mutual information, since $H(X\,|\,YZ) = H(XY\,|\,Z) - H(Y\,|\,Z)$ by statement 1.
5. Follows from statements 2 and 4.
6. $I(X;Z) + I(X;Y\,|\,Z) = H(X) - H(X\,|\,Z) + H(X\,|\,Z) - H(X\,|\,YZ) = I(X;YZ)$.
7. Follows from statements 5 and 6.

The proof of Proposition B.36 is finished.   □

Proposition B.37. *Let X, Y and Z be finite probability spaces with joint distribution XYZ. The following statements are equivalent:*

1. *X and Y are independent assuming z, i.e.,*
   $\text{prob}(x,y\,|\,z) = \text{prob}(x\,|\,z) \cdot \text{prob}(y\,|\,z)$, *for all $(x,y,z) \in XYZ$.*
2. *$H(XY\,|\,Z) = H(X\,|\,Z) + H(Y\,|\,Z)$.*
3. *$H(Y\,|\,Z) = H(Y\,|\,XZ)$.*
4. *$I(X;Y\,|\,Z) = 0$.*

Remark. We get the last three statements in Proposition B.32 from Proposition B.37 by taking $Z := \{z_0\}$ with $\text{prob}(z_0) := 1$ and $XYZ := XY \times Z$.

Proof. The equivalence of statements 1 and 2 follows from the computation in the proof of Proposition B.36, statement 2, by Lemma B.29. The equivalence of statements 2 and 3 is immediately derived from Proposition B.36, statement 1. The equivalence of statements 2 and 4 follows from Proposition B.36, statement 4.   □

Remark. All entropies considered above may in addition be conditioned on some event $\mathcal{E}$. We make use of this fact in our overview about some results on "unconditional security" in Section 9.6. There, mutual information of the form $I(X;Y\,|\,Z,\mathcal{E})$ appears. $\mathcal{E}$ is an event in XYZ (i.e., a subset of XYZ) and the mutual information is defined as usual, but the conditional probabilities assuming $\mathcal{E}$ have to be used. More precisely:

$$H(X\,|\,YZ,\mathcal{E}) := -\sum_{(x,y,z)\in\mathcal{E}} \text{prob}(y,z\,|\,\mathcal{E}) \cdot \text{prob}(x\,|\,y,z,\mathcal{E}) \cdot \log_2(\text{prob}(x\,|\,y,z,\mathcal{E})),$$

$$H(X\,|\,Z,\mathcal{E}) := -\sum_{(x,y,z)\in\mathcal{E}} \text{prob}(z\,|\,\mathcal{E}) \cdot \text{prob}(x\,|\,z,\mathcal{E}) \cdot \log_2(\text{prob}(x\,|\,z,\mathcal{E})),$$

$$I(X;Y\,|\,Z,\mathcal{E}) := H(X\,|\,Z,\mathcal{E}) - H(X\,|\,YZ,\mathcal{E}).$$

Here recall that, for example, $\mathrm{prob}(y,z\,|\,\mathcal{E}) = \mathrm{prob}(X \times \{y\} \times \{z\}\,|\,\mathcal{E})$ and $\mathrm{prob}(x\,|\,y,z,\mathcal{E}) = \mathrm{prob}\,(\{x\} \times Y \times Z\,|\,X \times \{y\} \times \{z\},\mathcal{E})$ The results on entropies from above remain valid if they are, in addition, conditioned on $\mathcal{E}$. The same proofs apply, but the conditional probabilities assuming $\mathcal{E}$ have to be used.

References

Textbooks

[Ash65] R.B. Ash: Information Theory. New York: John Wiley & Sons, 1965.

[BalDiaGab95] J.L. Balcázar, J. Díaz, J. Gabarró: Structural Complexity I. Berlin, Heidelberg, New York: Springer-Verlag, 1995.

[Bauer07] F.L. Bauer: Decrypted Secrets – Methods and Maxims of Cryptology. 4th ed. Berlin, Heidelberg, New York: Springer-Verlag, 2007.

[Bauer96] H. Bauer: Probability Theory. Berlin: de Gruyter, 1996.

[BerPer85] J. Berstel, D. Perrin: Theory of Codes. Orlando: Academic Press, 1985.

[Buchmann2000] J.A. Buchmann: Introduction to Cryptography. Berlin, Heidelberg, New York: Springer-Verlag, 2000.

[Cohen95] H. Cohen: A Course in Computational Algebraic Number Theory. Berlin, Heidelberg, New York: Springer-Verlag, 1995.

[CovTho92] T.M. Cover, J.A. Thomas: Elements of Information Theory. New York: John Wiley & Sons, 1992.

[DaeRij02] J. Daemen, V. Rijmen: The Design of Rijndael – AES – The Advanced Encryption Standard. Berlin, Heidelberg, New York: Springer-Verlag, 2002.

[Feller68] W. Feller: An Introduction to Probability Theory and Its Applications. 3rd ed. New York: John Wiley & Sons, 1968.

[Forster96] O. Forster: Algorithmische Zahlentheorie. Braunschweig, Wiesbaden: Vieweg, 1996.

[GanYlv67] R.A. Gangolli, D. Ylvisaker: Discrete Probability. New York: Harcourt, Brace & World, 1967.

[Goldreich99] O. Goldreich: Modern Cryptography, Probabilistic Proofs and Pseudorandomness. Berlin, Heidelberg, New York: Springer-Verlag, 1999.

[Goldreich01] O. Goldreich: Foundations of Cryptography – Basic Tools. Cambridge University Press, 2001.

[Goldreich04] O. Goldreich: Foundations of Cryptography. Volume II Basic Applications. Cambridge University Press, 2004.

[GolPeiSch94] S.W. Golomb, R.E. Peile, R.A. Scholtz: Basic Concepts in Information Theory and Coding. New York: Plenum Press, 1994.

[Gordon97] H. Gordon: Discrete Probability. Berlin, Heidelberg, New York: Springer-Verlag, 1997.

[Hamming86] R.W. Hamming: Coding and Information Theory. 2nd ed. Englewood Cliffs, NJ: Prentice Hall, 1986.

[HarWri79] G.H. Hardy, E.M. Wright: An Introduction to the Theory of Numbers. 5th ed. Oxford: Oxford University Press, 1979.

[HopUll79] J. Hopcroft, J. Ullman: Introduction to Automata Theory, Languages and Computation. Reading, MA: Addison-Wesley Publishing Company, 1979.

[IreRos82] K. Ireland, M.I. Rosen: A Classical Introduction to Modern Number Theory. Berlin, Heidelberg, New York: Springer-Verlag, 1982.

[Kahn67] D. Kahn: The Codebreakers: The Story of Secret Writing. New York: Macmillan Publishing Co. 1967.

[Knuth98] D.E. Knuth: The Art of Computer Programming. 3rd ed. Volume 2/Seminumerical Algorithms. Reading, MA: Addison-Wesley Publishing Company, 1998.

[Koblitz94] N. Koblitz: A Course in Number Theory and Cryptography. 2nd ed. Berlin, Heidelberg, New York: Springer-Verlag, 1994.

[Lang05] S. Lang: Algebra. 3rd ed. Berlin, Heidelberg, New York: Springer-Verlag, 2005.

[Luby96] M. Luby: Pseudorandomness and Cryptographic Applications. Princeton, NJ: Princeton University Press, 1996.

[MenOorVan96] A. Menezes, P.C. van Oorschot, S.A. Vanstone: Handbook of Applied Cryptography. Boca Raton, New York, London, Tokyo: CRC Press, 1996.

[MotRag95] R. Motwani, P. Raghavan: Randomized Algorithms. Cambridge, UK: Cambridge University Press, 1995.

[Osen74] L.M. Osen: Women in Mathematics. Cambridge, MA: MIT, 1974.

[Papadimitriou94] C.H. Papadimitriou: Computational Complexity. Reading, MA: Addison-Wesley Publishing Company, 1994.

[Rényi70] A. Rényi: Probability Theory. Amsterdam: North-Holland, 1970.

[Riesel94] H. Riesel: Prime Numbers and Computer Methods for Factorization. Boston, Basel: Birkhäuser, 1994.

[Rose94] H.E. Rose: A Course in Number Theory. 2nd ed. Oxford: Clarendon Press, 1994.

[Rosen2000] K.H. Rosen: Elementary Number Theory and Its Applications. 4th ed. Reading, MA: Addison-Wesley Publishing Company, 2000.

[Salomaa90] A. Salomaa: Public-Key Cryptography. Berlin, Heidelberg, New York: Springer-Verlag, 1990.

[Schneier96] B. Schneier: Applied Cryptography. New York: John Wiley & Sons, 1996.

[Simmons92] G.J. Simmons (ed.): Contemporary Cryptology. Piscataway, NJ: IEEE Press, 1992.

[Stinson95] D.R. Stinson: Cryptography – Theory and Practice. Boca Raton, New York, London, Tokyo: CRC Press, 1995.

Papers

[AleChoGolSch88] W.B. Alexi, B. Chor, O. Goldreich, C.P. Schnorr: RSA/Rabin functions: certain parts are as hard as the whole. SIAM Journal on Computing, 17(2): 194–209, April 1988.

[AumDinRab02] Y. Aumann, Y.Z. Ding, M.O. Rabin: Everlasting security in the bounded storage model. IEEE Transactions on Information Theory 48(6):1668-1680, 2002.

[AumRab99] Y. Aumann, M.O. Rabin: Information-theoretically secure communication in the limited storage space model. Advances in Cryptology - CRYPTO '99, Lecture Notes in Computer Science, 1666: 65–79, Springer-Verlag, 1999.

[Bach88] E. Bach: How to generate factored random numbers. SIAM Journal on Computing, 17(2): 179–193, April 1988.

[BarPfi97] N. Barić, B. Pfitzmann: Collision-free accumulators and fail-stop signature schemes without trees. Advances in Cryptology - EUROCRYPT '97, Lecture Notes in Computer Science, 1233: 480–494, Springer-Verlag, 1997.

[BatHor62] P. Bateman, R. Horn: A heuristic formula concerning the distribution of prime numbers. Mathematics of Computation, 16: 363–367, 1962.

[BatHor65] P. Bateman, R. Horn: Primes represented by irreducible polynomials in one variable. Proc. Symp. Pure Math., 8: 119–135, 1965.

[BeGrGwHåKiMiRo88] M. Ben-Or, O. Goldreich, S. Goldwasser, J. Håstad, J. Kilian, S. Micali, P. Rogaway: Everything provable is provable in zero-knowledge. Advances in Cryptology - CRYPTO '88, Lecture Notes in Computer Science, 403: 37–56, Springer-Verlag, 1990.

[Bellare99] M. Bellare: Practice oriented provable security. Lectures on Data Security. Lecture Notes in Computer Science, 1561: 1–15, Springer-Verlag, 1999.

[BelRog93] M. Bellare, P. Rogaway: Random oracles are practical: a paradigm for designing efficient protocols, Proc. First Annual Conf. Computer and Communications Security, ACM, New York, 1993:6273, 1993.

[BelRog94] M. Bellare, P. Rogaway: Optimal asymmetric encryption. Advances in Cryptology - EUROCRYPT '94, Lecture Notes in Computer Science, 950: 92–111, Springer-Verlag, 1995.

[BelRog96] M. Bellare, P. Rogaway: The exact security of digital signatures, how to sign with RSA and Rabin. Advances in Cryptology - EUROCRYPT '96, Lecture Notes in Computer Science, 1070: 399–416, Springer-Verlag, 1996.

[BelRog97] M. Bellare, P. Rogaway: Collision-resistant hashing: towards making UOWHF practical. Advances in Cryptology - CRYPTO '97, Lecture Notes in Computer Science, 1294: 470–484, Springer-Verlag, 1997.

[Bleichenbacher96] D. Bleichenbacher: Generating ElGamal signatures without knowing the secret key. Advances in Cryptology - EUROCRYPT '96, Lecture Notes in Computer Science, 1070: 10–18, Springer-Verlag, 1996.

[Bleichenbacher98] D. Bleichenbacher: A chosen ciphertext attack against protocols based on the RSA encryption standard PKCS #1. Advances in Cryptology - CRYPTO '98, Lecture Notes in Computer Science, 1462: 1–12, Springer-Verlag, 1998.

[BluBluShu86] L. Blum, M. Blum, M. Shub: A simple unpredictable pseudorandom number generator. SIAM Journal on Computing, 15(2): 364–383, 1986.

[BluGol85] M. Blum, S. Goldwasser: An efficient probabilistic public-key encryption scheme which hides all partial information. Advances in Cryptology - Proceedings of CRYPTO '84, Lecture Notes in Computer Science, 196: 289–299, Springer-Verlag, 1985.

[Blum82] M. Blum: Coin flipping by telephone: a protocol for solving impossible problems. Proceedings of the 24th IEEE Computer Conference, San Francisco, Calif., February 22–25, 1982: 133–137, 1982.

[Blum84] M. Blum: Independent unbiased coin flips from a correlated biased source. Proceedings of the IEEE 25th Annual Symposium on Foundations of Computer Science, Singer Island, Fla., October 24–26, 1984: 425–433, 1984.

[BluMic84] M. Blum, S. Micali: How to generate cryptographically strong sequences of pseudorandom bits. SIAM Journal on Computing, 13(4): 850–863, November 1984.

[Boer88] B. Den Boer: Diffie-Hellman is as strong as discrete log for certain primes. Advances in Cryptology - CRYPTO '88, Lecture Notes in Computer Science, 403: 530–539, Springer-Verlag, 1990.

[BoeBos93] B. den Boer, A. Bosselaers: Collisions for the compression function of MD5. Advances in Cryptology - EUROCRYPT '93, Lecture Notes in Computer Science, 765: 293–304, Springer-Verlag, 1994.

[Boneh01] D. Boneh: Simplified OAEP for the RSA and Rabin functions. Advances in Cryptology - CRYPTO 2001, Lecture Notes in Computer Science, 2139: 275–291, Springer-Verlag, 2001.

[BonVen96] D. Boneh, R. Venkatesan: Hardness of computing the most significant bits of secret keys in Diffie-Hellman and related schemes. Advances in Cryptology - CRYPTO '96, Lecture Notes in Computer Science, 1109: 129–142, Springer-Verlag, 1996.

[BonVen98] D. Boneh, R. Venkatesan: Breaking RSA may not be equivalent to factoring. Advances in Cryptology - EUROCRYPT '98, Lecture Notes in Computer Science, 1403: 59–71, Springer-Verlag, 1998.

[Brands93] S. Brands: An efficient off-line electronic cash system based on the representation problem. Technical Report CS-R9323. Amsterdam, NL: Centrum voor Wiskunde en Informatica (CWI), 1993.

[BraCre96] G. Brassard, C. Crépeau: 25 years of quantum cryptography. SIGACT News 27(3): 13–24, 1996.

[CachMau97] C. Cachin, U.M. Maurer: Unconditional security against memory-bounded adversaries. Advances in Cryptology - CRYPTO '97, Lecture Notes in Computer Science, 1294: 292–306, Springer-Verlag, 1997.

[CamMauSta96] J. Camenisch, U.M. Maurer, M. Stadler: Digital payment systems with passive anonymity revoking trustees. Proceedings of ESORICS '96, Lecture Notes in Computer Science, 1146: 33–43, Springer-Verlag, 1996.

[CamWie92] K.W. Campbell, M.J. Wiener: DES is not a group. Advances in Cryptology - CRYPTO '92, Lecture Notes in Computer Science, 740: 512–520, Springer-Verlag, 1993.

[CamPivSta94] J.L. Camenisch, J.M. Piveteau, M.A. Stadler: Blind signatures based on the discrete logarithm problem. Advances in Cryptology - EUROCRYPT '94, Lecture Notes in Computer Science, 950: 428–432, Springer-Verlag, 1995.

[CanGolHal98] R. Canetti, O. Goldreich, S. Halevi: The random oracle methodology, revisited. STOC'98, Dallas, Texas: 209–218, New York, NY: ACM, 1998.

[CanGolHal04] R. Canetti, O. Goldreich, S. Halevi: On the random-oracle methodology as applied to length-restricted signature schemes. First Theory of Cryptography Conference, TCC 2004, Lecture Notes in Computer Science, 2951: 40–57, Springer-Verlag, 2004.

[CarWeg79] J.L. Carter, M.N. Wegman: Universal classes of hash functions. Journal of Computer and System Sciences, 18: 143–154, 1979.

[ChaPed92] D. Chaum, T. Pedersen: Wallet databases with observers. Advances in Cryptology - CRYPTO '92, Lecture Notes in Computer Science, 740: 89–105, Springer-Verlag, 1993.

[Chaum82] D. Chaum: Blind signatures for untraceable payments. Advances in Cryptology - Proceedings of CRYPTO '82: 199–203, Plenum Press 1983.

[Coppersmith97] D. Coppersmith: Small solutions to polynomial equations and low exponent RSA vulnerabilities. Journal of Cryptology, 10(4): 233–260, 1997.

[CraDam96] R. Cramer, I. Damgård: New generation of secure and practical RSA-based signatures. Advances in Cryptology - CRYPTO '96, Lecture Notes in Computer Science, 1109: 173–185, Springer-Verlag, 1996.

[CraFraSchYun96] R. Cramer, M.K. Franklin, B. Schoenmakers, M. Yung: Multi-authority secret-ballot elections with linear work. Advances in Cryptology - EUROCRYPT '96, Lecture Notes in Computer Science, 1070: 72–83, Springer-Verlag, 1996.

[CraGenSch97] R. Cramer, R. Gennaro, B. Schoenmakers: A secure and optimally efficient multi-authority election scheme. Advances in Cryptology - EUROCRYPT '97, Lecture Notes in Computer Science, 1233: 103–118, Springer-Verlag, 1997.

[CraSho98] R. Cramer, V. Shoup: A practical public key cryptosystem provably secure against adaptive chosen ciphertext attack. Advances in Cryptology -

CRYPTO '98, Lecture Notes in Computer Science, 1462: 13–25, Springer-Verlag, 1998.

[CraSho2000] R. Cramer, V. Shoup: Signature schemes based on the strong RSA assumption. ACM Transactions on Information and System Security, 3(3): 161–185, 2000.

[Damgård87] I.B. Damgård: Collision-free hash functions and public-key signature schemes. Advances in Cryptology - EUROCRYPT '87, Lecture Notes in Computer Science, 304: 203–216, Springer-Verlag, 1988.

[Diffie88] W. Diffie: The first ten years of public key cryptology. In: G.J. Simmons (ed.): Contemporary Cryptology, 135–175, Piscataway, NJ: IEEE Press, 1992.

[DifHel76] W. Diffie, M.E. Hellman: New directions in cryptography. IEEE Transactions on Information Theory, IT-22: 644–654, 1976.

[DifHel77] W. Diffie, M. E. Hellman: Exhaustive cryptanalysis of the NBS data encryption standard. Computer, 10: 74–84, 1977.

[Ding01] Y.Z. Ding: Provable everlasting security in the bounded storage model. PhD Thesis, Cambridge, MA: Harvard University, May 2001.

[DinRab02] Y.Z. Ding, M.O. Rabin: Hyper-encryption and everlasting security. 19th Annual Symposium on Theoretical Aspects of Computer Science (STACS) 2002, Lecture Notes in Computer Science, 2285: 1–26, Springer-Verlag, 2002.

[Dobbertin96] H. Dobbertin: Welche Hash-Funktionen sind für digitale Signaturen geeignet? In: P. Horster (ed.): Digitale Signaturen, 81–92, Braunschweig, Wiesbaden: Vieweg 1996.

[Dobbertin96a] H. Dobbertin: Cryptanalysis of MD5. Presented at the rump session, Advances in Cryptology - EUROCRYPT '96.

[DwoNao94] C. Dwork, M. Naor: An efficient unforgeable signature scheme and its applications. Advances in Cryptology - CRYPTO '94, Lecture Notes in Computer Science, 839: 234–246, Springer-Verlag, 1994.

[DziMau02] S. Dziembowski, U. Maurer: Tight security proofs for the bounded-storage model. Proceedings of the 34th Annual ACM Symposium on Theory of Computing, Montréal, Québec, Canada, May 19–21, 2002: 341–350, 2002.

[DziMau04a] S. Dziembowski, U. Maurer: Optimal randomizer efficiency in the bounded-storage model. Journal of Cryptology, 17(1): 5–26, January 2004.

[DziMau04b] S. Dziembowski, U. Maurer: On generating the initial key in the bounded-storage model. Advances in Cryptology - EUROCRYPT 2004, Lecture Notes in Computer Science, 3027: 126–137, Springer-Verlag, 2004.

[ElGamal84] T. ElGamal: A public key cryptosystem and a signature scheme based on discrete logarithms. Advances in Cryptology - Proceedings of CRYPTO '84, Lecture Notes in Computer Science, 196: 10–18, Springer-Verlag, 1985.

[FiaSha86] A. Fiat, A. Shamir: How to prove yourself: practical solutions to identification and signature problems. Advances in Cryptology - CRYPTO '86, Lecture Notes in Computer Science, 263: 186–194, Springer-Verlag, 1987.

[FIPS46 1977] FIPS46: Data Encryption Standard. Federal Information Processing Standards Publication 46, U.S. Department of Commerce/National Bureau of Standards, National Technical Information Service, Springfield, Virginia, 1977.

[FIPS 113] FIPS 113: Computer data authentication. Federal Information Processing Standards Publication 113, U.S. Department of Commerce/National Bureau of Standards, http://www.itl.nist.gov/fipspubs/, 1985.

[FIPS 180-2] FIPS 180-2: Secure hash signature standard. Federal Information Processing Standards Publication 180-2, U.S. Department of Commerce/National Bureau of Standards, http://www.itl.nist.gov/fipspubs/, 2002.

[FIPS 198] FIPS 198: The keyed-hash message authentication code (HMAC). Federal Information Processing Standards Publication 198, U.S. Department of

Commerce/National Bureau of Standards, http://www.itl.nist.gov/fipspubs/, 2002.

[FisSch2000] R. Fischlin, C.P. Schnorr: Stronger security proofs for RSA and Rabin bits. Journal of Cryptology, 13(2): 221–244, 2000.

[FujOkaPoiSte2001] E. Fujisaki, T. Okamoto, D. Pointcheval, J. Stern: RSA-OAEP is secure under the RSA assumption. Advances in Cryptology - CRYPTO 2001, Lecture Notes in Computer Science, 2139: 260–274, Springer-Verlag, 2001.

[GenHalRab99] R. Gennaro, S. Halevi, T. Rabin: Secure hash-and-sign signatures without the random oracle. Advances in Cryptology - EUROCRYPT '99, Lecture Notes in Computer Science, 1592: 123–139, Springer-Verlag, 1999.

[GenJarKraRab99] R. Gennaro, S. Jarecki, H. Krawczyk, T. Rabin: Secure distributed key generation for discrete-log based cryptosystems. Advances in Cryptology - EUROCRYPT '99, Lecture Notes in Computer Science, 1592: 295–310, Springer-Verlag, 1999.

[Gill77] J. Gill: Computational complexity of probabilistic Turing machines. SIAM Journal on Computing, 6(4): 675–695, December 1977.

[GolLev89] O. Goldreich, L. Levin: A hard-core predicate for all one-way functions. Proceedings of the 21st Annual ACM Symposium on Theory of Computing, Seattle, Wash., May 15–17, 1989: 25–32, 1989.

[GolMic84] S. Goldwasser, S. Micali: Probabilistic encryption. Journal of Computer and System Sciences, 28(2): 270–299, 1984.

[GolMicRac89] S. Goldwasser, S. Micali, C. Rackoff: The knowledge complexity of interactive proof systems. SIAM Journal on Computing, 18: 185–208, 1989.

[GolMicRiv88] S. Goldwasser, S. Micali, R. Rivest: A digital signature scheme secure against chosen message attacks. SIAM Journal on Computing, 17(2): 281–308, 1988.

[GolMicTon82] S. Goldwasser, S. Micali, P. Tong: Why and how to establish a private code on a public network. Proceedings of the IEEE 23rd Annual Symposium on Foundations of Computer Science, Chicago, Ill., November 3–5, 1982: 134–144, 1982.

[GolMicWid86] O. Goldreich, S. Micali, A. Wigderson: Proofs that yield nothing but their validity and a methodology of cryptographic protocol design. Proceedings of the IEEE 27th Annual Symposium on Foundations of Computer Science, Toronto, October 27–29, 1986: 174–187, 1986.

[GolTau03] S. Goldwasser, Y. Tauman: On the (in)security of the Fiat-Shamir paradigm. Proceedings of the IEEE 44th Annual Symposium on Foundations of Computer Science, Cambridge, MA, USA, October 11–14, 2003: 102–113, 2003.

[GolTau03a] S. Goldwasser, Y. Tauman: On the (in)security of the Fiat-Shamir paradigm. Cryptology ePrint Archive, http://eprint.iacr.org, Report 034, 2003.

[Gordon84] J.A. Gordon: Strong primes are easy to find. Advances in Cryptology - EUROCRYPT '84, Lecture Notes in Computer Science, 209: 216–223, Springer-Verlag, 1985.

[HåsNäs98] J. Håstad, M. Näslund: The security of all RSA and Discrete Log bits. Proceedings of the IEEE 39th Annual Symposium on Foundations of Computer Science, Palo Alto, CA, November 8–11, 1998: 510–519, 1998.

[HåsNäs99] J. Håstad, M. Näslund: The security of all RSA and Discrete Log bits. Electronic Colloquium on Computational Complexity, http://eccc.hpi-web.de, ECCC Report TR99-037, 1999.

[ISO/IEC 9594-8] ISO/IEC 9594-8: Information technology - Open Systems Interconnection - The Directory: Authentication framework. International Organization for Standardization, Geneva, Switzerland, http://www.iso.org, 1995.

[ISO/IEC 9797-1] ISO/IEC 9797-1: Message Authentication Codes (MACs) – Part 1: Mechanisms using a block cipher. International Organization for Standardization, Geneva, Switzerland, http://www.iso.org, 1999.

[ISO/IEC 9797-2] ISO/IEC 9797-2: Message Authentication Codes (MACs) – Part 2: Mechanisms using a dedicated hash-function. International Organization for Standardization, Geneva, Switzerland, http://www.iso.org, 2002.

[ISO/IEC 10116] ISO/IEC 10116: Information processing - Modes of operation for an n-bit block cipher algorithm. International Organization for Standardization, Geneva, Switzerland, http://www.iso.org, 1991.

[ISO/IEC 10118-2] ISO/IEC 10118-2: Information technology - Security techniques - Hash-functions - Part 2: Hash-functions using an n-bit block cipher algorithm. International Organization for Standardization, Geneva, Switzerland, http://www.iso.org, 1994.

[Koblitz88] N. Koblitz: Primality of the number of points on an elliptic curve over a finite field. Pacific Journal of Mathematics, 131(1): 157–165, 1988.

[KobMen05] N. Koblitz, A.J. Menezes: Another look at "provable security". Journal of Cryptology, Online First: OF1–OF35, November 2005.

[Klima06] V. Klima: Tunnels in Hash Functions: MD5 Collisions Within a Minute. Cryptology ePrint Archive, http://eprint.iacr.org, Report 105, 2006.

[LeeMooShaSha55] K. de Leeuw, E.F. Moore, C.E. Shannon, N. Shapiro: Computability by probabilistic machines. In: C.E. Shannon, J. McCarthy (eds.): Automata Studies, 183–212, Princeton, NJ: Princeton University Press, 1955.

[Lu02] C. Lu: Hyper-encryption against space-bounded adversaries from on-line strong extractors. Advances in Cryptology - CRYPTO 2002, Lecture Notes in Computer Science, 2442: 257–271, Springer-Verlag, 2002.

[MatMeyOse85] S.M. Matyas, C.H. Meyer, J. Oseas: Generating strong one way functions with cryptographic algorithm. IBM Techn. Disclosure Bull., 27(10A), 1985.

[MatTakIma86] T. Matsumoto, Y. Takashima, H. Imai: On seeking smart public-key-distribution systems. The Transactions of the IECE of Japan, E69: 99–106, 1986.

[Maurer92] U.M. Maurer: Conditionally-perfect secrecy and a provably-secure randomized cipher. Journal of Cryptology, 5(1): 53–66, 1992.

[Maurer94] U.M. Maurer: Towards the equivalence of breaking the Diffie-Hellman protocol and computing discrete logarithms. Advances in Cryptology - CRYPTO '94, Lecture Notes in Computer Science, 839: 271–281, Springer-Verlag, 1994.

[Maurer95] U.M. Maurer: Fast generation of prime numbers and secure public-key cryptographic parameters. Journal of Cryptology, 8: 123–155, 1995.

[Maurer97] U.M. Maurer: Information-theoretically secure secret-key agreement by not authenticated public discussion. Advances in Cryptology - EUROCRYPT '92, Lecture Notes in Computer Science, 658: 209–225, Springer-Verlag, 1993.

[Maurer99] U.M. Maurer: Information-theoretic cryptography. Advances in Cryptology - CRYPTO '99, Lecture Notes in Computer Science, 1666: 47–65, Springer-Verlag, 1999.

[MauRenHol04] U. Maurer, R. Renner, C. Holenstein: Indifferentiability, impossibility results on reductions, and applications to the Random Oracle methodology. First Theory of Cryptography Conference, TCC 2004, Lecture Notes in Computer Science, 2951: 21–39, Springer-Verlag, 2004.

[MauWol96] U.M. Maurer, S. Wolf: Diffie-Hellman oracles. Advances in Cryptology - CRYPTO '96, Lecture Notes in Computer Science, 1109: 268–282, Springer-Verlag, 1996.

[MauWol97] U.M. Maurer, S. Wolf: Privacy amplification secure against active adversaries. Advances in Cryptology - CRYPTO '96, Lecture Notes in Computer Science, 1109: 307–321, Springer-Verlag, 1996.

[MauWol98] U.M. Maurer, S. Wolf: Diffie-Hellman, decision Diffie-Hellman, and discrete logarithms. Proceedings of ISIT '98, Cambridge, MA, August 16–21, 1998, IEEE Information Theory Society: 327, 1998.

[MauWol2000] U.M. Maurer, S. Wolf: The Diffie-Hellman protocol. Designs, Codes, and Cryptography, Special Issue Public Key Cryptography, 19: 147–171, Kluwer Academic Publishers, 2000.

[MicRacSlo88] S. Micali, C. Rackoff, B. Sloan: The notion of security for probabilistic cryptosystems. SIAM Journal on Computing, 17: 412–426, 1988.

[NaoYun89] M. Naor, M. Yung: Universal one-way hash functions and their cryptographic applications. Proceedings of the 21st Annual ACM Symposium on Theory of Computing, Seattle, Wash., May 15–17, 1989: 33–43, 1989.

[NaoYun90] M. Naor, M. Yung: Public-key cryptosystems provably secure against chosen ciphertext attack. Proceedings of the 22nd Annual ACM Symposium on Theory of Computing, Baltimore, MD, May 14–16, 1990: 427–437, 1990.

[NeeSch78] R.M. Needham, M.D. Schroeder: Using encryption for authentication in large networks of computers. Communications of the ACM, 21: 993–999, 1978.

[von Neumann63] J. von Neumann: Various techniques for use in connection with random digits. In: von Neumann's Collected Works, 768–770. New York: Pergamon, 1963.

[NeuTs'o94] B.C. Neuman, T. Ts'o: Kerberos: an authentication service for computer networks. IEEE Communications Magazine, 32: 33–38, 1994.

[Newman80] D.J. Newman: Simple analytic proof of the prime number theorem. Am. Math. Monthly 87: 693–696, 1980.

[NIST94] National Institute of Standards and Technology, NIST FIPS PUB 186, Digital Signature Standard, U.S. Department of Commerce, 1994.

[Okamoto92] T. Okamoto: Provably secure and practical identification schemes and corresponding signature schemes. Advances in Cryptology - CRYPTO '92, Lecture Notes in Computer Science, 740: 31–53, Springer-Verlag, 1993.

[OkaOht92] T. Okamoto, K. Ohta: Universal electronic cash. Advances in Cryptology - CRYPTO '91, Lecture Notes in Computer Science, 576: 324–337, Springer-Verlag, 1992.

[OngSchSha84] H. Ong, C.P. Schnorr, A. Shamir: Efficient signature schemes based on quadratic equations. Advances in Cryptology - Proceedings of CRYPTO '84, Lecture Notes in Computer Science, 196: 37–46, Springer-Verlag, 1985.

[Pedersen91] T. Pedersen: A threshold cryptosystem without a trusted party. Advances in Cryptology - EUROCRYPT '91, Lecture Notes in Computer Science, 547: 522–526, Springer-Verlag, 1991.

[Peralta92] R. Peralta: On the distribution of quadratic residues and nonresidues modulo a prime number. Mathematics of Computation, 58(197): 433–440, 1992.

[Pfitzmann96] B. Pfitzmann: Digital signature schemes - general framework and fail-stop signatures. Lecture Notes in Computer Science, 1100, Springer-Verlag, 1996.

[PohHel78] S.C. Pohlig, M.E. Hellman: An improved algorithm for computing logarithms over GF(p) and its cryptographic significance. IEEE Transactions on Information Theory, IT24: 106–110, January 1978.

[PoiSte2000] D. Pointcheval, J. Stern: Security arguments for digital signatures and blind signatures. Journal of Cryptology, 13(3): 361–396, 2000.

[PolSch87] J.M. Pollard, C.P. Schnorr: An efficient solution of the congruence $x^2 + ky^2 = m(\mathrm{mod}\, n)$. IEEE Transactions on Information Theory, 33(5): 702–709, 1987.

[Rabin63] M.O. Rabin: Probabilistic automata. Information and Control, 6: 230–245, 1963.

[Rabin79] M.O. Rabin: Digitalized signatures and public key functions as intractable as factorization. MIT/LCS/TR-212, MIT Laboratory for Computer Science, 1979.

[RacSim91] C. Rackoff, D.R. Simon: Non-interactive zero-knowledge proof of knowledge and chosen ciphertext attack. Advances in Cryptology - CRYPTO '91, Lecture Notes in Computer Science, 576: 433–444, Springer-Verlag, 1992.

[Rényi61] A. Rényi: On measures of entropy and information. Proc. 4th Berkeley Symposium on Mathematical Statistics and Probability, vol. 1: 547–561, Berkeley: Univ. of Calif. Press, 1961.

[RFC 1510] J. Kohl, C. Neuman: The Kerberos network authentication service (V5). Internet Request for Comments 1510 (RFC 1510), http://www.ietf.org, 1993.

[RFC 2104] H. Krawczyk, M. Bellare, R. Canetti: HMAC: Keyed-hashing for message authentication. Internet Request for Comments 2104 (RFC 2104), http://www.ietf.org, 1997.

[RFC 2246] The TLS protocol, Version 1.0. Internet Request for Comments 2246 (RFC 2246), http://www.ietf.org, 1999.

[RFC 2313] B. Kaliski: PKCS#1: RSA encryption, Version 1.5. Internet Request for Comments 2313 (RFC 2313), http://www.ietf.org, 1998.

[RFC 2409] The Internet Key Exchange (IKE). Internet Request for Comments 2409 (RFC 2409), http://www.ietf.org, 1998.

[RFC 3174] US Secure Hash Algorithm 1 (SHA1). Internet Request for Comments 3174 (RFC 3174), http://www.ietf.org, 2001.

[RFC 3369] R. Housley: Cryptographic Message Syntax (CMS). Internet Request for Comments 3369 (RFC 3369), http://www.ietf.org, 2002.

[RFC 3447] J. Jonsson, B. Kaliski: Public-Key Cryptography Standards (PKCS) #1: RSA cryptography specifications, Version 2.1. Internet Request for Comments 3447 (RFC 3447), http://www.ietf.org, 2003.

[RFC 4346] T. Dierks, E. Rescorla: The Transport Layer Security (TLS) protocol, Version 1.1. Internet Request for Comments 4346 (RFC 4346), http://www.ietf.org, 2006.

[Rivest90] R. Rivest: The MD4 message digest algorithm. Advances in Cryptology - CRYPTO '90, Lecture Notes in Computer Science, 537: 303–311, Springer-Verlag, 1991.

[RivShaAdl78] R. Rivest, A. Shamir, and L.M. Adleman: A method for obtaining digital signatures and public key cryptosystems. Communications of the ACM, 21(2): 120–126, 1978.

[RosSch62] J. Rosser, L. Schoenfield: Approximate formulas for some functions of prime numbers. Illinois J. Math. 6: 64–94, 1962.

[Santos69] E.S. Santos: Probabilistic Turing machines and computability. Proc. Amer. Math. Soc. 22: 704–710, 1969.

[SchnAle84] C.P. Schnorr, W. Alexi: RSA-bits are 0.5 + epsilon secure. Advances in Cryptology - EUROCRYPT '84, Lecture Notes in Computer Science, 209: 113–126, Springer-Verlag, 1985.

[Shannon48] C.E. Shannon: A mathematical theory of communication. Bell Systems Journal, 27: 379–423, 623–656, 1948.

[Shannon49] C.E. Shannon: Communication theory of secrecy systems. Bell Systems Journal, 28: 656–715, 1949.

[Shor94] P.W. Shor: Algorithms for quantum computation: discrete log and factoring. Proceedings of the IEEE 35th Annual Symposium on Foundations of

Computer Science, Santa Fe, New Mexico, November 20–22, 1994: 124–134, 1994.

[Shoup2001] V. Shoup: OAEP Reconsidered. Advances in Cryptology - CRYPTO 2001, Lecture Notes in Computer Science, 2139: 239–259, Springer-Verlag, 2001.

[Stinson92] D.R. Stinson: Universal hashing and authentication codes. Advances in Cryptology - CRYPTO '91, Lecture Notes in Computer Science, 576: 74–85, Springer-Verlag, 1992.

[Vadhan03] S. Vadhan: On constructing locally computable extractors and cryptosystems in the bounded storage model. Advances in Cryptology - CRYPTO 2003, Lecture Notes in Computer Science, 2729: 61–77, Springer-Verlag, 2003.

[Vazirani85] U.V. Vazirani: Towards a strong communication complexity, or generating quasi-random sequences from slightly random sources. Proceedings of the 17th Annual ACM Symposium on Theory of Computing, Providence, RI, May 6–8, 1985: 366–378, 1985.

[VazVaz84] U.V. Vazirani, V.V. Vazirani: Efficient and secure pseudorandom number generation. Proceedings of the IEEE 25th Annual Symposium on Foundations of Computer Science, Singer Island, Fla., October 24–26, 1984: 458–463, 1984.

[Vernam19] G.S. Vernam: Secret signaling system. U.S. Patent #1, 310, 719, 1919.

[Vernam26] G.S. Vernam: Cipher printing telegraph systems for secret wire and radio telegraphic communications. Journal of American Institute for Electrical Engineers, 45: 109–115, 1926.

[WaiPfi89] M. Waidner, B. Pfitzmann: The dining cryptographers in the disco: unconditional sender and recipient untraceability with computationally secure serviceability. Advances in Cryptology - EUROCRYPT '89, Lecture Notes in Computer Science, 434: 690, Springer-Verlag, 1990.

[WanFenLaiYu04] X. Wang, D. Feng, X. Lai, H. Yu: Collisions for hash functions MD4, MD5, HAVAL-128, RIPEMD. Rump Session, Advances in Cryptology - CRYPTO 2004. Cryptology ePrint Archive, http://eprint.iacr.org, Report 199, 2004.

[WatShiIma03] Y. Watanabe, J. Shikata, H. Imai: Equivalence between semantic security and indistinguishability against chosen ciphertext attacks. Proceedings Public Key Cryptography - PKC 2003, Lecture Notes in Computer Science, 2567: 71–84, Springer-Verlag, 2003.

[WegCar81] M.N. Wegman, J.L. Carter: New hash functions and their use in authentication and set equality. Journal of Computer and System Sciences, 22: 265–279, 1981.

[Wiener90] M.J. Wiener: Cryptanalysis of short RSA secret exponents. IEEE Transactions on Information Theory, 36: 553–558, 1990.

[Wolf98] S. Wolf: Unconditional security in cryptography. In: I. Damgård (ed.): Lectures on Data Security. Lecture Notes in Computer Science, 1561: 217–250, Springer-Verlag, 1998.

Internet

[BelDesJokRog97] M. Bellare, A. Desai, E. Jokipii, P. Rogaway: A concrete security treatment of symmetric encryption: analysis of the DES modes of operation. http://www-cse.ucsd.edu/users/mihir/papers/sym-enc.html, 1997.

[GolBel01] S. Goldwasser, M. Bellare: Lecture notes on cryptography. http://www-cse.ucsd.edu/users/mihir/papers/gb.html, 2001.

[DistributedNet] Distributed Net. http://www.distributed.net.

[NIST2000] National Institute of Standards and Technology. Advanced Encryption algorithm (AES) development effort. http://www.nist.gov/aes.

[RSALabs] RSA Laboratories. http://www.rsasecurity.com/rsalabs/.
[WanYinYu05] X. Wang, Y.L. Yin, H. Yu: Collision search attacks on SHA1. Preprint, http://theory.csail.mit.edu/~yiqun/shanote.pdf, February 2005.

Index